VOLUME NINETY ONE

ADVANCES IN VIRUS RESEARCH

Control of Plant Virus Diseases: Vegetatively-Propagated Crops

VOLUME NINETY ONE

ADVANCES IN VIRUS RESEARCH

Control of Plant Virus Diseases: Vegetatively-Propagated Crops

Edited by

GAD LOEBENSTEIN

Department of Plant Pathology
Volcani Center Bet Dagan,
Bet Dagan, Israel

NIKOLAOS I. KATIS

Faculty of agriculture, Forestry and Natural Environment, School of Agriculture,
Plant Pathology Lab,
Aristotle University of Thessaloniki,
Thessaloniki, Greece

ELSEVIER

AMSTERDAM • BOSTON • HEIDELBERG • LONDON
NEW YORK • OXFORD • PARIS • SAN DIEGO
SAN FRANCISCO • SINGAPORE • SYDNEY • TOKYO
Academic Press is an imprint of Elsevier

Academic Press is an imprint of Elsevier
225 Wyman Street, Waltham, MA 02451, USA
525 B Street, Suite 1800, San Diego, CA 92101-4495, USA
125 London Wall, London, EC2Y 5AS, UK
The Boulevard, Langford Lane, Kidlington, Oxford OX5 1GB, UK

First edition 2015

ISBN: 978-0-12-802762-2
ISSN: 0065-3527

For information on all Academic Press publications
visit our website at store.elsevier.com

CONTENTS

CONTRIBUTORS

Marina Barba
Consiglio per la Ricerca e Sperimentazione in agricoltura, Centro di Ricerca per la Patologia Vegetale, Rome, Italy

Matthieu Chabannes
CIRAD, UMR BGPI, F-34398, Montpellier, France

Wilmer Cuellar
Centro Internacional de Agricultura Tropical (CIAT), Cali, Colombia

Morag Ferguson
International Institute of Tropical Agriculture (IITA), Nairobi, Kenya

Marc Fuchs
Department of Plant Pathology and Plant–Microbe Biology, Cornell University, New York State Agricultural Experiment Station, Geneva, New York, USA

Rachid Hanna
IITA, P.O Box. 2008 (Messa), Yaoundé, Cameroon

Vincenza Ilardi
Consiglio per la Ricerca e Sperimentazione in agricoltura, Centro di Ricerca per la Patologia Vegetale, Rome, Italy

Marie-Line Iskra-Caruana
CIRAD, UMR BGPI, F-34398, Montpellier, France

Edward Kanju
International Institute of Tropical Agriculture (IITA), Dar es Salaam, Tanzania

Theodora Kapari Isaia
Agricultural Research Institute, Plant Protection Section, Nicosia, Cyprus

Nikolaos I. Katis
Faculty of agriculture, Forestry and Natural Environment, School of Agriculture, Plant Pathology Lab, Aristotle University of Thessaloniki, Thessaloniki, Greece

P. Lava Kumar
International Institute of Tropical Agriculture (IITA), Oyo Road, PMB 5320, Ibadan, Nigeria

Richard F. Lee
USDA ARS National Clonal Germplasm Repository for Citrus and Dates, Riverside, California, USA

James P. Legg
International Institute of Tropical Agriculture (IITA), Dar es Salaam, Tanzania

Gad Loebenstein
Department of Plant Pathology, Agricultural Research Organization, Bet Dagan, Israel

T. Makeshkumar
Central Tuber Crops Research Institute (CTCRI), Thiruvananthapuram, India

Varvara I. Maliogka
Faculty of agriculture, Forestry and Natural Environment, School of Agriculture, Plant Pathology Lab, Aristotle University of Thessaloniki, Thessaloniki, Greece

Giovanni P. Martelli
Università degli Studi di Bari "Aldo Moro", Bari, Italy

Robert R. Martin
USDA-ARS Horticultural Crops Research Unit, Corvallis, Oregon, USA

Pheneas Ntawuruhunga
International Institute of Tropical Agriculture (IITA), Lusaka, Zambia

Graziella Pasquini
Consiglio per la Ricerca e Sperimentazione in agricoltura, Centro di Ricerca per la Patologia Vegetale, Rome, Italy

Ramasamy Selvarajan
National Research Center for Banana, Tiruchirapalli, PIN # 620102, TN, India

Leena Tripathi
International Institute of Tropical Agriculture (IITA), Nairobi, Kenya

Ioannis E. Tzanetakis
Department of Plant Pathology, Division of Agriculture, University of Arkansas, Fayetteville, Arkansas, USA

Christina Varveri
Laboratory of Virology, Benaki Phytopathological Institute, Kifissia, Greece

PREFACE

This volume covers control of vegetative propagated crops and follows the previous one on control in seed-propagated ones. While in many seed-propagated plants the virus is not transmitted through seed, plantings from vines, tubers, cuttings from a virus-carrying plant will almost in all cases be infected by the virus. This therefore will require first to eliminate the virus from a mother plant through meristem cultures, combined by thermo-therapy, or cryotherapy and later propagating them under controlled conditions.

The main crops that are covered in this volume include cassava, citrus, sweetpotato, stone and pome fruits, grape vines, small fruits, and bananas and plantain, and a general chapter on principles of supplying virus-tested material. Each chapter deals with a specific crop (e.g., Citrus) or a number of related crops (e.g., Berries) and summarizes the most important viruses that infect the crop(s), their epidemiology and ecology, and finally the measures that can be adopted for their management. Unfortunately, we did no locate a qualified author for writing the chapter on potato. The interested reader might look up the following book for relevant information: Loebenstein, Berger, Brunt, and Lawson (2001).

We would like to express our gratitude to Prof. Karl Maramorosch who encouraged us to proceed with the subject and Ms. Helene Kabes and her staff who helped us with the technical details. We also express our thanks for their excellent collaboration to all colleagues who contributed to this book.

Finally, we hope that this volume will be of interest and serve as an excellent practical resource and help to plant virologists, horticulturists, and practitioners and will further contribute to the control of viral diseases of vegetative propagated crops plants and minimize the losses they cause to the crops.

Gad Loebenstein and Nikolaos I. Katis

November 2014

REFERENCE

Loebenstein, G., Berger, P. H., Brunt, A. A., & Lawson, R. H. (2001). *Virus and virus-like diseases of potatoes and production of seed-potatoes.* New York: Springer Science & Business Media.

CHAPTER ONE

Principles for Supplying Virus-Tested Material

Christina Varveri*,[1], Varvara I. Maliogka[†], Theodora Kapari-Isaia[‡]

*Laboratory of Virology, Benaki Phytopathological Institute, Kifissia, Greece

[†]Faculty of agriculture, Forestry and Natural Environment, School of Agriculture, Plant Pathology Lab, Aristotle University of Thessaloniki, Thessaloniki, Greece

[‡]Agricultural Research Institute, Plant Protection Section, Nicosia, Cyprus

[1]Corresponding author: e-mail address: c.varveri@bpi.gr

Contents

Abstract

Production of virus-tested material of vegetatively propagated crops through national certification schemes has been implemented in many developed countries for more than 60 years and its importance for being the best virus control means is well acknowledged by growers worldwide. The two most important elements of certification schemes are the use of sensitive, reliable, and rapid detection techniques to check the health status of the material produced and effective and simple sanitation procedures for the elimination of viruses if present in candidate material before it enters the scheme. New technologies such as next-generation sequencing platforms are expected to further enhance the efficiency of certification and production of virus-tested material, through the clarification of the unknown etiology of several graft-transmissible diseases.

Advances in Virus Research, Volume 91
ISSN 0065-3527
http://dx.doi.org/10.1016/bs.aivir.2014.10.004

The successful production of virus-tested material is a demanding procedure relying on the close collaboration of researchers, official services, and the private sector. Moreover, considerable efforts have been made by regional plant protection organizations such as the European and Mediterranean Plant Protection Organization (EPPO), the North American Plant Protection Organization (NAPPO), and the European Union and the USA to harmonize procedures, methodologies, and techniques in order to assure the quality, safety, and movement of the vegetatively propagated material produced around the world.

1. INTRODUCTION

It is well known that plant species which are propagated in a vegetative manner, especially perennial woody plants, suffer from multiple virus and virus-like diseases, which, over time, induce a severe deterioration in their health status. Trading and the use of infected propagation material are major factors that have contributed to this situation. Therefore, one of the most important, successful, and acknowledged ways to control viruses in these plant species is by the identification, development, and propagation of virus-free (VF) clones (Hadidi, Khetarpal, & Koganezawa, 1998; Hull, 2002). The identification of such clones requires the use of reliable detection methods in order to ensure the health status with respect to diseases of known and unknown etiology. New sequencing technologies are expected to greatly contribute to and facilitate the detection of nonidentified viruses (Kreuze et al., 2009). When a plant variety is 100% virus infected, different technologies can be used for sanitation, that is, to free a plant or plant part from the virus. It is important to select the desirable clones before sanitation according to their horticultural qualities. Once suitable, VF material has been obtained it must be multiplied in such a way as to prevent reinfection and to allow the evaluation of the horticultural value (trueness to type).

Certification schemes aim at the production of certified plant propagation material using appropriate procedures which safeguard the health status and ensure the horticultural characteristics. The production and distribution of VF/virus-tested (VT) propagation material to growers through national certification programs have been implemented in many countries for more than 60 years (Cutting & Montgomery, 1973; Németh, 1984; Reed & Foster, 2011; Waterworth, 1998). To date, VF plant material has been widely used in many vegetatively propagated horticultural crops, such as tuber (Faccioli, 2001; Loebenstein, Fuentes, Cohen, & Salazar, 2003; Slack & Singh, 1998), fruit tree (Barba, 1998; Mink, 1998), grapevine

(Martelli & Walter, 1998), citrus (Roistacher, 1998), strawberry (Spiegel, 1998), and ornamentals (Krczal, 1998).

2. VIRUS DETECTION

The use of reliable, sensitive virus detection methods is essential in all certification schemes (EPPO, 2008a, 2008b; Rowhani, Uyemoto, Golino, & Martelli, 2005). The candidate nuclear stock plants are tested to assure that they are free from viruses or virus-like agents, according to the directives of the plant protection organizations. In the case that sanitation treatments have been applied to eliminate viral infections from these plants, the resulting material has to be retested for viruses before it can be further propagated to generate "virus-free" plantlets.

Three main groups of diagnostic methods, i.e., bioassays, serological, and molecular methods, are applied. Of these, bioassays are mainly used for testing the candidate nuclear stock plants, while serological and molecular techniques are used in other steps of the certification procedure:

a. Biological testing includes grafting of the plant material onto indicator plants. Symptoms develop usually after a long period of time and are compared with those of virus positive and negative control plants. Even though this method of virus indexing is time consuming and requires greenhouse facilities and personnel to look after the plants, it is still a compulsory step in several certification schemes, such as that of grapevine, mainly because some diseases cannot be otherwise identified reliably (EPPO, 2008b; Martelli, 1999). Bioassays also include the mechanical inoculation of herbaceous hosts which are used for detecting sap-transmitted viruses, such as *Nepo-*, *Poty-*, *Carla-*, and *Potexvirus*es (EPPO, 2008a, 2008b; Grousset & Smith, 1998; Rowhani et al., 2005). Biological indexing is normally applied in combination with serological or molecular techniques.

b. Serological methods, which are based on the specific interaction between the viral antigens and homologous antibodies, are routinely used for monitoring the sanitary status of plant material. Even though different serology-based techniques have been developed, the enzyme-linked immunosorbent assay (ELISA) is the most widely applied. ELISA has significantly affected plant virus diagnosis (Boonham et al., 2014; Torrance & Jones, 1981) by shortening the time needed to get results, simplifying detection, and allowing the testing of large numbers of samples. Different types of ELISA (direct and indirect) using polyclonal and/or monoclonal antibodies have been developed and successfully applied for routine virus indexing of most vegetatively propagated agricultural crops. The production of monoclonal antibodies against viruses like *Citrus tristeza virus* (CTV) solved the problems of specificity and increased sensitivity of ELISA tests (Cambra, Gorris, Marroquín, et al., 2000).

It should be mentioned that high-quality antiserum is required to achieve specific and sensitive detection of the tested virus and this necessitates virus isolation. Nevertheless, when it comes to fruit tree and grapevine viruses, this is not always an easy task since most of the viruses affecting these crop plants are phloem-limited, appear in low concentrations, and usually occur in mixed infections (Choueri, Boscia, Digiaro, Castellano, & Martelli, 1996). Furthermore, the reliability of ELISA in woody species is significantly affected by the uneven distribution of viruses in the infected plant and the seasonal fluctuation of their titer (Constable, Connellan, Nicholas, & Rodoni, 2012; Mekuria et al., 2003; Monis & Bestwick, 1996; Salem, Mansour, Al-Musa, & Al-Nsour, 2003; Tsai, Daugherty, & Almeida, 2012; Uyemoto, Luhn, Asai, Beede, & Beutel, 1989; Varveri, Holeva, & Bem, 1997). Therefore, it is important to use an ELISA material from the part of the plant and collected at the time of the year in which virus concentrations are highest.

More recently, monoclonal antibodies have been successfully used in serological techniques like tissue print-ELISA, which allows the sensitive analysis of thousands of samples in an easy way, without the need of extract preparation, by directly printing sections of plant material on nitrocellulose membranes (Cambra, Gorris, Román, et al., 2000). Another serological technique that has been developed lately is the lateral flow immune test. A kit is used for the rapid on-site detection of viral pathogens. It is widely applied during potato seed certification where inspectors are searching for virus-infected plants in the field. The assay needs only a small amount of leaf extract and results are obtained within 15–30 min. Lateral-flow tests for the six most important potato viruses have been developed which have a sensitivity comparable to that of DAS-ELISA (Drygin et al., 2012).

c. Molecular methods focus on the detection of viral nucleic acids. Among the various techniques that have been developed, polymerase chain reaction (PCR) is the most widely applied. The incorporation of a first, reverse transcription (RT) step is also often necessary because most of the plant viruses have an RNA genome. The major advantage of PCR is its higher sensitivity, as compared to the serological techniques. This is especially important for the detection of viruses which infect woody hosts since they often occur in low concentrations throughout the year (Kinard, Scott, & Barnett, 1996; Uyemoto et al., 1989). On the other hand, the reliability of PCR as a virus detection method is affected by the genetic diversity of the viruses, which, especially in the case of RNA viruses infecting vegetatively propagated, woody, plant species, can be high (Alabi, Martin, & Naidu, 2010; Mathioudakis, Maliogka, Katsiani, & Katis, 2009; Šubr & Glasa, 2013). Therefore, in order to circumvent this problem, attention should be paid to the primers used in the RT-PCR assays. Primer design should be based on conserved regions of the genome so as to be able to detect the whole range of virus variants. Another factor that also affects the performance of the RT-PCR is the template used in the assay. This is especially crucial in woody plants which contain high quantities of polysaccharides and phenolic compounds which inhibit the enzymes used in the assay (Nassuth, Pollari, Helmeczy, Stewart, & Kofalvi, 2000; Rowhani, Chay, Golino, & Falk, 1993). Various approaches have been taken in order to avoid these substances, among which total RNA extraction is the most widely used.

Different variants of the conventional RT-PCR have been developed and successfully applied for virus indexing of vegetatively propagated plant species, for example, immunocapture RT-PCR, nested RT-PCR, or multiplex PCR assays (Gambino & Gribaudo, 2006; Hassan, Myrta, & Polak, 2006, Olmos, Cambra, Esteban, Gorris, & Terrada, 1999; Sefc, Leonhardt, & Steinkellner, 2000). Especially the last variant, which has the ability to detect simultaneously different viral species, can be valuable in certification schemes by reducing the time, cost, and labor needed for virus indexing. Recently, real-time quantitative PCR (qPCR) has been developed and this is now widely applied by many laboratories worldwide. This technique, which is based on the detection of a fluorescent signal generated during DNA amplification in a closed tube, achieves high sensitivity and has lower contamination risks than conventional PCR. A range of single reaction or multiplex qPCR assays have been developed for the detection of major viral pathogens infecting vegetatively propagated hosts (Lopez-Fabuel et al., 2013; Mortimer-Jones, Jones, Jones, Thomson, & Dwyer, 2009; Osman, Hodzic, Omanska-Klusek, Olineka, & Rowhani, 2013; Ruiz-Ruiz, Moreno, Guerri, & Ambrós, 2007).

Finally, a great breakthrough in plant virus discovery was achieved with the recent development of high-throughput or next-generation sequencing (NGS) technologies. NGS is performed using different platforms and template preparations (for review, see Boonham et al., 2014; Sebastien, Olmos, Jijakli, & Candresse, 2014) and allows the rapid, simultaneous detection of all known or unknown viral sequences present in a sample. Thus, this technology has significantly advanced the clarification of disease etiology (Al Rwahnih et al., 2013; Vives et al., 2013; Zhang, Singh, Kaur, & Qiu, 2011) and has the potential to be used in the future for routine plant virus diagnosis. So far, several limitations, such as the high cost of the method (Sebastien et al., 2014), have prevented its wider adoption in diagnostics. Nevertheless, it could be used in combination with the currently applied techniques, for a broad-spectrum screening of valuable plant material of vegetatively propagated species, before its use in a certification scheme.

3. VIRUS ELIMINATION

The elimination of viruses, viroids, and phytoplasmas from infected initial (mother) propagation material is a prerequisite for the production of healthy, vegetatively propagated crop material. Methods used are thermotherapy, meristem tissue culture, *in vitro* micrografting, *in vitro* chemotherapy, and cryotherapy of shoot tips, followed by shoot-tip tissue culture or *in vitro* micrografting.

3.1. Thermotherapy

Thermotherapy or heat therapy is the oldest method used for the elimination of viruses, viroids, and phytoplasmas from vegetatively propagated plants and has been used since the end of the nineteenth century. The first reference to the use of heat therapy of plants dates back to 1869 when Scottish gardeners immersed bulbs in hot water before planting. In 1936, Kunkel reported the use of dry heat or hot water treatment for the cure of peach yellows diseases (Nyland & Goheen, 1969). Walkey and Cooper (1975) have also studied the effect of temperature on virus eradication.

Thermotherapy has been used for virus elimination in grapevines (Leonhardt, Wawrosch, Auer, & Kopp, 1998), stone fruits (Manganaris, Economou, Boubourakas, & Katis, 2003; Mink, Wample, & Howell, 1998), citrus (Muhammad, Muhammad, Attique, & Sher, 2005), pome fruits (Howell, Burgess, Mink, Skrzeczkowski, & Zhang, 1998; Tan, Wang, & Hong, 2010), potatoes (Salazar, 1996), and strawberries and others

(Nyland & Goheen, 1969; Panattoni, Luvisi, & Triolo, 2013). The infected plant is incubated in a controlled temperature cabinet at 30–40 °C for a period of 6–12 weeks. Elimination of viruses from the whole plant is usually impossible without causing severe damage to or even death of the plant. In most cases, after thermotherapy, cuttings or bud grafts are removed from the heat-treated plant and propagated *in vivo* or *in vitro* using meristem tips to produce new VF plants.

Alternatively, thermotherapy can be applied *in vitro* with the advantage of reducing procedure time. Shoots from infected plants are cultured *in vitro* and then incubated in a controlled temperature cabinet at 30–40 °C for several days. *In vitro* thermotherapy has been used successfully to eliminate *Apple chlorotic leaf spot virus* (ACLSV) from apricot (*Prunus armeniaca*) shoot cultures; *Prunus necrotic ring spot virus* (PNRSV) and ACLSV from peach (*Prunus persica*) shoot cultures; and *Prune dwarf virus* (PDV) and ACLSV from sour cherry (*Prunus avium*) shoot cultures (Gella & Errea, 1998).

3.2. Low-temperature therapy

Thermotherapy at elevated temperatures (about 37 °C) does not eliminate most viroids and therefore low-temperature treatments have been used instead. Lizarraga, Salazar, Roca, and Schilde-Rentscheler (1980) observed that *Potato spindle tuber viroid* (PSTVd) concentration was lower in potato plants grown at low temperatures (8 °C) and successfully eliminated PSTVd using meristem cultures from potato plants grown at low temperature. Similarly, Paduch-Cichal and Kryczyński (1987) have also reported the elimination of PSTVd from infected potato plants and also the elimination of *Chrysanthemum stunt* (ChSVd), *Chrysanthemum chlorotic mottle* (ChCMVd), and *Cucumber palefruit viroids* (CPFVd) from infected chrysanthemum plants after 6 months therapy in a growth chamber at 5 °C. Low-temperature treatment and meristem culture have also been used for the elimination of *Apple scar skin viroid* (ASSVd) from pears (Postman & Hadidi, 1995) and the eradication of *Hop latent viroid* (HSVd) from hops (Adams, Barbara, Morton, & Darby, 1996).

3.3. Meristem culture *in vitro*

Meristem culture *in vitro* is used for the elimination of viruses and related pathogens from a large number of vegetatively propagated plants and it is the main method used in plant virus elimination programs. The shoot apical meristem is comprised of the leaf primordia, which turn into leaves, and the apical dome, where the stem elongates and it is usually free of pathogens. Limmaset and Cornuet (1949) observed that virus titers declined toward the apical meristem. As early as 1952, Morel and Martin produced VF dahlia plants by meristem culture (Faccioli & Marani, 1998).

The isolation of meristems and virus elimination involves the excision of a suitable explant from the infected parent plant, the aseptic culture of the explant in a nutrient medium, which in most cases is based on MS (Murashige & Skoog, 1962) and finally the establishment of the new plantlet in soil (Figs. 1 and 2). Other tissue culture techniques used for virus elimination are callus culture, protoplast culture, and the culture of reproductive tissues (George, 1993b).

A considerable number of publications report on the use of meristem culture *in vitro* for the production of VF plants, for example, for shallots sanitized from *Onion yellow dwarf* (OYDV) and *Shallot latent viruses* (SLV; Fletcher, Fletcher, & Lewthwaite, 1980), grapevine freed from leafroll, yellow speckle, fleck, and summer mottle diseases (Barlass, Skene, Woodham, & Krake, 1982), red clover freed from *White clover mosaic virus* (WCMV) and *Red clover necrotic mosaic virus* (RCNMV), white clover freed from phyllody and clover red leaf diseases (Dale & Cheyne, 1993), and cassava from Cassava mosaic disease (Kartha & Gamborg, 1975). Further examples include the elimination of *Sugarcane yellow leaf virus* (SYLV) and sugarcane yellows phytoplasma from sugarcanes (Parmessur, Aljanabi, Saumtally, & Dookun-Saumtally, 2002) and the production of healthy potato minitubers from the varieties Spunta, Nicola, and Cara (Gregoriou, Minas, Kapari, Seraphides, & Papayiannis, 2007).

In many cases, meristem culture is used in combination with other methods including thermotherapy, chemotherapy, and cryotherapy in order to improve the efficacy of pathogen elimination.

3.4. Micrografting *in vitro*

For some plants, such as citrus, stone fruits, and other woody species, meristem cultures are not successful (George, 1993b; Navarro, 1988). In these cases, the meristem tip is grafted onto a VF rootstock (Fig. 3). The micrografting technique was first used for the elimination of viruses and viroids in citrus by Navarro, Roistacher, and Murashige (1975). Several scientists thereafter adopted this method to produce VF plant material in citrus (Ioannou, Kyriakou, & Ioannou, 1991; Kapari-Isaia, Kyriakou, Ioannides, Papayiannis, & Minas, 2007; Kapari-Isaia et al., 2002, 2011;

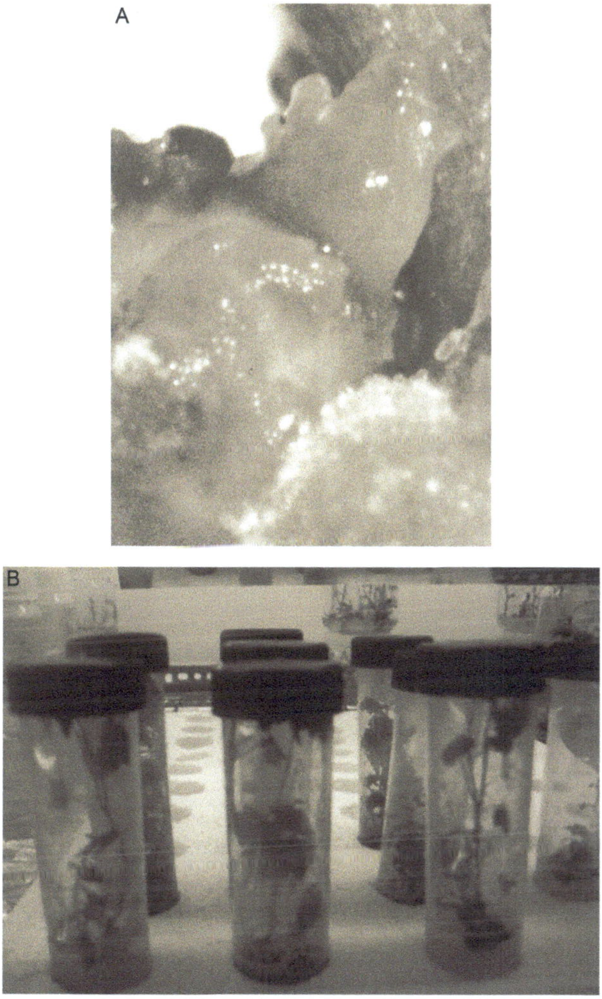

Figure 1 Grapevine meristem culture *in vitro*. (A) Grapevine meristem and (B) rooted *in vitro* grapevine microplants in transparent polycarbonate tubes. *Minas, Papayiannis, and Kapari-Isaia (2007).* (See the color plate.)

Mukhopadhyay et al., 1997; Navarro, Civerolo, Juarez, & Garnsey, 1991; Navarro, Juarez, & Pina, 2001; Navarro et al., 1975; Navarro, Roistacher, & Murashige, 1976; Singh, 2001). Shoot-tip grafting has also been used for the elimination of specific pathogens from fruit trees such as peach (Fig. 4), almond (Rizgi, Zemzami, & Spiegel, 2001; Yildirim et al., 2010) and other prunus species (Conejero et al., 2013; Kapari-Isaia, Kyriakou, Samouel, Papayiannis, & Gabriel, 2009; Navarro, Llacer,

Figure 2 Potato meristem culture *in vitro*. (A) Aseptically prepared microcuttings for *in vitro* mass micropropagation and (B) culture with grown up and rooted microcuttings ready to be transferred for acclimatization *ex vitro*. *Minas, Gregoriou, Kapari-Isaia, and Papayiannis (2007).* (See the color plate.)

Cambra, Arregui, & Juarez, 1982), apple (Huang & Millikan, 1980), pear (Faggioli, Martino, & Barba, 1997; Zilka et al., 2002), avocado (Raharjo & Litz, 2005), and cashew (Mneney & Mantell, 2001).

3.5. Chemotherapy

In chemotherapy, antiviral chemicals such as acycloguanosine, azidothymidine, acyclovir, ribavirin, or 2-thiouracil are added to the tissue culture medium to prevent virus replication or movement from infected to healthy

Figure 3 (See legend on next page.)

tissues. The use of these chemicals, combined with meristem tip culture or micrografting *in vitro*, results in the successful elimination of certain viruses. When chemotherapy is applied to a meristem culture *in vitro*, the chemicals are directly added to the nutrient medium in various dosages which are optimized for different plants and viruses (Long & Cassells, 1986; Sanjeev et al., 2007; Weiland, Cantos, Troncoso, & Perez-Camacho, 2004).

Virazole has been used for the elimination of *Cucumber mosaic virus* (CMV) and *Potato virus Y* (PVY) and in the regeneration of VF tobacco plants (Cassells & Long, 1980). PVY and *Potato virus S* (PVS) were eliminated from potato by chemotherapy of cultured axillary bud tips (Wambugu, 1985). Hansen and Lane (1985) have reported the elimination of ACLSV from apple shoot cultures using ribavirin. More recently, chemotherapy was successfully used in the eradication of serious viral pathogens from grapevine explants (Panattoni, D'Anna, & Triolo, 2007; Skiada, Maliogka, Katis, & Eleftheriou, 2013).

3.6. Cryotherapy

Cryotherapy of shoot tips is a novel application of the cryopreservation technique (process where shoot tips are exposed to the ultra-low temperature of liquid nitrogen, stored, and regenerated for multiplication) and is used for plant pathogen elimination. It is based on the fact that meristematic cells have a lower water content compared to more differentiated cells and generally do not contain viruses. As a result, meristematic cells are not killed by the formation of ice crystals during application of cryopreservation and it is possible to produce VF plants by culturing them *in vitro*. When compared with traditional methods, such as meristem culture, cryotherapy of shoot tips facilitates the treatment of large numbers of samples, produces higher rates of pathogen-free plants, and is independent of shoot-tip size and cryogenic methods (Feng et al., 2013; Wang, Cuellar, Rajamaki, Hirata, & Valkonen, 2008; Wang, Panis, Engelmann, Lambardi, & Valkonen, 2009).

The application of cryotherapy of shoot tips for a certain plant depends on the availability of tissue culture and cryopreservation protocols suitable for it. Many cryopreservation protocols are genotype specific and this is the main limitation of applying cryotherapy (Wang et al., 2009). Modern

Figure 3—Cont'd Citrus micrografing *in vitro*. (A and B) Meristems 0.2–0.5 mm long are aseptically isolated in a laminar flow cabinet using a binocular stereomicroscope to visualize the tissues. (C and D) Micrografted citrus plantlets in culture tubes. *Kapari-Isaia et al. (2002)*. (See the color plate.)

Figure 4 Peach micrografting *in vitro*. (A) Peach meristems 0.2–0.5 mm long micrografted *in vitro*. (B) Micrografted peach plantlets in culture tubes ready to be transferred for acclimatization *ex-vitro*. *Kapari-Isaia, Kyriakou, Samouel, Papayiannis, and Gabriel (2009)*. (See the color plate.)

dehydration techniques are based on vitrification that is the solidification of liquids without crystallization (Benson, 2008). Some of the basic steps for cryotherapy include production of *in vitro* plantlets, preconditioning and preculture conditions for the shoot tips, osmoprotection, cryoprotection, rewarming protocol, and regrowth conditions (Reed, 2008). To achieve a high efficiency in cryotherapy, it may be necessary to adjust the established cryopreservation protocol to increase mortality of nonmeristematic cells (Wang et al., 2009).

Cryotherapy of shoot tips was first used by Brison, de Boucaud, Pierronnet, and Dosba (1997) for the elimination of *Plum pox virus* (PPV) from a prunus rootstock. A number of more recent articles report on the use of cryotherapy for obtaining VF plants from infected mother plants. CMV and *Banana streak virus* (BSV) were eliminated from banana (*Musa* spp.; Helliot, Panis, Poumay, & Swennen, 2002). *Grapevine virus A* (GVA; Wang et al., 2003), *Potato leafroll virus* (PLRV), and PVY (Wang, Liu, He, Xie, & You, 2006; Wang, Wang, Hong, Tang, & Teng, 2006) are further examples of viruses eliminated by cryopreservation of *in vitro*-grown shoot tips. Recently, Volk, Bonnart, Krueger, and Lee (2012) reported the cryopreservation of citrus shoot tips in combination with micrografting, while Beneli, Carlo, and Engelmann (2013) referred to the advantages in the cryopreservation of shoot-derived germplasm of *Actinidia*, *Diospyros*, *Malus*, *Olea*, *Prunus*, *Pyrus*, and *Vitis* species.

3.7. Combination of methods

Combinations of the above methods may be used to eliminate specific pathogens from plants. In all cases, thermotherapy, chemotherapy, or cryotherapy is used in combination with meristem culture or shoot-tip grafting *in vitro*. A considerable number of publications deal with this subject.

3.7.1 Thermotherapy and apical meristem culture or shoot-tip grafting

Mulin, Smith, Frazier, Schlegel, and McCall (1974) reported the elimination of *Strawberry mild yellow-edge virus* (SMYEV), *Strawberry mottle virus* (SMoV), and pallidosis disease-associated viruses from strawberry plants by thermotherapy and meristem culture; similarly, VF prunus, malus, banana, and garlic plants were produced by heat treatment and meristem culture (Conci & Nome, 1991; Knapp et al., 1995; Lloyd & Bustamante, 1974; Manganaris et al., 2003; Ucman, Žel, & Ravnikar, 1998; Wang, Liu, et al., 2006; Wang, Wang, et al., 2006). Postman and Hadidi (1995) eliminated ASSVd from pears by *in vitro* thermotherapy and apical meristem culture. Furthermore, seed potato, sweet potato, taro, and grapevine plants were freed from viral infection using the same methodology (Li, Xu, & Chen, 2002; Mervat & Ashoub, 2009; Minas, Gregoriou, et al., 2007; Minas, Papayiannis, et al., 2007) and the combination of thermotherapy and shoot-tip grafting was reported by Koizumi (1984) to be successful in the elimination of Tatter leaf-citrange stunt virus complex from Satsuma mandarin.

3.7.2 *Chemotherapy and tissue culture or shoot-tip grafting*

Chemotherapy and tissue culture were used for the elimination of OYDV and SLV in shallots (Cassells & Long, 1982), *Raspberry bushy dwarf virus* (RBDV) from axillary bud cultures of red raspberry cv. Lloyd George (Kudell & Buchenauer, 1989), PVY and PVS from potato (Wambugu, 1985), as well as *Cymbidium mosaic virus* (CyMV) and *Odontoglossum ringspot virus* (ORV) from orchids (Lim, Wong, & Goh, 1993). Recently, Sedlak, Paprstein, and Talacko (2011) used chemotherapy with ribavirin and meristem culture for the elimination of *Apple stem pitting virus* (ASPV) from *in vitro*-grown pear plants of the cultivars "Alexander Lucas," "Bohemica," "Elektra," and "Rote Williams."

A combination of chemotherapy and shoot-tip grafting was successfully applied for the elimination of *Citrus ringspot virus* (CRV) from citrus kinnow plants (Sanjeev et al., 2007).

3.7.3 *Chemotherapy, thermotherapy, and meristem* in vitro *culture*

The combination of chemotherapy, thermotherapy, and *in vitro* culture was reported to achieve the elimination of PDV, PNRSV, and ACLSV from sweet cherries (Deogratias, Dosba, & Lutz, 1989), ACLSV from pears (Cieslinka, 2002), ACLSV and PNRSV from *Prunus* spp. (Cieslinka, 2007), and PNRSV from *Begonia* spp. plants (Neeraj, Raja, & Zaidi, 2005). The effect of chemotherapy and thermotherapy on virus concentrations in potato plantlets cultured *in vitro* and the production of VF potato plants has also been reported (Awan, Mughal, Iftikhar, & Khan, 2007; Griffiths, Slack, & Dodds, 1990).

4. CERTIFICATION SCHEMES AND PROGRAMS

Important and reputable certification programs and organizations in developed countries operate, some since the 1960s, delivering high-quality VT plant propagation material preventing the introduction of viruses in the fields. NCPN-FT (formerly NRSP5 or IR-2), NCPP, and FPS in the USA, CTIFL and SOC in France, EMLA and SASA in the UK, Naktuinbouw in the Netherlands, and CVIPS in Spain are only some examples to show the investment various countries have made to promote their respective agricultural economy (Boyé & Desvignes, 1984; Cutting & Montgomery, 1973; Ebbels, 1979; Navarro et al., 2002; NCPN-FT, 2012; Reed & Foster, 2011; Rowhani et al., 2005).

At the end of the 1990s, Regional Plant Protection Organizations (RPPOs) such as the European and Mediterranean Plant Protection Organization (EPPO, members: 50 countries) and the North American Plant Protection Organization (NAPPO, members: the USA, Canada, and Mexico) began drafting technical guidelines for their member countries concerning the production of certified plant propagation material. According to EPPO, a certification scheme is a "System for the production of vegetatively propagated plants for planting, intended for further propagation or for sale, obtained from nuclear stock after several propagation stages under conditions ensuring that stated health standards are met. The filiations of the material are recorded throughout the scheme." In line with the above, NAPPO defines that a "virus-certified stock refers to plants for planting and propagation produced under an official virus testing and certification program" (NAPPO, 2013). Consequently, certification schemes are essentially quality control systems for propagating and planting material, which will be officially certified by the officially delegated authorities by the issuing of a certificate or label. Through certification, propagating material is assured to be free not only from quarantine organisms but from important indigenous nonquarantine pathogens as well, in compliance with each country's requirements as dictated by the local and international markets.

Not all pathogens or viruses are or can be excluded by a certification scheme. In the past, two types of categories, VF and VT, have been used. The first corresponds to individual plants tested for all virus and virus-like pathogens known to infect the host in a specific region, while the second one focuses on the most important pathogens. Although the first term is older and more popular to nurserymen, the preferred term nowadays is pathogen tested and covers only organisms particularly mentioned in the published scheme (usually viruses, viroids, and phytoplasmas). All recent EPPO certification schemes refer only to the latter category (EPPO, 2006).

Furthermore, different approaches for the production of healthy planting material of a certain cultivated plant, certification or classification, may apply. In a typical EPPO certification scheme, the certified material is descended by not more than a fixed number of steps from individual plants each of which is tested and found free from pests, and is then maintained and propagated under rigorous conditions excluding recontamination. In a classification scheme, the classified material is descended by one or more steps from material which, as a population, meets certain health standards and is maintained and propagated under conditions minimizing recontamination. Which of the approaches is appropriate for a given cultivated plant depends on considerations of cost and resources, health status required, practical possibilities for testing, rate of recontamination, and value of the final material.

Normally, the operation of certification schemes is run by official governmental authorities or officially recognized private organizations, although there are differences between countries and continents. Private companies and nurseries' associations are not precluded from participating in an official scheme (e.g., MIVA in Italy and AVASA in Spain) (Pina, Chomé, Vives, & Navarro, 2012; Savino, 1992) or from running their own schemes (e.g., companies producing ornamentals) (Waterworth, 1998).

4.1. Principles

Although differences exist among national or regional certification schemes and among crops certain steps are common and are used to classify propagation material into categories or levels:

1. *Primary source selection (candidate material)*: Individual plants are selected for their agronomical characteristics but they are usually of unknown health status.
2. *Nuclear stock production and maintenance (prebasic material, certification level: G1)*: Plants free from all known harmful viruses and professionally identified for true-to-type phenotype are produced. Sanitation methods can be used if necessary. Plants are maintained under very strict isolation conditions to prevent any (re)infection and are individually tested. They are usually put in greenhouses or screenhouses in order to avoid contact with soil, entry of insects, etc.
3. *Nuclear stock propagation (basic, certification level: G2)*: Plants from the previous category are propagated and maintained under very strict isolation conditions to prevent any (re)infection by root contact, pollen, or aerial vectors and are retested as appropriate. Nuclear stock plants are multiplied in as few steps as possible under conditions ensuring freedom from infection (i.e., isolation, insecticide treatment, etc.).
4. *Production of propagation stock (foundation or propagation block, certification level: G3)*: Nurseries obtain basic material from foundation sources and use it to start their propagation of nursery stock. This material must successfully meet testing regulations for known viruses and graft-transmissible pathogens. Any vegetatively propagated material used for the production of a mother plant must originate from a plant of the previous level(s). Budding, regrafting, or top working is not permitted,

unless the propagative material is from the previous level(s) and with prior approval by the certifying agency.

5. *Increase blocks of certified material (normally certification level: G4)*: Material which is produced from basic stocks and is often the material distributed for sale. Appropriate growing conditions must be followed, the field must be kept clean cultivated, and the surrounding buffer zone must be planted in a manner approved by the certifying agency.
6. *Certification and labeling*: The certifying authority issues official certificates on the basis of procedures, inspections, and tests have been closely followed during the production process. The overall health status of the plants is visually checked. This material cannot be used for further multiplication in an official scheme.

An example of the whole process adapted from OEPP/EPPO (2006) for the production of pathogen-tested olive trees and rootstocks is shown in Fig. 5.

4.2. Harmonization

The structure and operation of certification schemes vary among countries and continents resulting in trading and use of propagating material of unequal health status and quality. Regional Plant Protection Organizations and FAO (Food and Agricultural Organization of the United Nations) have drafted standards and recommendations, and the European Union has drafted laws in an attempt to harmonize certification schemes and facilitate plant movement.

Since 1994, EPPO has produced certification schemes which focus on the phytosanitary aspects of certification giving thorough details on the selection, growth, and maintenance of the candidate material and on the propagation of this material in several stages under conditions ensuring that stated health standards are met (Roy, 2011). Appropriate checks on specified pests are specified throughout the scheme. Information is provided, as necessary, on relevant pests, cultural practices, inspection, and testing methods. Up to now, EPPO has approved and published 34 certification/classification schemes (Table 1) using the same format and terminology (http://archives.eppo.int/EPPOStandards/certification.htm).

Recent developments within the European Union have brought, at the time of writing, a proposal for a new Directive for the production and making available on the market of plant reproductive material (PRM law) for both vegetatively and seed propagated crops with the aim of harmonizing and replacing the existing 12 Directives with one Directive (EU, 2013). The proposed Regulation Proposal sets out rules for the certification of

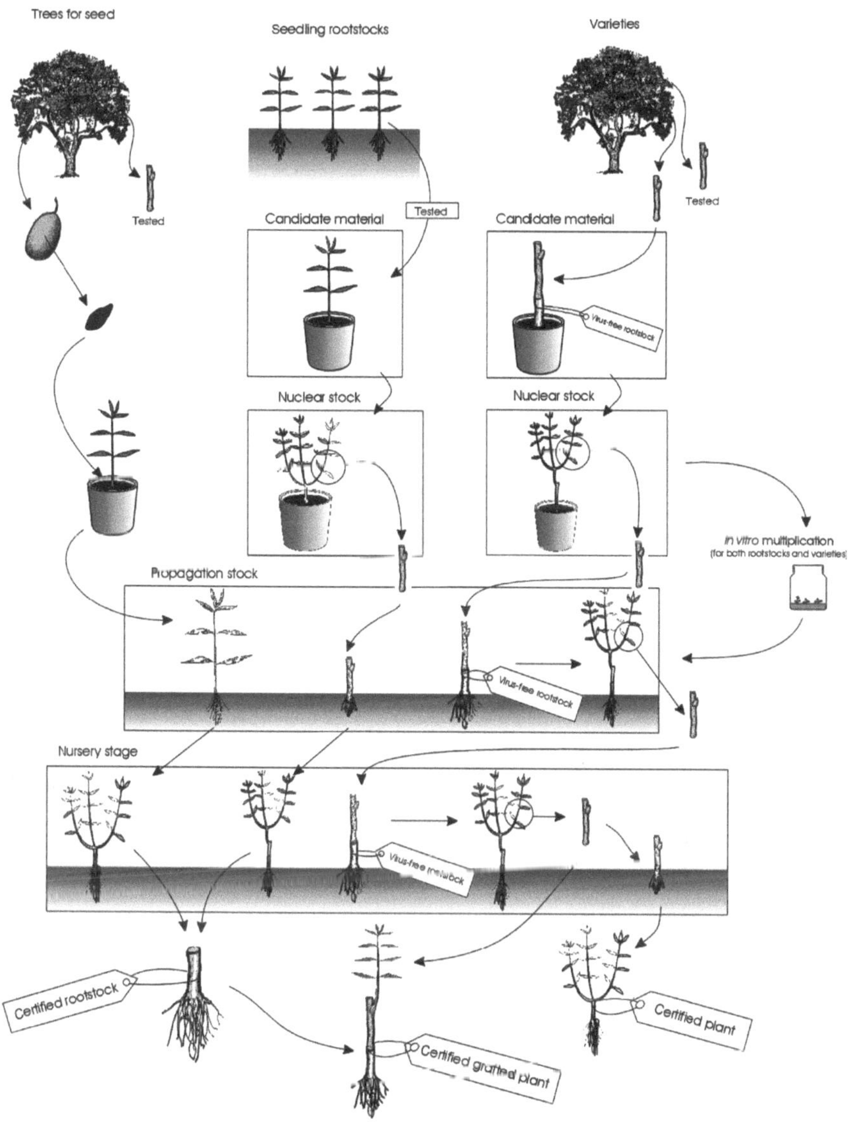

Figure 5 Certification scheme for the production of pathogen-tested olive trees and rootstocks. *EPPO (2006) with permission from Wiley.*

PRM as well as for the activities aiming at verifying the reliability of certification at the postcertification stage, to ensure the fulfillment of the applicable quality requirements. Coherence and synergies with the Plant Health Law concerning the plant health checks which are part of the PRM certification process or integration of general principles relating to official

Table 1 Plants covered by EPPO certification/classification schemes

Target crop	Standard	Year	EPPO bulletin
Fruit/small fruit crops			
Almond, apricot, peach, and plum	PM 4/30(1)	2001	*31*, 463–478
Cherry	PM 4/29(1)	2001	*31*, 447–462
Citrus trees and rootstocks	PM 4/12(1)	1995	*25*, 737–755
Grapevine varieties and rootstocks	PM 4/8(2)	1994	*24*, 347–367
Hazelnut	PM 4/31(1)	2004	*34*, 145–147
Hop	PM 4/16(2)	1997	*27*, 175–184
Malus, *Pyrus*, and *Cydonia* spp.	PM 4/27(1)	1999 2001	*29*, 239–252 *31*, 445–446
Olive trees and rootstocks	PM 4/17(2)	2006	*36*, 77–83
Ribes spp.	PM 4/9(2)	2008	*38*, 14–18
Rubus spp.	PM 4/10(2)	1994	*24*, 865–873
Sambucus spp.	PM 4/32(1)	2008	*38*, 19–24
Strawberry	PM 4/11(2)	1994	*24*, 875–889
Vaccinium spp.	PM 4/18(1)	1994	*27*, 195–204
Ornamentals			
Begonia spp.	PM 4/19(2)	2002	*32*, 135–146
Bulbous iris	PM 4/15(2)	2002	*32*, 129–134
Carnation	PM 4/2(2)	2002	*32*, 55–66
Chrysanthemum	PM 4/6(2)	2002	*32*, 105–114
Crocus	PM 4/14(2)	2002	*32*, 123–128
Freesia	PM 4/22(2)	2002	*32*, 179–184
Herbaceous ornamentals	PM 4/34(1)	2008	*38*, 31–52
Hyacinth	PM 4/23(2)	2002	*32*, 185–190
Kalanchoë spp.	PM 4/25(2)	2002	*32*, 199–210
Lily	PM 4/4(2)	2002	*32*, 79–90
Narcissus	PM 4/5(2)	2002	*32*, 91–104
New Guinea hybrids of impatiens	PM 4/20(2)	2002	*32*, 147–158

Table 1 Plants covered by EPPO certification/classification schemes—cont'd

Target crop	Standard	Year	EPPO bulletin
Pelargonium	PM 4/3(3)	2002	*32*, 67–78
Petunia spp.	PM 4/26(2)	2002	*32*, 211–221
Rose	PM 4/21(2)	2002	*32*, 159–178
Tulip	PM 4/13(2)	2002	*32*, 115–122
Other			
Seed potatoes	PM 4/28(1)	1999	*29*, 253–267
Poplar and willow	PM 4/33(1)	2008	*38*, 25–30

controls embedded in Regulation (EC) No. 882/2004 on official controls are foreseen: "Pre-basic, basic or certified material shall be certified and identified through an official label ('official label') certifying that the material complies with the relevant production and quality requirements. Standard material shall be identified through an operator's label ('operator's label'). The Commission shall be empowered to adopt delegated acts supplementing the certification schemes and taking into account the applicable international technical and scientific standard recommendations of EPPO. Each EU Member State shall adopt its own enforcement and implementation policies using the EU Directive as a guide. Propagating material and plants from Countries outside of the European Union (EU) may only be marketed within the Community if they offer the same guarantees as materials produced in the Community" (EU, 2013).

NAPPO aims at developing "science-based Regional Standards for Phytosanitary Measures (RSPM) which are intended to protect agricultural, forest and other plant resources against regulated plant pests, while facilitating trade" (www.nappo.org). NAPPO standards combine in a single document quarantine issues relevant to trade with production issues relevant to certification and define some basic guidelines, such as that the certification program should clearly define certification levels. These levels represent successive generations of propagation from the original, tested material, and may have additional phytosanitary measures applied depending on the generation. Certification levels are a categorical measure of the health status of certified plants. Eligibility criteria should be established at each level, including nomenclature, propagation, and pest management measures, and the number of generations removed from the original tested material. NAPPO has issued standards for all important crops (grapevine, fruit trees, citrus, and potato) and they refer to the pests, their status in each area (presence and absence) and appropriate detection methods. Detailed certification programs are delivered by the member countries. A State Level Model Regulatory Standard for "Virus-Tested Certification Program for Prunus, Malus, Pyrus, Chaenomeles, and Cydonia Nursery Stock Production Systems" was, for example, recently issued (October 2012) by the National Clean Plant Network (fruit tree group) FT in the USA to harmonize procedures among states (NCPN-FT, 2012).

FAO has also developed protocols for the safe multiplication of pathogen-free materials of vegetatively propagated crops belonging to tropical or subtropical agricultural systems (i.e., yam, cassava, sweet potato, Andean tubers, bananas, plantains, and other species of Musaceae, cocoyam, garlic, Hausa potato, konjac, and taro) which are staple foods in many developing countries. In collaboration with the International Potato Centre (CIP) and a team of international experts, protocols have been established to improve the quality and availability of planting materials, particularly for the small farmer. The crops described varied greatly in their reproductive means and therefore on the types of processes needed to obtain quality materials. However, in general terms, the technical papers focused on quality and sanitary aspects of the production of planting materials (FAO, 2010).

4.3. Effectiveness

Although the importance of certification schemes is well acknowledged by growers worldwide, few studies analyze the economic impact of using certified material. Cembali, Folwell, Wandschneider, Eastwell, and Howell (2003) examined the economic effects of NRSP5 in New York State and calculated the benefits from avoiding yield losses and quality declines for apples, sweet cherries, and clingstone peaches, as a result of that center's program of testing and clean plant provision. They estimated a total gross annual benefit of approximately $227 million or 420 times the cost of the program. In a more recent study, the costs and benefits of a virus testing and certification program for *Grapevine leafroll-associated virus 3* in the North Coast region of California were examined (Fuller, Alston, & Golino, 2013). The authors showed that Californian grapevine growers will receive a very large benefit—over 100 times the cost—from using certified grapevine material at current costs. They compared grower costs and benefits from using certified stock, and they extrapolated to the industry as a whole to estimate costs and benefits to consumers and producers of wine grapes, in the North Coast Californian region. They found that the benefits from certification are large—in excess of $50 million per year for the region—and that they substantially outweighed the costs. They also found large potential benefits from rogueing and replacing diseased vines rather than leaving them in the vineyard where they can spread the disease.

It is unfortunate that recently some growers and regulators have become complacent and that the investment of some countries in certification schemes has decreased (Reed & Foster, 2011). An example comes from Greece, with an EU "Protected Zone" status for *Citrus tristeza virus*, where in 2001 certified citrus material imported from Spain was found CTV positive (although at a low level of 0.67%; Dimou, Drosopoulou, Moschos, Spanou, & Dermatas, 2002). Since then, actions have been taken by the exporting country to make procedures stricter and strengthen controls (M. Cambra, personal communication). The free trade of certified propagation material within EU member states makes it absolutely imperative that the traded material fully fulfills all health requirements foreseen, so that pathogens do not disperse from one country to another.

5. PROSPECTS

Nowadays in most countries, certification schemes are well in place and actions for harmonization are being taken (EPPOs, EU, and the USA). The development of highly efficient detection techniques targeting multiple pathogens and allowing the use of composite samples is believed to facilitate considerably controls needed in all steps of the production of VT material. A revolution is anticipated to arise from the next-generation sequence technologies for the identification of virus pathogens of unknown etiology for which laborious and time-consuming biological tests have to be applied. A recent example is the identification of the causal agent of Citrus vein enation/woody gall disease through deep sequencing of small RNAs under the provisional name Citrus vein enation virus (CVEV; Vives et al., 2013). Screening of candidate material to enter a certification scheme for diseases hitherto of unknown etiology and securing freedom from them at later certification steps will be greatly simplified by using the new technology available (NGS). Furthermore, there is a lot of experience worldwide on the different sanitation methods which are combined to produce VF material. Cryotherapy is a technology producing high rates of pathogen-free plants, independently of shoot-tip size. It is also believed that it will assist particular certification schemes based on sanitation of candidate material. In all cases, it is certain that certification is a quite demanding procedure in expertise, facilities, and manpower, and that the close collaboration of all implicated parties from pure researchers up to civil servants and nursery men is a prerequisite for its success.

ACKNOWLEDGMENTS

The authors thank Dr Catherine-Margaret Cook, National Agricultural Research Foundation (Agricultural Research Centre of Macedonia-Thrace, Thermi, Greece) for correcting the chapter. Her review improved it substantially.

REFERENCES

Adams, A. N., Barbara, D. J., Morton, A., & Darby, P. (1996). The experimental transmission of hop latent viroid and its elimination by low temperature treatment and meristem culture. *Annals of Applied Biology, 128*, 37–44.

Alabi, O. J., Martin, R. R., & Naidu, R. A. (2010). Sequence diversity, population genetics and potential recombination events in Grapevine rupestris stem pitting-associated virus in Pacific North-West vineyards. *Journal of General Virology, 91*, 265–276.

Al Rwahnih, M., Dave, A., Anderson, M., Rowhani, A., Uyemoto, J. K., & Sudarshana, M. R. (2013). Association of a DNA virus with grapevines affected by red blotch disease in California. *Phytopathology, 103*, 1069–1076.

Awan, A. R., Mughal, S. M., Iftikhar, Y., & Khan, H. Z. (2007). *In vitro* elimination of potato leaf roll polerovirus from potato varieties. *European Journal of Scientific Research, 18*, 155–164.

Barba, M. (1998). Virus certification of fruit tree propagative material in Western Europe. In A. Hadidi, R. K. Khaterpal, & H. Koganezawa (Eds.), *Plant virus disease control* (pp. 288–293). St. Paul, MN: APS Press.

Barlass, M., Skene, K. G. M., Woodham, R. C., & Krake, L. R. (1982). Regeneration of virus-free grapevines using *in vitro* apical culture. *Annals of Applied Biology, 101*, 291–295.

Beneli, C., Carlo, A., & Engelmann, F. (2013). Recent advances in the cryopreservation of shoot-derived germplasm of economically important fruit trees of Actinidia, Diospyros, Malus, Olea, Prunus, Pyrus and Vitis. *Biotechnology Advances, 31*, 175–185.

Benson, E. (2008). Cryopreservation theory. In B. M. Reed (Ed.), *Plant cryopreservation: A practical guide* (pp. 15–32). New York: Springer.

Boonham, N., Kreuze, J., Winter, S., Van der Vlugt, R., Bergervoet, J., Tomlinson, J., et al. (2014). Methods in virus diagnostics: From ELISA to next generation sequencing. *Virus Research, 186*, 20–31.

Boyé, R., & Desvignes, J. C. (1984). Bilan des quinze années de sélection conservatrice du matérielvégétale fruitier. *Fruits, 39*, 637–645.

Brison, M., de Boucaud, M. T., Pierronnet, A., & Dosba, F. (1997). Effect of cryopreservation on the sanitary state of a cv *Prunus* rootstock experimentally contaminated *with Plum pox potyvirus*. *Plant Science, 123*, 189–196.

Cambra, M., Gorris, M. T., Marroquín, C., Román, M. P., Olmos, A., Martínez, M. C., et al. (2000). Incidence and epidemiology of *Citrus tristeza virus* in the Valencian Community of Spain. *Virus Research, 71*, 85–95.

Cambra, M., Gorris, M. T., Román, M. P., Terrada, E., Garnsey, S. M., Camarasa, E., et al. (2000). Routine detection of *Citrus tristeza virus* by direct immunoprinting-ELISA method using specific monoclonal and recombinant antibodies. In *Proceedings fourteenth IOCV conference* (pp. 34–41). Riverside, CA: IOCV.

Cassells, A. C., & Long, R. D. (1980). The regeneration of virus-free plants from cucumber mosaic virus- and potato virus Y-infected tobacco explants cultured in the presence of virazole. *Zeitschrift fur Naturforschung, 35*, 350–351.

Cassells, A. C., & Long, R. D. (1982). The elimination of potato viruses X, Y, S and M in meristem and explant cultures of potato in the presence of virazole. *Potato Research, 25*, 165–173.

Cembali, T., Folwell, R. J., Wandschneider, P., Eastwell, K. C., & Howell, W. E. (2003). Economic implications of a virus prevention program in deciduous tree fruits in the US. *Crop Protection, 22*, 1149–1156.

Choueri, E., Boscia, D., Digiaro, M., Castellano, M. A., & Martelli, G. P. (1996). Some properties of a hitherto undescribed filamentous virus of the grapevine. *Vitis, 35*, 91–93.

Cieslinka, M. (2002). Elimination of *Apple chlorotic leaf spot virus* (ACLSV) from pear by *in vitro* thermotherapy and chemotherapy. *Acta Horticulturae, 596*, 481–484.

Cieslinka, M. (2007). Application of thermo- and chemotherapy *in vitro* for eliminating some viruses infecting *Prunus* sp. fruit trees. *Journal of Fruit and Ornamental Plant Research, 15*, 117–124.

Conci, V. C., & Nome, S. F. (1991). Virus free garlic (Allium sativum L.) plants obtained by thermotherapy and meristem tip culture. *Journal of Phytopathology, 132*, 186–192.

Conejero, A., Romero, C., Cunill, M., Mestre, M. A., Martínez-Calvo, J., Badenes, M. L., et al. (2013). *In vitro* shoot-tip grafting for safe *Prunus* budwood exchange. *Scientia Horticulturae, 150*, 365–370.

Constable, F. E., Connellan, J., Nicholas, P., & Rodoni, B. C. (2012). Comparison of enzyme-linked immunosorbent assays and reverse transcription-polymerase chain reaction for the reliable detection of Australian grapevine viruses in two climates during three growing seasons. *Australian Journal of Grape and Wine Research, 18*, 239–244.

Cutting, C. V., & Montgomery, H. B. S. (1973). *More and better fruit with EMLA*. East Malling/Long Ashton: East Malling/Long Ashton Research Station.

Dale, P. J., & Cheyne, V. A. (1993). The elimination of clover diseases by shoot tip culture. *Annals of Applied Biology, 123*, 25–32.

Deogratias, J. M., Dosba, F., & Lutz, A. (1989). Eradication of *Prune dwarf virus*, *Prunus necrotic ringspot virus* and *Apple chlorotic leaf spot virus* in sweet cherries by a combination of chemotherapy, thermotherapy and *in vitro* culture. *Canadian Journal of Plant Pathology, 11*, 337–342.

Dimou, D., Drosopoulou, J., Moschos, E., Spanou, C., & Dermatas, P. (2002). First report of *Citrus tristeza virus* (CTV) in Greece. In *Proceedings fifteenth IOCV conference* (pp. 78–82). Riverside, CA: IOCV.

Drygin, Y. F., Blintsov, A. N., Grigorenko, V. G., Andreeva, I. P., Osipov, A. P., Varitzev, Y. A., et al. (2012). Highly sensitive field test lateral flow immunodiagnostics of PVX infection. *Applied Microbiology and Biotechnology, 93*, 179–189.

E. U. (2013). *Proposal for a Regulation of the European Parliament and of the Council on the production and making available on the market of plant reproductive material (plant reproductive material law)*. http://ec.europa.eu/dgs/health_consumer/pressroom/docs/proposal_aphp_en.pdf.

Ebbels, D. L. (1979). A historical review of certification schemes for vegetatively-propagated crops in England and Wales. *ADAS Quarterly Review, 32*, 21–58.

EPPO. (2006). Certification scheme for the production of pathogen-tested olive trees and rootstocks. *Bulletin OEPP/EPPO Bulletin, 6*, 77–83.

EPPO. (2008a). Certification scheme for strawberry. *Bulletin OEPP/EPPO Bulletin, 38*, 430–437.
EPPO. (2008b). Certification scheme. Pathogen-tested material of grapevine varieties and rootstocks. *Bulletin OEPP/EPPO Bulletin, 38*, 422–429.
Faccioli, G. (2001). Control of potato viruses using meristem culture and stem-cutting cultures, thermotherapy and chemotherapy. In G. Loebenstein, P. H. Berger, A. Brunt, & R. H. Lawson (Eds.), *Virus and virus-like diseases of potatoes and production of seed-potatoes* (pp. 365–390). Dordrecht: Kluwer.
Faccioli, G., & Marani, F. (1998). Virus elimination by meristem tip culture and tip micrografting. In A. Hadidi, R. H. Khetarpal, & H. Koganezawa (Eds.), *Plant virus disease control* (pp. 346–380). St. Paul, MN: APS Press.
Faggioli, F., Martino, L., & Barba, M. (1997). *In vitro* micrografting of *Pyrus communis* shoot tips. *Advances in Horticultural Science, 11*, 25–29.
FAO. (2010). Quality declared planting material. Protocols and standards for vegetatively propagated crops. In J. Fajardo, N. Lutaladio, M. Larinde, C. Rosell, I. Barker, W. Roca, & E. Chujoy (Eds.), *Plant Production and Protection Paper 195*. Rome, Italy: FAO.
Feng, C., Wang, R., Li, J., Wang, B., Yin, Z., Cui, Z., et al. (2013). Production of pathogen-free horticultural crops by cryotherapy of *in vitro*-grown shoot tips. In M. Lambardi, E. A. Ozudogru, & S. M. Jain (Eds.), *Protocols for micropropagation of selected economically-important horticultural plants. Methods in molecular biology* (pp. 463–482).New York City: Humana Press, Copyright Holder: Springer Science+Business Media New York.
Fletcher, P. J., Fletcher, J. D., & Lewthwaite, S. L. (1980). In vitro elimination of onion yellow dwarf and shallot latent viruses in shallots (Allium cepa var. ascalonicum L.). *New Zealand Journal of Crop and Horticultural Science, 26*, 23–26.
Fuller, K. B., Alston, J. M., & Golino, D. A. (2013). *The benefits from certified virus-free nursery stock:* A case study of grapevine leafroll-3 in the North Coast Region of California: RMI-CWE Working Paper Number 1306. Davis, CA: Robert Mondavi Institute Center For Wine Economics.
Gambino, G., & Gribaudo, I. (2006). Simultaneous detection of nine grapevine viruses by multiplex reverse transcription-polymerase chain reaction with co-amplification of a plant RNA as internal control. *Phytopathology, 96*, 1223–1229.
Gella, R., & Errea, P. (1998). Application of *in vitro* therapy for ilarvirus elimination in three *Prunus* species. *Journal of Phytopathology, 146*, 445–449.
George, E. F. (1993a). Micrografting. In E. F. George, M. A. Hall, & Geet-Jan De Klerk (Eds.), *Plant propagation by tissue culture, part I. The technology* (pp. 64–66). Westbury, England: Exegetics Ltd.
George, E. F. (1993b). Methods of elimination. In E. F. George, M. A. Hall, & Geet-Jan De Klerk (Eds.), *Plant propagation by tissue culture, part I. The technology* (pp. 146–151). Westbury, England: Exegetics Ltd.
Gregoriou, S., Minas, G., Kapari, Th., Seraphides, N., & Papayiannis, L. (2007). Production of minitubers of the varieties Spunta, Nicola and Cara from healthy micro propagated material in Cyprus. In A. J. Haverkort & B. V. Anisimov (Eds.), *Potato production and innovative technologies* (pp. 160–168). Wageningen, Netherlands: Wageningen Academic Publishers.
Griffiths, H. M., Slack,, St. A., & Dodds, J. H. (1990). Effect of chemical and heat therapy on virus concentrations in *in vitro* potato plantlets. *Canadian Journal of Botany, 68*, 1515–1521.
Grousset, F., & Smith, I. M. (1998). EPPO certification scheme for seed potatoes. *Bulletin OEPP/EPPO Bulletin, 28*, 561–567.

Hadidi, A., Khetarpal, R. K., & Koganezawa, H. (1998). *Plant virus disease control.* St. Paul, MN: APS Press.

Hansen, A. J., & Lane, W. D. (1985). Elimination of *Apple chlorotic leafspot virus* from apple shoot cultures by ribavirin. *Plant Disease, 69,* 134–135.

Hassan, M., Myrta, A., & Polak, J. (2006). Simultaneous detection and identification of four pome fruit viruses by one-tube pentaplex RT-PCR. *Journal of Virological Methods, 133,* 124–129.

Helliot, B., Panis, B., Poumay, Y., & Swennen, R. (2002). Cryopreservation for the elimination of cucumber mosaic virus (CMV) and *banana* streak virus (BSV) from Banana (*Musa* spp.). *Plant Cell Reports, 20,* 1117–1122.

Howell, W. E., Burgess, J., Mink, G. I., Skrzeczkowski, L. J., & Zhang, Y. P. (1998). Elimination of apple fruit and bark deforming agents by heat therapy. *Acta Horticulturae, 472,* 641–648.

Huang, S., & Millikan, D. F. (1980). *In vitro* micrografting of apple shoot-tips. *HortScience, 15,* 741–743.

Hull, R. (2002). *Matthew's plant virology* (4th ed.). New York: Academic (Chapter 16).

Ioannou, M., Kyriakou, A., & Ioannou, N. (1991). *Production of healthy Lapithos lemon plants by shoot-tip grafting* in vitro*: Tech. Bulletin 139.* Nicosia, Cyprus: Agricultural Research Institute.

Kapari-Isaia, Th., Kyriakou, A., Ioannides, I., Papayiannis, L., & Minas, G. (2007). Sanitation of local citrus clones and/or varieties in Cyprus. *Acta Horticulturae, 741,* 301–306.

Kapari-Isaia, Th., Kyriakou, A., Samouel, S., Papayiannis, L. C., & Gabriel, I. (2009). Sanitation of stone fruits in Cyprus. *Phytopathologia Mediterranea, 48,* 323.

Kapari-Isaia, T. H., Minas, G. J., Polykarpou, D., Iosephidou, E., Arseni, Sp, & Kyriakou, A. (2002). Shoot-tip grafting *in vitro* for elimination of viroids and citrus psorosis virus in the local "Arakapas" mandarin in Cyprus. In *Proceedings fifteenth IOCV conference* (pp. 417–419), Riverside, CA: IOCV.

Kapari-Isaia, Th., Voloudakis, A. E., Kyriakou, A., Ioannides, I., Papayiannis, L., Samouel, S., et al. (2011). Sanitation of citrus varieties and/or clones by in vitro micrografting in Cyprus and Greece. *Acta Horticulturae, 892,* 279–285.

Kartha, K. K., & Gamborg, O. L. (1975). Elimination of Cassava mosaic disease by meristem culture. *Phytopathology, 65,* 826–828.

Kinard, G. R., Scott, S. W., & Barnett, O. W. (1996). Detection of apple chlorotic leaf spot and apple stem grooving viruses using RT-PCR. *Plant Disease, 80,* 616–621.

Knapp, E., Hanzer, V., Weiss, H., da Câmara Machado, A., Weiss, B., Wang, Q., et al. (1995). New aspects of virus elimination in fruit trees. *Acta Horticulturae, 386,* 409–418.

Koizumi, M. (1984). Elimination of Tatter leaf-citrange stunt virus from Satsuma mandarin by shoot-tip grafting following pre-heat-treatment. In *Proceedings ninth IOCV conference* (pp. 229–233). Riverside, CA: IOCV.

Krczal, G. (1998). Virus certification of ornamental plants—The Europe strategy. In A. Hadidi, R. K. Khetarpal, & H. Koganezawa (Eds.), *Plant virus disease control* (pp. 277–287). St. Paul, MN: APS Press.

Kreuze, J., Perez, A., Untiveros, M., Quispe, D., Fuentes, S., Barker, I., et al. (2009). Complete viral genome sequence and discovery of novel viruses by deep sequencing of small RNAs: A generic method for diagnosis, discovery and sequencing of viruses. *Virology, 388,* 1–7.

Kudell, A. R., & Buchenauer, H. (1989). Elimination of Raspberry bushy dwarf virus from axillary bud cultures of red Raspberry cv. Lloyd George by antiviral compounds. *Journal of Phytopathology, 124,* 332–336.

Leonhardt, W., Wawrosch, Ch., Auer, A., & Kopp, B. (1998). Monitoring of virus diseases in Austrian grapevine varieties and virus elimination using *in vitro* thermotherapy. *Plant Cell, Tissue and Organ Culture, 52,* 71–74.

Li, Y., Xu, C., & Chen, J. (2002). Establishment of virus free taro (Colosia esulenta cv. Fenghuayunaitou) by meristem-tip culture with thermotherapy. *Pakistan Journal of Plant Pathology, 1*, 40–43.

Lim, S., Wong, S., & Goh, C. (1993). Elimination of Cymbidium mosaic virus and Odontoglossum ringspot virus from orchids by meristem culture and thin section culture with chemotherapy. *Annals of Applied Biology, 122*, 289–297.

Limmaset, P., & Cornuet, P. (1949). Recherche de virus de la mosaïque du Tabac (Marmor Tabaci Holmes) dans les méristèmes des plantes infectées. *Comptes Rendus Hebdomadaires des Séances de l'Académie des Sciences, 228*, 1971–1972.

Lizarraga, R. E., Salazar, L. F., Roca, W. H., & Schilde-Rentscheler, L. (1980). Elimination of potato spindle tuber viroid by low temperature and meristem culture. *Phytopathology, 70*, 754–755.

Lloyd, A. B., & Bustamante, M. (1974). Heat treatment and meristem culture for the production of virus-free bananas. *Phytopathology, 64*, 320–322.

Loebenstein, G., Fuentes, S., Cohen, J., & Salazar, L. F. (2003). Sweet potato. In G. Loebenstein, & G. Thottappilly (Eds.), *Virus and virus-like diseases of major crops in developing countries* (pp. 223–248). Dordrecht: Kluwer Academic.

Long, R. D., & Cassells, A. C. (1986). Elimination of viruses from tissue cultures in the presence of antiviral chemicals. In L. A. Withers, & P. G. Alderson (Eds.), *Plant tissue culture and its agricultural applications* (pp. 239–248). UK: Butterworths.

Lopez-Fabuel, I., Wetzel, T., Bertolini, E., Bassler, A., Vidal, E., Torres, L. B., et al. (2013). Real-time multiplex RT-PCR for the simultaneous detection of the five main grapevine viruses. *Journal of Virological Methods, 188*, 21–24.

Manganaris, G. A., Economou, A. S., Boubourakas, I. N., & Katis, N. I. (2003). Elimination of PPV and PNRSV through thermotherapy and meristem tip culture in nectarine. *Plant Cell Reports, 22*, 195–200.

Martelli, G. P. (1999). Infectious diseases and certification of grapevines. In G. P. Martelli & M. Digiaro (Eds.), *Options Mediterraneennes, Series B. Etudes et Recherches: 29, Proceedings of the Mediterranean network on grapevine closteroviruses 1992–1997 and the viroses and virus-like diseases of the grapevine a bibliographic report 1985–1997* (pp. 47–64)

Martelli, G. P., & Walter, B. (1998). Virus certification of grapevines. In A. Hadidi, R. K. Khetarpal, & H. Koganezawa (Eds.), *Plant virus disease control* (pp. 261–276). St. Paul, MN: APS Press.

Mathioudakis, M. M., Maliogka, V. I., Katsiani, A. T., & Katis, N. I. (2009). Incidence and molecular variability of Apple stem pitting and Apple chlorotic leaf spot viruses in apple and pear orchards in Greece. *Journal of Plant Pathology, 92*, 139–147.

Mekuria, G., Ramesh, S. A., Alberts, E., Bertozzi, T., Wirthensohn, M., Collins, G., et al. (2003). Comparison of ELISA and RT-PCR for the detection of *Prunus necrotic ring spot virus* and *Prune dwarf virus* in almond (*Prunus dulcis*). *Journal of Virological Methods, 114*, 65–69.

Mervat, M. M. El Far, & Ashoub, A. (2009). Utility of thermotherapy and Meristem tip for freeing sweet potato from viral infection. *Australian Journal of Basic and Applied Sciences, 3*, 153–159.

Minas, G., Gregoriou, S., Kapari-Isaia, Th., & Papayiannis, L. (2007). Seed potatoes production in *Cyprus starting* from *in vitro* apical meristem. *Acta Horticulturae, 741*, 283–288.

Minas, G., Papayiannis, L., & Kapari-Isaia, Th. (2007). A rapid protocol for in vitro micro propagation of grapevine, cultivars and root stocks, starting from apical meristem. *Acta Horticulturae, 741*, 283–288.

Mink, G. I. (1998). Virus certification of deciduous fruit trees in the United States and Canada. In A. Hadidi, R. K. Khetarpal, & H. Koganezacbewa (Eds.), *Plant virus disease control* (pp. 294–300). St. Paul, MN: APS Press.

Mink, G. I., Wample, R., & Howell, W. E. (1998). Heat treatment of perennial plants to eliminate phytoplasmas, viruses and viroids while maintaining plant survival.

In A. Hadidi, R. H. Khetarpal, & H. Koganezawa (Eds.), *Plant virus disease control* (pp. 332–345). St. Paul, MN: APS Press.

Mneney, E. E., & Mantell, S. H. (2001). In vitro micrografting of cashew. *Plant Cell, Tissue and Organ Culture, 66*, 49–58.

Monis, J., & Bestwick, R. K. (1996). Detection and localization of grapevine leafroll associated closteroviruses in greenhouse and tissue culture grown plants. *American Journal of Enology and Viticulture, 47*, 199–205.

Mortimer-Jones, S. M., Jones, M. G. K., Jones, R. A. C., Thomson, G., & Dwyer, G. I. (2009). A single tube, quantitative real-time RT-PCR assay that detects four potato viruses simultaneously. *Journal of Virological Methods, 161*, 289–296.

Muhammad, I., Muhammad, A., Attique, A., & Sher, H. (2005). Elimination of citrus tristeza closterovirus from bud-wood through thermotherapy. *Pakistan Journal of Botany, 37*, 423–430.

Mukhopadhyay, S., Jaishree, R., Sharma, B. S., Gurung, A., Sengupta, R. K., & Nath, P. S. (1997). Micropropagation of Darjeeling orange (*Citrus reticulata* Blanco) by shoot-tip grafting. *Journal of Horticultural Science, 72*, 493–499.

Mulin, R. H., Smith, S. H., Frazier, N. W., Schlegel, D. E., & McCall, S. R. (1974). Meristem culture frees strawberries of Mild yellow edge, Pallidosis and Mottle diseases. *Phytopathology, 64*, 1425–1429.

Murashige, T., & Skoog, F. (1962). A revised medium for rapid growth and bioassays with tobacco tissue culture. *Physiologia Plantarum, 15*, 473–497.

NAPPO. (2013). *RSPM 5. Glossary of phytosanitary terms*. Ottawa, Canada: The Secretariat of the North American Plant Protection Organization.

Nassuth, A., Pollari, E., Helmeczy, K., Stewart, S., & Kofalvi, S. A. (2000). Improved RNA extraction and one-tube RT-PCR assay for simultaneous detection of control plant RNA plus several viruses in plant extracts. *Journal of Virological Methods, 90*, 37–49.

Navarro, L. (1988). Application of shoot-tip grafting *in vitro* to woody species. *Acta Horticulturae, 227*, 43–56.

Navarro, L., Civerolo, E. L., Juarez, J., & Garney,, S. M. (1991). Improving therapy methods for citrus germplasm exchange. In *Proceedings eleventh IOCV conference* (pp. 400–408). Riverside, CA: IOCV.

Navarro, L., Juarez, J., & Pina, J. A. (2001). Strategies and problems for the production of high quality nursery citrus trees worldwide. In *Proceedings of the sixteenth world congress of the international society of Citrus Nurserymen, Brazil* (pp. 1–10).

Navarro, L., Llacer, C., Cambra, M., Arregui, J., & Juarez, J. (1982). Shoot-tip grafting *in vitro* for elimination of viruses in peach plants (*Prunus persica* Batsch). *Acta Horticulturae, 130*, 185–192.

Navarro, L., Pina, J. A., Juárez, J., Ballester-Olmos, J. F., Arregui, J. M., Ortega, C., et al. (2002). The citrus variety improvement program in Spain in the period 1975–2001. In *Proceedings of the fifteenth IOCV conference* (pp. 306–316). Riverside, CA: IOCV.

Navarro, L., Roistacher, C. N., & Murashige, T. (1975). Improvement of shoot-tip grafting *in vitro* for virus free citrus. *Journal of the American Society for Horticultural Science, 100*, 471–479.

Navarro, L., Roistacher, C. N., & Murashige, T. (1976). Effect of size of shoot tips on psorosis-A and exocortis content of navel orange plants obtained by shoot-tip grafting *in vitro*. In *Proceedings of the seventh IOCV conference* (pp. 194–197), Riverside, CA: IOCV.

NCPN-FT. (2012). *State level model regulatory standard: Virus-tested certification program for* Prunus, Malus, Pyrus, Chaenomeles, *and* Cydonia *nursery stock production systems*. Prosser, Washington: NCPN-FT.

Neeraj, V., Raja, R., & Zaidi, A. A. (2005). *In vitro* production of *Prunus necrotic ringspot virus*-free begonias through chemo- and thermotherapy. *Scientia Horticulturae, 103*, 239–247.

Németh, M. (1984). *Virus, mycoplasma and rickettsia diseases of fruit trees*. Dordrecht: Martinus Nijhoff.
Nyland, G., & Goheen, A. C. (1969). Heat therapy of virus diseases of perennial plants. *Annual Review of Phytopathology*, *7*, 331–354.
OEPP/EPPO. (2006). EPPO Standards PM 4/17 (2) Schemes for the production of healthy plants for planting. Pathogen-tested olive trees and rootstocks. *Bulletin OEPP/EPPO Bulletin*, *36*, 77–83.
Olmos, A., Cambra, M., Esteban, O., Gorris, M. T., & Terrada, E. (1999). New device and method for capture, reverse transcription and nested PCR in a single closed-tube. *Nucleic Acids Research*, *27*, 1564–1565.
Osman, F., Hodzic, E., Omanska-Klusek, A., Olineka, T., & Rowhani, A. (2013). Development and validation of a multiplex quantitative PCR assay for the rapid detection of *Grapevine virus A, B and D*. *Journal of Virological Methods*, *194*, 138–145.
Paduch-Cichal, E., & Kryczyński, S. (1987). A low temperature therapy and meristem-tip culture for eliminating four viroids from infected plants. *Journal of Phytopathology*, *118*, 341–346.
Panattoni, F., D'Anna, E., & Triolo, A. (2007). Antiviral activity of tiazofurin and mycophenolic acid against *Grapevine Leafroll-associated Virus 3* in *Vitis vinifera* explants. *Antiviral Research*, *73*, 206–211.
Panattoni, A., Luvisi, A., & Triolo, E. (2013). Elimination of viruses in plants: Twenty years of progress. *Spanish Journal of Agricultural Research*, *11*, 173–188.
Parmessur, Y., Aljanabi, S., Saumtally, S., & Dookun-Saumtally, A. (2002). *Sugarcane yellow leaf virus* and sugarcane yellows phytoplasma: Elimination by tissue culture. *Plant Pathology*, *51*, 561–566.
Pina, J. A., Chomé, P., Vives, M. C., & Navarro, L. (2012). The citrus nursery tree certification program in Spain. In *Book of abstracts XII international citrus congress, Valencia, Spain* (p. 250).
Postman, J. D., & Hadidi, A. (1995). Elimination of *Apple scar skin viroid* from pears by *in vitro* thermotherapy and apical meristem culture. *Acta Horticulturae*, *386*, 536–543.
Raharjo, S. H. T., & Litz, R. E. (2005). Micrografting and *ex vitro* grafting for somatic embryo rescue and plant recovery in avocado (*Persea americana*). *Plant Cell, Tissue and Organ Culture*, *82*, 1–9.
Reed, B. M. (2008). Cryopreservation-practical considerations. In B. M. Reed (Ed.), *Plant cryopreservation: A practical guide* (pp. 3–13). New York: Springer.
Reed, P. J., & Foster, J. A. (2011). Exclusion of pome and stone fruit viruses, viroids and phytoplasmas by certification and quarantine. In A. Hadidi, M. Barba, T. Candresse, & W. Jelkmann (Eds.), *Virus and virus like diseases of pome and stone fruits* (pp. 381–388). St. Paul, MN: APS Press.
Rizgi, A., Zemzami, M., & Spiegel, S. (2001). Recovery of virus-free almond plants by improved *in vitro* shoot-tip grafting. *Acta Horticulturae*, *550*, 447–453.
Roistacher, C. N. (1998). Indexing for viruses in citrus. In A. Hadidi, R. K. Khetarpal, & H. Koganezawa (Eds.), *Plant virus disease control* (pp. 301–319). St. Paul, MN: APS Press.
Rowhani, A., Chay, C., Golino, D. A., & Falk, B. W. (1993). Development of a polymerase chain reaction technique for the detection of grapevine fanleaf virus in grapevine tissue. *Phytopathology*, *83*, 749–753.
Rowhani, A., Uyemoto, J. K., Golino, D. A., & Martelli, G. P. (2005). Pathogen testing and certification of *Vitis* and *Prunus* species. *Annual Review of Phytopathology*, *43*, 261–278.
Roy, A. S. (2011). Control measures of pome and stone fruit viruses, viroids and phytoplasmas: Role of international organizations. In A. Hadidi, M. Barba, T. Candresse, & W. Jelkmann (Eds.), *Virus and virus like diseases of pome and stone fruits* (pp. 407–413). St. Paul, MN: APS Press.

Ruiz-Ruiz, S., Moreno, P., Guerri, J., & Ambrós, S. (2007). A real-time RT-PCR assay for detection and absolute quantitation of *Citrus tristeza virus* in different plant tissues. *Journal of Virological Methods, 145*, 96–105.

Salazar, L. F. (1996). *Potato viruses and their control.* Lima, Peru: International Potato Center.

Salem, N., Mansour, A., Al-Musa, A., & Al-Nsour, A. (2003). Seasonal variation of *Prunus necrotic ringspot virus* concentration in almond, peach, and plum cultivars. *Phytopathologia Mediterranea, 42*, 155–160.

Sanjeev, S., Balwinder, S., Gita, R., Aijar Asghar, Z., Vipin, H., Avinash, N., et al. (2007). Production of Indian Citrus ringspot virus free plants of kinnow employing chemotherapy coupled with shoot tip grafting. *Journal of Central European Agriculture, 8*, 1–8.

Savino, V. (1992). Certification of grapevine in Italy. In G. P. Martelli (Ed.), *Grapevine viruses and certification in EEC countries: State of the art* (pp. 55–65).Bari, Italy: CIHEAM, Quaderno n. 3.

Sebastien, M., Olmos, A., Jijakli, H., & Candresse, T. (2014). Current impact and future directions of high throughput sequencing in plant virus diagnostics. *Virus Research, 188*, 90–96.

Sedlak, J., Paprstein, F., & Talacko, L. (2011). Elimination of *Apple stem pitting virus* from pear cultivars by *in vitro* chemotherapy. *Acta Horticulturae, 923*, 111–115.

Sefc, K. M., Leonhardt, W., & Steinkellner, H. (2000). Partial sequence identification of Grapevine-leafroll-associated virus-1 and development of a highly sensitive IC-RT-PCR detection method. *Journal of Virological Methods, 86*, 101–106.

Singh, I. P. (2001). Shoot-tip grafting of Mosambi (*C. sinensis* L. Osbeck) in India. In *Proceedings sixteenth world congress of the international society of Citrus Nurserymen, Brazil* (pp. 128–130).

Skiada, F. G., Maliogka, V. I., Katis, N. I., & Eleftheriou, E. P. (2013). Elimination of *Grapevine rupestris stem pitting-associated virus* (GRSPaV) from two *Vitis vinifera* cultivars by *in vitro* chemotherapy. *European Journal of Plant Pathology, 135*, 407–414.

Slack, S. A., & Singh, R. P. (1998). Control of viruses affecting potatoes through seed potato certification programs. In A. Hadidi, R. K. Khetarpal, & H. Koganezawa (Eds.), *Plant virus disease control* (pp. 249–260). St. Paul, MN: APS Press.

Spiegel, S. (1998). Virus certification of strawberries. In A. Hadidi, R. K. Khetarpal, & H. Koganezawa (Eds.), *Plant virus disease control* (pp. 320–324). St. Paul, MN: APS Press.

Šubr, Z., & Glasa, M. (2013). Unfolding the secrets of *Plum pox virus*: From epidemiology to genomics. *Acta Virologica, 57*, 217–228.

Tan, R., Wang, L., & Hong, N. (2010). Enhanced efficiency of virus eradication following thermotherapy of shoot-tip cultures of pear. *Plant Cell, Tissue and Organ Culture, 101*, 229–235.

Torrance, L., & Jones, R. A. C. (1981). Recent developments in serological methods suited for use in routine testing for plant viruses. *Plant Pathology, 30*, 1–24.

Tsai, C. W., Daugherty, M. P., & Almeida, R. P. P. (2012). Seasonal dynamics and virus translocation of *Grapevine leafroll associated virus 3* in grapevine cultivars. *Plant Pathology, 61*, 977–985.

Ucman, R., Žel, J., & Ravnikar, M. (1998). Thermotherapy in virus elimination from garlic: Influences on shoot multiplication from meristems and bulb formation *in vitro*. *Scientia Horticulturae, 73*, 193–202.

Uyemoto, J. K., Luhn, C. F., Asai, W. K., Beede, R., & Beutel, J. A. (1989). Incidence of ilarviruses in young peach trees in California. *Plant Disease, 73*, 217–220.

Varveri, C., Holeva, R., & Bem, F. P. (1997). Effect of sampling time and plant part on the detection of two viruses in apricot and one in almond by ELISA. *Annales de l' Institut Phytopathologique Benaki, 18*, 25–33.

Vives, M. C., Velázquez, K., Pina, J. A., Moreno, P., Guerri, J., & Navarro, L. (2013). Identification of a new enamovirus associated with citrus vein enation disease by deep sequencing of small RNAs. *Phytopathology, 103*, 1077–1086.

Volk, G. M., Bonnart, R., Krueger, R., & Lee, R. (2012). Cryopreservation of citrus shoot tips using micrografting and recovery. *Cryoletters, 33*, 418–426.

Walkey, D. G. A., & Cooper, V. C. (1975). Effect of temperature on virus eradication and growth of infected tissue cultures. *Annals of Applied Biology, 80*, 185–190.

Wambugu, F. M. (1985). Eradication of potato virus Y and S from potato by chemotherapy of cultured axillary bud tips. *American Potato Journal, 62*, 667–672.

Wang, Q., Cuellar, W., Rajamaki, M. L., Hirata, Y., & Valkonen, J. P. T. (2008). Combined thermotherapy and cryotherapy for efficient virus eradication: Relation of virus distribution, subcellular changes, cell survival and viral RNA degradation in shoot tips. *Molecular Plant Pathology, 9*, 237–250.

Wang, Q. C., Liu, Y., He, W., Xie, Y. H., & You, M. S. (2006). Cryotherapy of potato shoot tips for efficient elimination of *Potato Leafroll Virus* (PLRV) and *Potato Virus Y* (PVY). *Potato Research, 49*, 119–129.

Wang, Q., Mawassi, M., Li, P., Gafny, R., Sela, I., & Tanne, E. (2003). Elimination of *Grapevine virus A (GVA)* by cryopreservation of *in vitro*-grown shoot tips of *Vitis vinifera* L. *Plant Science, 165*, 321–327.

Wang, Q., Panis, B., Engelmann, F., Lambardi, M., & Valkonen, J. P. T. (2009). Cryotherapy of shoot tips: A technique for pathogen elimination to produce healthy planting materials and prepare healthy plant genetic resources for cryopreservation. *Annals of Applied Biology, 154*, 351–363.

Wang, L., Wang, W., Hong, N., Tang, R., & Teng, X. (2006). Effect of thermotherapy on elimination of *Apple stem grooming virus* and *Apple chlorotic leaf spot virus* for *in vitro* cultured pear shoot-tips. *HortScience, 41*, 729–732.

Waterworth, H. E. (1998). Certification for plant viruses—An overview. In A. Hadidi, R. K. Khetarpal, & H. Koganezawa (Eds.), *Plant virus disease control* (pp. 325–331). St. Paul, MN: APS Press.

Weiland, C. M., Cantos, M., Troncoso, A., & Perez-Camacho, F. (2004). Regeneration of virus-free plants by *in vitro* chemotherapy of GFLV (*Grapevine Fanleaf Virus*) infected explants of *Vitis vinifera* cv "Zalema". *Acta Horticulturae, 652*, 463–466.

Yildirim, H., Onay, A., Suzere, I. V., Tilkat, E., Ozden-Tokatli, Y., & Akdemir, H. (2010). Micrografting of almonds (*Prunus dulcis* Mill) cultivars "Ferragnes" and "Ferraduel". *Scientia Horticulturae, 125*, 361–367.

Zhang, Y., Singh, K., Kaur, R., & Qiu, W. (2011). Association of a novel DNA virus with the grapevine vein-clearing and vine decline syndrome. *Phytopathology, 101*, 1081–1090.

Zilka, S., Faingersh, E., Rotbaum, A., Tam, Y., Spiegel, S., & Malca, N. (2002). *In vitro* production of virus-free pear plants. *Acta Horticulturae, 596*, 477–479.

CHAPTER TWO

Control of Sweet Potato Virus Diseases

Gad Loebenstein*
Department of Plant Pathology, Agricultural Research Organization, Bet Dagan, Israel
*Corresponding author: e-mail address: gad-talma@barak.net.il

Contents

Abstract

Sweet potato (*Ipomoea batatas*) is ranked seventh in global food crop production and is the third most important root crop after potato and cassava. Sweet potatoes are vegetative propagated from vines, root slips (sprouts), or tubers. Therefore, virus diseases can be a major constrain, reducing yields markedly, often more than 50%. The main viruses worldwide are *Sweet potato feathery mottle virus* (SPFMV) and *Sweet potato chlorotic stunt virus* (SPCSV). Effects on yields by SPFMV or SPCSV alone are minor, or but in complex infection by the two or other viruses yield losses of 50%. The orthodox way of controlling viruses in vegetative propagated crops is by supplying the growers with virus-tested planting material. High-yielding plants are tested for freedom of viruses by PCR, serology, and grafting to sweet potato virus indicator plants. After this, meristem tips are taken from those plants that reacted negative. The meristems were grown into plants which were kept under insect proof conditions and away from other sweet potato material for distribution to farmers after another cycle of reproduction.

Advances in Virus Research, Volume 91
ISSN 0065-3527
http://dx.doi.org/10.1016/bs.aivir.2014.10.005

1. INTRODUCTION

Sweet potato (*Ipomoea batatas*) is ranked seventh in global food crop production and is the third most important root crop after potato and cassava. They are grown on about 8.1 million hectares, yielding ca. 131 million tons, with an average yield of about 15 t/ha (FAOSTAT, 2011). They are mainly grown in developing countries, which account for over 95% of world output. The cultivated area of sweet potato in China, about 3.7 million hectares, accounted for 70% of the total area of sweet potato cultivation in the world. China produces about 80 million tons, ca. 46% of the total world production. Vietnam is the second largest producer. Sweet potato is a "poor man's crop," with most of the production done on a small or subsistence level. Sweet potato produces more biomass and nutrients per hectare than any other food crop in the world. Thus, for example, across East Africa's semiarid, densely populated plains, thousands of villages depend on sweet potato for food security and the Japanese used it when typhoons demolished their rice fields. Sweet potato is grown for both the leaves, which are used as greens, and the tubers, for a high carbohydrate and beta-carotene source. Yields differ greatly in different areas or even fields in the same location. Thus, the average yield in African countries is about 4.7 t/ha, with yields of 9.1, 4.5, 1.9, and 2.9 t/ha in Kenya, Uganda, Sierra Leone, and Nigeria, respectively. The yields in Asia are significantly higher, averaging 20.0 t/ha. China, Japan, Korea, and Israel have the highest yields with about 22.0, 21.7, 15.6, and 33.3 t/ha, respectively. In South America, the average yield is 12.3 t/ha, with Argentina, Peru, and Uruguay in the lead with 14, 16.8, and 10.9 t/ha, respectively. For comparison, the average yield in the United States is 22.8 t/ha (FAOSTAT, 2012).

These differences in yields are mainly due to variation in quality of the propagation material. Sweet potatoes are vegetative propagated from vines, root slips (sprouts), or tubers, and farmers in African and other countries often take vines for propagation from their own fields year after year. Thus, if virus diseases are present in the field they will inevitable be transmitted with the propagation material to the newly planted field, resulting in a decreased yield. Often these fields are infected with several viruses, thereby compounding the effect on yields. In China, on average, losses of over 20% due to sweet potato virus diseases (SPVDs) are observed (Gao, Gong, & Zhang, 2000), mainly due to *Sweet potato feathery mottle virus* (SPFMV) and *Sweet potato latent virus* (SPLV). The infection rate in the Shandong province reaches 5–41% (Shang et al., 1999). In countries were care is taken to provide virus-tested planting material as, among others in the United States and Israel, yields increase markedly, up to seven times and more.

2. THE MAIN VIRUSES

2.1. *Sweet potato feathery mottle virus* Genus *Potyvirus*

Sweet potato feathery mottle virus Genus *Potyvirus* (SPFMV) is the most common sweet potato virus worldwide. Certain isolates in the United States cause much economic damage by inducing cracking or internal corkiness in some cultivars. In Africa, SPFMV causes a SPVD in a complex infection with the whitefly-transmitted *Sweet potato sunken vein virus* Genus *Crinivirus* (SPSVV) [synonym: *Sweet potato chlorotic stunt* Genus *Crinivirus* (SPCSV)].

Most sweet potato cultivars infected by SPFMV alone show only mild circular spots on their leaves or light green patterns along veins. However, when infected together with the whitefly-transmitted SPSVV stunting of the plants, feathery vein clearing, and yellowing of the plants are observed. In controlled experiments, SPFMV-infection alone did not reduce yields compared to virus-free controls, while the complex infection with SPCSV reduced yields by 50% or more SPFMV is transmitted in a nonpersistent manner by aphids, including *Aphis gossypii*, *Myzus persicae*, *A. craccivora*, and *Lipaphis erysimi*. The virus can be transmitted mechanically to various *Ipomoea sp.*, as *I. batatas*, *I. setosa*, *I. nil*, *I. incarnata*, and *I. purpurea*, and *Nicotiana benthamiana* and *Chenopodium amaranticolor* (for some strains). The virus is transmitted by grafting but not by seed or pollen or by contact between plants. The virus can best be diagnosed by grafting on *I. setosa*, causing vein clearing or on *I. incarnata* and *I. nil* inducing systemic vein clearing, vein banding, and ringspots. SPFMV can be diagnosed by ELISA, and antisera are commercially available. However, ELISA reliably detects SPFMV only in leaves with symptoms. It is best to sample several leaves from a plant, as the virus seems to be unevenly distributed. SPFMV can also be detected together with three other sweet potato viruses by multiple one-step reverse transcription-PCR (Li et al., 2012).

East Africa (EA) appears as a hotspot for evolution and diversification of SPFMV (Tugume, Settumba Mukasa, & Omongo, 2010).

Virions are filamentous, not enveloped, usually flexuous, and with a modal length of 830–850 nm. The genome consists of single-stranded linear RNA, with a poly(A) region. Many strains of SPFMV (Moyer, 1986), isolates, variants, and serotypes of SPFMV have been reported By comparing coat protein (CP) gene sequences of isolates, it was shown that isolates from EA form a separate cluster (Kreuze, Karyeija, Gibson, & Valkonen, 2000). The complete nucleotide sequence of a sweet potato feathery mottle virus severe strain (SPFMV-S) genomic RNA was determined from overlapping cDNA clones and by directly sequencing viral RNA. The viral RNA genome is 10,820 nucleotides long, excluding the poly(A) tail and contains one open reading frame starting at nucleotide 118 and ending at 10,599, potentially encoding a polyprotein of 3493 amino acids (Mr 393,800) (Sakai et al., 1997). Though SPFMV alone generally causes only minor damage, its control is imperative as in combination with other viruses its effect on plant growth and yields may become substantial.

2.2. *Sweet potato chlorotic stunt virus* Genus *Crinivirus*

Infection of sweet potato by *Sweet potato chlorotic stunt virus* Genus *Crinivirus* (SPCSV; possible synonym: *Sweet potato sunken vein virus*, SPSVV) alone produced on cv. Georgia Jet mild symptoms consisting of slight yellowing of veins, with some sunken secondary veins on the upper sides of the leaves and swollen veins on their lower sides. Upward rolling of the three to five distal leaves was also observed. Effects on yields by SPSVV or SPCSV alone are minor or but in complex infection with SPFMV or other viruses yield losses of 50% and more are observed (Milgram, Cohen, & Loebenstein, 1996; Untiveros, Fuentes, & Salazar, 2007). This was concurrent with a significant reduction in chlorophyll content (Njeru et al., 2004).

SPCSV and/or SPSVV are transmitted by the whitefly *Bemisia tabaci* biotype B, *Trialeurodes abutilonea*, and *B. afer* (Gamarra et al., 2010; Ng & Falk, 2006; Schaefers & Terry, 1976; Sheffield, 1957; Sim, Valverde, & Clark, 2000; Valverde, Sim, & Lotrakul, 2004) in a semipersistent manner, requiring at least 1 h for acquisition and infection feeding and reaching a maximum after 24 h for both of them. The virus is graft transmissible, but not by mechanical inoculation. The virus was transmitted by whiteflies to *I. setosa*, *N. clevelandii*, *N. benthamiana*, and *Amaranthus palmeri* from *I. setosa* and by grafting to other Ipomoeas.

The virus is best being diagnosed on a pair of sweet potato plants—one healthy, the other infected by SPFMV. On the healthy plants hardly any symptoms will become apparent, while (if carrying SPFMV) severe symptoms of SPVD will appear. Diagnosing SPSVV (or probably also SPCSV) by PCR can be erratic as the virus is not distributed evenly in the plant, and sampling only a limited number of small tissue pieces may not enable detection of the virus.

The complete nucleotide sequences of genomic RNA1 (9407 nts) and RNA2 (8223 nts) were determined, revealing that SPCSV possesses the second largest identified positive-strand single-stranded RNA genome among plant viruses after *Citrus tristeza virus* (Kreuze, Savenkov, & Valkonen, 2002). The complete genome sequences of two SPCSV isolates from China was also determined (Qin et al., 2013) The virus has been reported from East and Southern Africa, Nigeria, Niger, Indonesia, Israel, Egypt, Spain, Argentina, Brazil, Peru (Karyeija, Gibson, & Valkonen, 1998), and from the United States (Abad et al., 2007).

SPVD is caused by the interaction of SPFMV and SPCSV/SPSVV. Characteristic symptoms of the disease include vein clearing, chlorosis, and stunting. The disease was described by Schaefers and Terry (1976) in Nigeria and is the most important virus (complex) disease in EA, where sweet potato is often the main food staple (Karyeija et al., 1998). The disease was described in Israel by (Loebenstein & Harpaz,1960), the United States (Abad et al., 2007), Spain (Trenado, Lozano, Valverde, & Navas-Castillo, 2007), and occurs probably in Italy (Parrella, De Stradis, & Giorgini, 2006).

It can cause losses over 50%, especially in Uganda and Kenya, though in another study from Uganda, losses were much smaller, probably due to relatively high levels of virus resistance in their landraces (Gibson, Mawanga, Kasule, Mpembe, & Carey, 1997).

The sequence of infections of the causal agents of SPVD effects symptom severity and virus titters (Mcgregor et al., 2009). Symptoms were significantly more severe in plants infected with SPCSV followed by SPFMV compared to plants infected with SPFMV followed by SPCSV. Virus titers were not significantly different for SPCSV, but SPFMV titers, in plants infected with SPCSV followed by SPFMV, were significantly higher than all other treatments.

2.3. *Sweet potato mild mottle virus* Genus *Ipomovirus*

Sweet potato mild mottle virus Genus *Ipomovirus* (SPMMV, synonym: Sweet potato B virus, Sheffield, 1957) has so far been reported *inter alia* from West- and South Africa, Indonesia, China, Philippines, India, New Zealand, and Egypt.

SPMMV can cause leaf mottling, stunting, and loss of yields. Cultivars differ greatly in their reaction to the virus, some being symptomlessly infected, others apparently are immune. The virus is transmitted semi-persistently by *B. tabaci*, by grafting and by mechanical inoculation. It is not transmitted by seed or by contact between plants. The virus was transmitted to plants in 14 families (Mcgregor et al., 2009) Diagnostic species:

N. tabacum, *N. glutinosa*—vein clearing, leaf puckering, mottling, and distortion; *C. quinoa*—local lesions, not systemic; *I. setosa*—conspicuous systemic vein chlorosis. Diagnosis can be confirmed by serological tests of *I. setosa*. Commercial antisera are available. After 3–4 weeks, new growth is almost symptomless. Sap transmission from sweet potato to test plants is often difficult. The virus is best maintained in *N. glutinosa*, *N. clevelandii, and N. tabacum*, and can be assayed quantitatively on *C. quinoa*, where SPMMV induces local lesions.

The virus can be purified from systemically infected *N. tabacum* (Hollings, Stone, & Bock, 1976). Virions are flexous rod-shaped particles, 800–950 nm in length, containing 5% RNA and 95% protein. The genome consists of single-stranded RNA. The viral RNA was cloned and the assembled genomic sequence was 10,818 nts in length with a polyadenylated tract at the 3-terminus. The sequence accession code is Z73124. Almost all known potyvirus motifs are present in the polyprotein of SPMMV, except some motifs in the putative helper-component and CP, which are incomplete or missing. This may account for its vector relations (Colinet, Nguyen, Kummert, Lepoivre, & Xia, 1998). The CP has a MW of 37,700. A synergism was observed in sweet potato doubly infected by SPMMV and SPCSV (but not by SPFMV) (Untiveros et al., 2007). Mukasa, Rubaihayo, & Valkonen (2006) showed that SPMMV titers increased approximately 1000-fold.

2.4. *Sweet potato latent virus* Genus *Potyvirus*

Sweet potato latent virus Genus *Potyvirus* (SPLV) is widespread in China and has been reported also from Egypt (research.cip.cgiar.org—without published records). SPLV may cause mild chlorosis but in most cultivars the infection is symptomless.

Symptoms often disappear after infection, but the plants remain infected. Crystal inclusions are observed in the nucleus and pinwheels in the cytoplasm. SPLV isolates from Japan and China were transmitted by the aphid *Myzus persicae* (Usugi, Nakano, Akira, & Hayashi, 1991) and the virus can be transmitted by mechanical inoculation and by grafting. It is not transmitted by seed. SPLV in the presence of SPCSV causes a synergistic disease (Untiveros et al., 2007).

Diagnostic species: *N. benthamiana*—systemic mosaic and stunting; *N. clevelandii*—systemic pin-prick chlorotic lesions; *C. quinoa*, *C. amaranticolor*—brown necrotic local lesions, no systemic infection; *I. setosa*—systemic mottle. Diagnosis can be confirmed by serological tests of *I. setosa*.

The virus is best maintained in *N. benthamiana* or *N. clevelandii* and can be assayed on *C. quinoa* or *C. amaranticolor.*

The virus can be purified according to Liao, Chien, Chung, Chiu, and Han (1979). Virus particles are flexuous rods, 750–790 nm in length. The capsid protein has a MW of 36,000. By using MAbs and polyclonal antibodies some epitopes common to SPLV and SPFMV are found. These can easily be differentiated when potyvirus cross-reactive MAbs are used, indicating a distant relationship (Hammond, Jordan, Larsen, & Moyer, 1992).

Combining RT-PCR with degenerate oligonucleotide primers derived from the conserved regions of potyviruses, it was possible to identify SPLV, as well as SPFMV and *Sweet potato G virus* (Colinet et al., 1998), and to differentiate between two strains of SPLV (Colinet, Kummert, & Lepoivre, 1997).

Sequence data accession codes are: X84011 SPLV (Chinese) mRNA for CP; X84012 SPLV (Taiwan) mRNA for CP. According to Nishiguchi et al. (2001), SPLV has 58% homology to SPFMV-S.

The best way to control this virus, as well as other viruses infecting sweet potato is by establishing propagation nurseries derived from virus-tested mother plants.

2.5. *Sweet potato leaf curl virus* Genus *Begomovirus*

Sweet potato leaf curl virus Genus *Begomovirus* (SPLCV) has been reported *inter alia* from the United States, Taiwan, Korea, Argentina, and Japan. Infected plants show upward leaf curling. The disease can be diagnosed on *I. nil* and *I. muricata* and by real-time PCR. Comparative sequence analysis from 11 isolates sampled from the United States, Brazil, China, Korea, and Spain were between 65% and 93.6% identical (Pardina et al., 2012). The virus is transmitted by *B. tabaci* in a persistent manner (ICTVdB Management, 2006).

3. TRANSGENIC APPROACHES TO CONTROL THE VIRUSES IN SWEET POTATO

The orthodox way of controlling viruses in vegetative propagated crops is by supplying the growers with virus tested planting material (see below). However, in African countries this is almost impractical because most sweet potatoes are grown by smallholder's poor farmers. Some varieties of sweet potato are resistant to some viruses. Thus, the landrace sweet potato variety "Huachano" is extremely resistant to SPFMV (Kreuze et al., 2008).

However, no cultivar is resistant to SPVD. Several attempts were made to induce resistance by transgenic approaches.

An attempt to transform sweet potato for obtaining resistance to SPFMV using the rice cysteine-inhibitor gene, which inhibits the proteolysis of a polyprotein in potyviruses, thus interfering with their replication, was reported by Cipriani et al. (2001). Improved resistance to SPFMV was observed in 18 of the 25 transgenic lines.

In Japan, sweet potato were transformed with the CP-encoding sequence of SPFMV and found to be resistant to SPFMV following experimental inoculation by grafting (Okada et al., 2001, 2002). These transgenic plants were also resistant when graft inoculated with different field isolates of SPFMV (Okada & Saito, 2008). The *CP* gene was inherited by the next generation confirming the stability of viral resistance for at least one generation (Okada et al., 2006). However, resistance that works under controlled conditions may not necessarily work in the field when inoculated by vectors.

With financial assistance from USAID/ABSP, a collaborative research project between Kenya Agricultural Research Institute (KARI) and Monsanto was launched in 1991 to develop a virus SPFMV-resistant sweet potato, using pathogen-derived resistance (PDR). However, their resistance broke down in EA (New Scientist, February 7, 2004, p. 7) possibly because the transgene was not from a locally prevalent SPFMV strain or because the transgenes still carried a small amount of virus, or because the plants became infected with SPCSV. For these reasons, the commonly encountered mixed virus infections in the field and the genetic variability of sweet potato viruses possess an important challenge that needs to be met before sustainable virus resistance can be obtained (Tairo et al., 2005).

When a landrace sweet potato variety Huachano that is extremely resistant to SPFMV was genetically engineered for resistance to SPCSV (Kreuze et al., 2008). In this cultivar as in many others (Karyeija, Kreuze, Gibson, & Valkonen, 2000) the high levels of resistance to SPFMV breaks down following infection with SPCSV and the plants succumb to the severe SPVD. This exemplifies how important the resistance to SPCSV would be in order to protect sweet potatoes against SPVD and other severe synergistic diseases induced by SPCSV with other viruses. The transgene was designed to express an SPCSV-homologous transcript that forms a double-stranded structure and hence efficiently primes virus-specific resistance. Many transgenic lines accumulated only low concentrations of SPCSV following infection and no symptoms developed.

These results showed that sweet potato could be protected against the disease caused by SPCSV using PDR. However, the low concentrations of SPCSV in the transgenic plants were still sufficient to break down the natural high levels of resistance to SPFMV. Apparently, complete immunity to SPCSV seems to be required for prevention of the severe virus diseases in sweet potato.

3.1. The orthodox approach for control

The orthodox way of controlling viruses in vegetative propagated crops is by supplying the growers with virus tested planting material. High yielding plants are tested for freedom of viruses by PCR, serology, and grafting to sweet potato virus indicator plants as *I. setosa*. After this meristem tips are taken from those plants that reacted negative. The meristems were grown into plants which were kept under insect proof conditions and away from other sweet potato material. The resulting plants have to be tested carefully. It is not enough to test them by PCR as the virus often is not distributed evenly in the plant. We preferred to test SPFMV by grafting on *I. setosa*, causing vein clearing (Fig. 1) followed by remission, or on *I. incarnata*- and *I. nil*-inducing systemic vein clearing, vein banding, and ringspots. SPFMV can be diagnosed by ELISA, and antisera are commercially available. However, ELISA reliably detects SPFMV only in leaves with symptoms and when co-infected with SPCSV (Gutiérrez, Fuentes, & Salazar, 2003). SPCSV can best be diagnosed on a pair of sweet potato plants—one healthy, the other infected by SPFMV. On the healthy plants hardly any symptoms will become apparent, while (if carrying SPFMV) severe symptoms of SPVD will appear. The virus can also be diagnosed by immunosorbent electron microscopy and by ELISA. The plants are then grown under insect-proof

Figure 1 *I. setosa* infected with *Sweet potato feathery mottle virus* (SPFMV), showing vein clearing. (See the color plate.)

Figure 2 Nursery of virus-tested stocks. (See the color plate.)

(Fig. 2) conditions and propagation is continued by cuttings. The farmers plant a certain number of vines in the open field and continue to propagate until the stand of the field is complete. The following year they will again start from virus-tested plants. Such programs are operating in Israel and in the Shandong province of China (Gao et al., 2000). This scheme when rigorously applied in Israel resulted during 1985–2000 in yields of 45–60 t/ha, a yield increase of at least 100%, while in China increases ranged between 22% and 92%. The payoff to the farmers has been high and in China the use of pathogen-free material is being extended. In Israel, use of certified material was common practice until 2000. However, afterward some farmers started to use planting material from their own fields and nurseries did not supply high-quality plantlets, both resulting in lower yields.

REFERENCES

Abad, J. A., Parks, E. J., New, S. L., Fuentes, S., Jesper, W., & Moyer, J. W. (2007). First report of *sweetpotato chlorotic stunt virus*, a component of sweetpotato virus disease, in North Carolina. *Plant Disease, 91*, 327.

Cipriani, S., Fuentes, G., Bello, V., Salazar, L. F., Ghislain, M., & Zhang, D. P. (2001). *Transgene expression of rice cysteine proteinase inhibitors for the development of resistance against* sweetpotato feathery mottle virus: *CIP program report 1999–2000* (pp. 267–271).

Colinet, D., Kummert, J., & Lepoivre, P. (1997). Evidence for the assignment of two strains of SPLV to the genus potyvirus based on coat protein and 3' non-coding region sequence data. *Virus Research, 49*, 91–100.

Colinet, D., Nguyen, M., Kummert, J., Lepoivre, P., & Xia, F. Z. (1998). Differentiation among potyviruses infecting sweetpotato based on genus- and virus-specific reverse transcription polymerase chain reaction. *Plant Disease, 82*, 223–229.

FAOSTAT. (2011). *FAO Statistical Databases*. http://apps.fao.org/.

FAOSTAT. (2012). *FAO Statistical Databases*. http://apps.fao.org/.

Gamarra, H. A., Fuentes, S., Morales, F. J., Glover, R., Malumphy, C., & Barker, I. (2010). *Bemisia afer* sensu lato, a vector of sweet potato chlorotic stunt virus. *Plant Disease, 94*, 510–514.

Gao, F., Gong, Y. F., & Zhang, P. B. (2000). Production and development of virus-free sweet potato in China. *Crop Protection, 19*, 105–111.

Gibson, R. W., Mawanga, R. O. M., Kasule, S., Mpembe, I., & Carey, E. E. (1997). Apparent absence of viruses in most symptomless field-grown sweetpotato in Uganda. *Annals of Applied Biology, 130*, 481–490.

Gutiérrez, D. L., Fuentes, S., & Salazar, L. F. (2003). Sweetpotato virus disease (SPVD): Distribution, incidence, and effect on sweetpotato yield in Peru. *Plant Disease, 87*, 297–302.

Hammond, J., Jordan, R. L., Larsen, R. C., & Moyer, J. W. (1992). Use of monoclonal antisera and monoclonal antibodies to examine serological relationships among three filamentous viruses of sweetpotato. *Phytopathology, 82*, 713–717.

Hollings, M., Stone, O. M., & Bock, K. R. (1976). Purification and properties of sweetpotato mild mottle virus, a whitefly-borne virus from sweetpotato (Ipomoea batatas) in East Africa. *Annals of Applied Biology, 82*, 511–528.

ICTVdB Management (2006). Sweet potato leaf curl virus. In C. Büchen-Osmond (Ed.), *ICTVdB—The universal virus database, version 4*. New York, USA: Columbia University.

Karyeija, R. F., Gibson, R. W., & Valkonen, J. P. T. (1998). Resistance in sweetpotato virus disease (SPVD) in wild East African Ipomoea. *Annals of Applied Biology, 133*, 39–44.

Karyeija, R. F., Kreuze, J. F., Gibson, R. W., & Valkonen, J. P. T. (2000). Two serotypes of *sweetpotato feathery mottle virus* in Uganda and their interaction with resistant sweetpotato cultivars. *Phytopathology, 90*, 1250–1255.

Kreuze, J. F., Karyeija, R. F., Gibson, R. W., & Valkonen, J. P. T. (2000). Comparisons of coat protein gene sequences show that East African isolates of sweetpotato feathery mottle virus form a genetically distinct group. *Archives of Virology, 145*, 567–574.

Kreuze, J. F., Samolski, I., Untiveros, M., Cuellar, W. J., Lajo, G., Cipriani, P. G., et al. (2008). RNA silencing mediated resistance to a crinivirus (*closteroviridae*) in cultivated sweetpotato (*Ipomoea batatas* L.) and development of sweetpotato virus disease following co-infection with a potyvirus. *Molecular Plant Pathology, 9*, 589–598.

Kreuze, J. F., Savenkov, E. I., & Valkonen, J. P. T. (2002). Complete genome sequence and analyses of the subgenomic RNAs of sweetpotato chlorotic stunt virus reveal several new features for the genus Crinivirus. *Journal of Virology, 76*, 9260–9270.

Li, F., Zuo, R., Abad, J., Xu, D., Bao, G., & Li, R. (2012). Simultaneous detection and differentiation of four closely related sweet potato potyviruses by a multiplex one-step RT-PCR. *Journal of Virological Methods, 186*, 161–166.

Liao, C. H., Chien, K., Chung, M. L., Chiu, R. J., & Han, Y. H. (1979). A study of a sweetpotato virus disease in Taiwan. I. Sweetpotato yellow spot virus disease. *Journal of Agricultural Research of China, 28*, 127–137.

Loebenstein, G., & Harpaz, I. (1960). Virus diseases of sweetpotatoes in Israel. *Phytopathology, 50*, 100–104.

McGregor, C. E., Miano, D. W., LaBonte, D. R., Hoy, M., Clark, C. A., & Rosa, G. J. M. (2009). Differential gene expression of resistant and susceptible sweetpotato plants after infection with the causal agents of sweetpotato virus disease. *Journal of the American Society of Horticultural Science, 134*, 658–666.

Milgram, M., Cohen, J., & Loebenstein, G. (1996). Effects of sweetpotato feathery mottle virus and sweetpotato sunken vein virus on sweetpotato yields and rate of reinfection on virus-free planting material in Israel. *Phytoparasitica, 24*, 189–193.

Moyer, J. F. (1986). Variability among strains of sweetpotato feathery mottle virus. *Phytopathology, 76*, 1126 (Abstr.).
Mukasa, S. B., Rubaihayo, P. R., & Valkonen, J. P. T. (2006). Interactions between a crinivirus, an ipomovirus and a potyvirus in co-infected sweet potato plants. *Plant Pathology, 55*, 458–467.
Ng, J. C. K., & Falk, B. W. (2006). Virus-vector interactions mediating nonpersistent and semipersistent transmission of plant viruses. *Annual Review of Phytopathology, 44*, 183–212.
Nishiguchi, M., Okada, Y., Sonoda, S., Mori, M., Kimura, T., Hanada, K., et al. (2001). Sweetpotato feathery mottle virus derived resistance: CP mediated resistance and gene silencing. In *International workshop on sweetpotato cultivar decline study September 8–9, 2000* (pp. 120–124). Miyakonojo, Miyazaki, Japan: Kyushu National Agricultural Experiment Station.
Njeru, R. W., Mburu, M. W. K., Cheramgoi, E., Gibson, R. W., Kiburi, Z. M., Obudho, E., et al. (2004). Studies on the physiological effects of viruses on sweet potato yield in Kenya. *Annals of Applied Biology, 145*, 71–76.
Okada, Y., Nishiguchi, M., Saito, A., Kimura, T., Mori, M., Hanada, K., et al. (2002). Inheritance and stability of the virus-resistant gene in the progeny of transgenic sweet potato. *Plant Breeding, 121*, 249–253.
Okada, Y., Nishiguchi, M., Saito, A., Kimura, T., Mori, M., Hanada, K., et al. (2006). Inheritance and stability of the virus-resistant gene in the progeny of transgenic sweet potato. *Plant Breeding, 121*, 249–253.
Okada, Y., & Saito, A. (2008). Evaluation of resistance to complex infection of SPFMV in transgenic sweet potato. *Breeding Science, 58*, 243–250.
Okada, Y., Saito, A., Nishiguchi, M., Kimura, T., Mori, M., Hanada, K., et al. (2001). Virus resistance in transgenic sweetpotato [*Ipomoea batatas* L. (Lam)] expressing the coat protein gene of sweetpotato feathery mottle virus. *Theoretical and Applied Genetics, 103*, 743–751.
Pardina, P. R., Luque, A., Nome, C., Colomba, E. L., Delgado, S. F., & Di Feo, L. (2012). First report of sweet potato leaf curl virus infecting sweet potato in Argentina. *Australasian Plant Disease Notes*, 7, 157–160.
Parrella, G., De Stradis, A., & Giorgini, M. (2006). Sweetpotato feathery mottle virus is the casual agent of sweetpotato virus disease (SPVD) in Italy. *Plant Pathology, 55*, 818.
Qin, Y., Zhang, Z., Qiao, Q., Zhang, D., Tian, Y., Wang, Y., et al. (2013). Complete genome sequences of two sweet potato chlorotic stunt virus isolates from China. *Genome Announcements, 1*, 218–313.
Sakai, J., Mori, M., Morshita, M., Tanaka, K., Hanada, K., Usugi, T., et al. (1997). Complete nucleotide sequence and genome organization of sweet potato feathery mottle virus (S strain) genomic RNA: The large coding region of the P1 gene. *Archives of Virology, 142*, 1553–1562.
Schaefers, G. A., & Terry, E. R. (1976). Insect transmission of sweetpotato agents in Nigeria. *Phytopathology, 66*, 642–645.
Shang, Y. F., Yang, C. L., Zhao, J. H., Li, C. S., Lu, X. B., Wang, S. J., et al. (1999). Progress in the researches and application of virus-free sweetpotato in Shandong province. *Research progress in plant protection and plant nutrition.* (pp. 104–108). Beijing, China: China Agric. Press.
Sheffield, F. M. L. (1957). Virus diseases of sweetpotato in East Africa. I. Identification of the viruses and their insect vectors. *Phytopathology, 47*, 582–590.
Sim, J., Valverde, R. A., & Clark, C. A. (2000). Whitefly transmission of *sweetpotato chlorotic stunt virus*. *Plant Disease, 84*, 1250.
Tairo, F., Musaka, S. B., Jones, R. A. C., Kullaia, A., Rubaihayo, P. R., & Valkonen, J. P. T. (2005). Unravelling the genetic diversity of the three main viruses involved in sweetpotato virus disease (SPVD), and its practical implications. *Molecular Plant Pathology, 6*, 199–211.

Trenado, H. P., Lozano, G., Valverde, R. A., & Navas-Castillo, J. (2007). First report of *sweetpotato virus G* and sweetpotato virus 2 infecting sweetpotato in Spain. *Plant Disease, 91*, 1687 (Abstr.).

Tugume, A. K., Settumba Mukasa, B., & Omongo, C. A. (2010). Unraveling the vector transmission biology of the ipomovirus *sweet potato mild mottle virus* (*potyviridae*) in sweetpotato (Lam.). In *Third RUFORUM Biennial Meeting 24–28 September 2012, Entebbe, Uganda* (pp. 319–324).

Untiveros, M., Fuentes, S., & Salazar, L. F. (2007). Synergistic interaction of sweet potato chlorotic stunt virus (crinivirus) with Carla-, cucumo-, ipomo-, and potyviruses infecting sweet potato. *Plant Disease, 91*, 669–676.

Usugi, T., Nakano, M., Akira, A., & Hayashi, T. (1991). Three filamentous viruses from sweet potato in Japan. *Annals of the Phytopathological Society of Japan, 57*, 512–521.

Valverde, R. A., Sim, J., & Lotrakul, P. (2004). Whitefly transmission of sweetpotato viruses. *Virus Research, 100*, 123–128.

CHAPTER THREE

Control of Pome and Stone Fruit Virus Diseases

Marina Barba[1], Vincenza Ilardi, Graziella Pasquini
Consiglio per la Ricerca e Sperimentazione in agricoltura, Centro di Ricerca per la Patologia Vegetale, Rome, Italy
[1]Corresponding author: e-mail address: marina.barba@entecra.it

Contents

Abstract

Many different systemic pathogens, including viruses, affect pome and stone fruits causing diseases with adverse effects in orchards worldwide. The significance of diseases caused by these pathogens on tree health and fruit shape and quality has resulted in the imposition of control measures both nationally and internationally

Control measures depend on the identification of diseases and their etiological agents. Diagnosis is the most important aspect of controlling fruit plant viruses. Early detection of viruses in fruit trees or in the propagative material is a prerequisite for their control and to guarantee a sustainable agriculture.

Many quarantine programs are in place to reduce spread of viruses among countries during international exchange of germplasm. All these phytosanitary measures are overseen by governments based on agreements produced by international organizations. Also certification schemes applied to fruit trees allow the production of planting material of known variety and plant health status for local growers by controlling the

Advances in Virus Research, Volume 91
ISSN 0065-3527
http://dx.doi.org/10.1016/bs.aivir.2014.11.001

propagation of pathogen-tested mother plants. They ensure to obtain propagative material not only free of "quarantine" organisms under the national legislation but also of important "nonquarantine" pathogens. The control of insect vectors plays an important role in the systemic diseases management, but it must be used together with other control measures as eradication of infected plants and use of certified propagation material.

Apart from the control of the virus vector and the use of virus-free material, the development of virus-resistant cultivars appears to be the most effective approach to achieve control of plant viruses, especially for perennial crops that are more exposed to infection during their long life span. The use of resistant or tolerant cultivars and/or rootstocks could be potentially the most important aspect of virus disease management, especially in areas in which virus infections are endemic. The conventional breeding for virus-tolerant or resistant fruit tree cultivars using available germplasm is a long-term strategy, and the development and production of these cultivars may take decades, if successful.

Genetic engineering allows the introduction of specific DNA sequences offering the opportunity to obtain existing fruit tree cultivars improved for the desired resistance trait. Unfortunately, genetic transformation of pome and stone fruits is still limited to few commercial genotypes. Research carried out and the new emerging biotechnological approaches to obtain fruit tree plants resistant or tolerant to viruses are discussed.

1. INTRODUCTION: THE IMPORTANCE OF TEMPERATE FRUIT TREES WORLDWIDE

Temperate fruit trees are members of the family *Rosaceae* whose species include those of pome (apple, quince, and pear) and stone fruits (apricot, peach, plum, almond, and cherry). Their fruits contribute significantly to human foods as their daily consumption may reduce the risk of cardiovascular diseases and certain cancer.

Nearly 50% of the total global volume of temperate tree fruits is produced by China, Turkey, USA, Brazil, Italy, and Spain (www.mapsofworld.com). China is the main country for producing apple, pear, and plum, Turkey is the main producer of cherry and apricot, while the USA is the largest producer of almonds (FAOSTAT, 2012).

The geographical distribution of temperate fruit trees is highly influenced by environmental conditions. Climate is the most significant variable affecting their production: they need low temperatures to break dormancy. The chilling requirement is different according to the species; for example, apples have the highest chilling requirement of all temperate fruit trees followed by apricots and lastly by peaches. Cultivars of different species can show different responses to different climatic conditions which result in better adaptability to local temperatures.

Climate condition and change have an important impact on the qualitative and quantitative aspects of fruit industry. Presence of pathogens and their vectors is influenced by climate which could increase the risk of introduction of exotic diseases that may modify the map of pests' distribution.

This aspect is evident also for systemic pathogens such as viruses of pome and stone fruits which are widely distributed. Losses due to viruses in pome and stone fruits are often insidious, frequently unnoticed, and untreated. Some viruses are largely latent (produce no visible symptoms) in some of their hosts, and their infection often results in plants that grow slowly, attain a smaller size, produce fewer fruits, have altered product composition, and have a reduced life span or other impacts that often go unnoticed (Hadidi & Barba, 2011).

2. MAJOR VIRUSES AFFECTING TEMPERATE FRUIT TREES

Pome and stone fruits are affected by many viruses which cause diseases with adverse effects in orchards worldwide. They cause a wide range of symptoms, ranging from symptomless (latent) to a general decline in vigor and productivity. Leaf symptoms include distortion or twisting, mottling, rolling, necrotic spots, shot holes, and unusual color patterns. Fruits may show reductions in size and quality, distortions in shape, and alterations such as ringspots, mottling, and line patterns.

Virus and virus-like agents of stone and pome fruit crops induce high economic losses to all sectors of the production chain. For example, *Plum pox virus* (PPV), the etiological agent of sharka disease, causes severe damages and has enormous economic and social impacts. Tens of millions of euros and dollars have been spent for controlling this pathogen without success. Other viruses like *Ilarviruses* induce mild but significant losses in orchards reducing the sustainability of many orchard operations. The major viruses identified on pome and stone fruits belong to several families and genera (Table 1).

Below is a brief description of the major economically important pome and stone fruit viruses.

Table 1 Plant hosts of the major viruses affecting temperate fruit trees

Virus	Host
Family: *Betaflexiviridae* Genus: *Trichovirus*	
Apple chlorotic leaf spot virus	Apple, pear, quince, peach, apricot, almond, sour and sweet cherry, and European and Japanese plum
Cherry mottle leaf virus	Sweet and sour cherry, apricot, and peach
Family: *Betaflexiviridae* Genus: *Capillovirus*	
Apple stem grooving virus	Apple, nashi, and other pomaceous species
Family: *Betaflexiviridae* Genus: *Foveavirus*	
Apple stem pitting virus	Most commercial pome fruit cultivars
Family: *Betaflexiviridae* Genus: *unassigned*	
Cherry green ring mottle virus	Cherry and other cultivated *Prunus* species
Family: *Bromoviridae* Genus: *Ilarvirus*	
Prunus necrotic ringspot virus	All stone fruit species
Apple mosaic virus	Apple, pear, almond, apricot, cherry, peach, and plum. Hazelnut
Prune dwarf virus	Sweet and sour cherry, plum and peach, apricot, and almond
Family: *Closteroviridae* Genus: *Ampelovirus*	
Little cherry virus 1 and *Little cherry virus* 2	Sweet cherry is the most affected. Limited to *Prunus* species
Family: *Potyviridae* Genus: *Potyvirus*	
Plum pox virus	Apricot, peach, European plum, Japanese plum, myrobalan, sweet, and sour cherry

2.1. Family: *Betaflexiviridae*

2.1.1 Genus: Trichovirus

2.1.1.1 *Apple chlorotic leaf spot virus*

Apple chlorotic leaf spot virus (ACLSV) was isolated for the first time from apple trees in the USA after transmission to *Malus platycarpa* (Mink & Shay, 1959).

In most commercial apple cultivars the infection generally is latent, but it causes an important disease of apple trees grown on Maruba kaido (*Malus prunifolia* cv. Ringo) rootstocks in Japan (Yaegashi, Yoshikawa, & Candresse, 2011; Yanase, 1974). In some stone fruit species ACLSV may be symptomless and causes bark splitting or it may be associated with severe fruit deformations, yield reduction, graft incompatibility, and bud necrosis (Myrta, Matic, Malinowski, Pasquini, & Candresse, 2011). Some virus isolates induce a severe disease in apricot and plum characterized by depressions and protuberances that deform the fruit, often confused with the "sharka" disease due to PPV, and named for this reason as "pseudopox."

ACLSV is a filamentous, flexuous particle virus, 680–780 nm long and 12 nm in width. It shows high molecular variability and a number of virus isolates differing in pathogenicity have been described.

In apple trees, ACLSV frequently is detected in coinfection with *Apple stem grooving virus* and/or *Apple stem pitting virus*.

ACLSV is mainly transmitted by grafting. No natural virus vectors are currently known (Yoshikawa, 2001) and ACLSV is not known to be seed or pollen transmitted.

2.1.1.2 *Cherry mottle leaf virus*

The virus was first observed in Oregon, USA, in 1920 (McLarty, 1935).

The virus induces chlorotic mottling and distortion of the leaves. It affects fruit marketable value as the infected trees produce smaller and flavorless fruits, and their ripening may be delayed. In some regions, *Cherry mottle leaf virus* (CMLV) represents the most severe disease on some cherry cultivars (James, 2011).

CMLV is a flexuous, filamentous virus, 760 nm in length and 10.2 nm in width, and with striations 3.1 nm in pitch.

CMLV can be transmitted by grafting (Nemeth, 1986) and also efficiently by the bud/scale mite *Eriophyes inaequalis* (Oldfield, 1970).

2.1.2 Genus: Capillovirus

2.1.2.1 *Apple stem grooving virus*

The virus was first reported in *Malus sylvestris* cv. Virginia Crab from the USA in the 1960s, and strain of the virus, named *Citrus tatter leaf virus* (CTLV), was described on citrus from Meyer lemon tree in California. A strain of *Apple stem grooving virus* (ASGV) also was recently reported in New Zealand in kiwifruit (*Actinidia deliciosa*) cuttings imported from China (Massart, Jijakli, & Kummert, 2011).

The virus is latent on apple cultivars. It causes stem grooving, brown line, and graft union abnormalities only when an infected cultivar is grafted on a sensitive rootstock such as *Malus pumila* cv. Virginia Crab and symptoms are due to the hypersensitive resistance of Virginia Crab to the virus (Massart et al., 2011). In nurseries, infected plants appear less vigorous than the healthy ones.

ASGV causes topworking disease of apple trees grown on Mitsuba kaido (*Malus sieboldii*) in Japan (Yanase, 1974, 1983).

ASGV particles are flexuous filaments, 620 nm long (apple strain), 650 nm long (citrus strain), and 680 nm long (*Actinidia* isolate), 12 nm in width, with obvious cross-banding and helical symmetry.

ASGV is only transmitted by infected propagative material and by grafting. No vector is currently known.

2.1.3 Genus: Foveavirus

2.1.3.1 *Apple stem pitting virus*

Apple stem-pitting disease was first described as incompatibility between certain apple cultivars and the rootstock *M. sylvestris* 'Virginia crab' in the USA in the 1940s.

Apple stem pitting virus (ASPV) is common in commercial apple cultivars but is mostly latent. In susceptible cultivars of different pome species and indicator plants ASPV causes a variety of symptoms. Xylem pitting in the stem appears on susceptible apple cultivars Charden and Reinette Clochard (Desvignes et al., 1999) and on woody indicator plants *M. pumila* 'Virginia crab.' ASPV induces epinasty and decline in apple indicator plants *M. pumila* 'Spy 227', causes a severe topworking disease in Japan and lethal decline in apple trees grown on sensitive *M. sieboldii* rootstocks.

Many commercial pear cultivars develop leaf symptoms only during the first few years of growth, whereas later on the affected trees are symptomless (Jelkmann & Paunovic, 2011). In quince, it causes diffused leaf spots and severe fruit deformation (Mathioudakis, Maliogka, Dovas, Paunovic, & Katis, 2009).

ASPV consists of flexuous, filamentous particles, approximately 800 nm in length and 12–15 nm in width, with helical symmetry.

ASPV is transmitted by grafting and through infected propagative material (Jelkmann & Paunovic, 2011). It has no known vector.

2.1.4 Genus: Unassigned

2.1.4.1 *Cherry green ring mottle virus*

The disease was first reported in 1937 affecting sour cherry (*Prunus cerasus*) trees in Michigan, USA, and described as a viral disease in 1951 (Rasmussen et al., 1951).

Latent infections of *Cherry green ring mottle virus* (CGRMV) are common in most *Prunus* species. Sometimes, it induces leaf symptoms consisting of green islands and rings cast against a yellow mottle background. Fruits are misshapen, fruit flesh is pitted, necrotic, and bitter in taste. Sour cherry cultivar Montmorency infected with the virus often produces unmarketable fruits (Jelkmann, Rott, & Uyemoto, 2011).

CGRMV has flexuous, filamentous particles, approximately 800 nm.

CGRMV is transmitted during vegetative propagation, such as grafting, and by natural root grafting between healthy and diseased trees (Barksdale, 1959; Milbrath, 1966). CGRMV has no known vector and is not seed-transmitted (Fridlund, 1966; Gilmer & Brase, 1962).

2.2. Family: *Bromoviridae*

2.2.1 Genus: Ilarvirus

2.2.1.1 *Prunus necrotic ringspot virus*

The disease was first reported in the USA in peach orchards by Cochran and Hutchins (1941).

It causes serious diseases in nurseries. Symptom severity is determined by the virus isolate and host cultivar (Nyland, Gilmer, & Moore, 1976), and synergistic interactions of *Prunus necrotic ringspot virus* (PNRSV) with other viruses, for example *Prune dwarf virus* (PDV), result in more severe disease symptoms.

PNRSV has nonenveloped isometric and quasi-isometric (bacilliform) particles of 23, 25, and 27 nm. The virus particles are relatively unstable in tissue extracts, a feature of *Ilarviruses*. The virus occurs as numerous strains, isolates, or biotypes that vary widely in their pathological, biophysical, and serological properties (Hammond, 2011).

PNRSV is easily transmitted by routine plant propagation methods (Cole, Mink, & Regev, 1982; George & Davidson, 1963; Mink, 1992; Nyland et al., 1976) and by root grafting in orchards. PNRSV is also carried on and in pollen grains and is readily transmitted through seeds (Amari, Burgos, Pallás, & Sánchez-Pina, 2009; Aparicio, Myrta, Di Terlizzi, & Pallás, 1999). Also, it is transmitted by different thrips species carrying infected pollen to healthy plants.

2.2.1.2 *Apple mosaic virus*

The virus was described for the first time in apple (Bradford & Joly, 1933).

The severity of symptoms in apple depends on cultivar susceptibility. Fruits do not develop diagnostic symptoms. Apple trees infected with the virus show pale yellow to bright cream irregular spots or bands along major veins on spring leaves. The spots may become necrotic on severely affected leaves after exposure to summer sun and heat. The symptomatic leaves drop prematurely. The distribution of symptomatic leaves may be erratic throughout the tree or limited to a single limb.

In pear, *Apple mosaic virus* (ApMV) infection is usually symptomless (Petrzik & Lenz, 2011).

In stone fruits, it causes typical yellow line pattern, bright yellow blotches, rings, bright yellow vein clearing, and/or oak-leaf pattern (Diekmann & Putter, 1996; Nemeth, 1986; Posnette & Ellenberger, 1957). Symptoms generally appear at the beginning of summer and, in some cases, are present only on a limited number of leaves randomly distributed on the plants (Paunovic, Pasquini, & Barba, 2011).

ApMV is an isometric virus (virion diameter of about 30 nm) with relatively unstable virions. ApMV is serologically related to PNRSV and it may have been erroneously identified as PNRSV when using serological tests (Petrzik & Lenz, 2011).

ApMV infects a large number of woody hosts and is frequently found in mixed infections with PNRSV, and PDV on their common stone fruit hosts. In apple, it often occurs together with ACLSV, ASPV, ASGV, and other apple-infecting viruses.

ApMV is only transmitted by vegetative propagation and by grafting. Also, slow natural spread in nurseries via root grafting occurs (Dhingra, 1972; Hunter, Chamberlain, & Atkinson, 1958). No insect vector is known for the virus. ApMV is not pollen-transmissible (Barba, Pasquini, & Quacquarelli, 1986; Digiaro, Savino, & Di Terlizzi, 1992; Sweet, 1980).

2.2.1.3 *Prune dwarf virus*

PDV was first described by Thomas and Hildebrand (1936) on 'Fellenberg prune' (*Prunus domestica*) showing stunting and leaf malformation symptoms.

In plum, PDV causes stunting and leaf malformation and shortened internodes. In Italian prune, it decreases the length of shoots, their diameter, number of leaves, and the photosynthetic total area (Hadidi & Barba, 2011). In cherry, PDV may cause leaf chlorotic spots, rings and diffuse mottling, and possibly stem pitting and flat limb. Fruits can be malformed and their production is reduced. In some apricot cultivars, PDV has been reported to induce gummosis on the trunk. In most peach cultivars, PDV induces mild stunting while leaves become dark green and more erect than those of noninfected trees, but infection by severe isolates can cause important yield reduction and poor quality of fruits.

Peachs infected with both PDV and PNRSV (peach stunt disease, PSD), display bark splitting, increased sucker production and yield is reduced by up to 60%.

The virus causes economic losses on stone fruit trees, especially in sour and sweet cherry, almond, and peach (Nolasco, Neves, & Faria, 1991; Rampitsch, Eastwell, & Hall, 1995; Uyemoto & Scott, 1992). PDV frequently occurs in mixed infections with other ilarviruses.

PDV is a multicomponent virus with five types of particles differing in size. Unenveloped virions vary from quasi-isometric, about 19–20 nm in diameter, to bacilliform with length up to 73 nm (Caglayan, Ulubas Serce, Gazel, & Varveri, 2011). Several strains of PDV have been described.

PDV is transmitted by grafting (buds, scions), pollen and seed. Pollen transmission depends on many factors such as fruit tree species and circumstances affecting pollination. Pollen transmission in sweet and sour cherry shows the highest transmission rates (George & Davidson, 1964; Gilmer, 1965). Seed transmission occurs in sweet cherry, sour cherry, mahaleb, and myrobalan, but infection rates vary with the species (Caglayan et al., 2011).

PDV infection causes yield reduction and is responsible for significant losses in almond (Nolasco et al., 1991), peach, and sweet cherry (Rampitsch et al., 1995; Uyemoto & Scott, 1992).

2.3. Family: *Closteroviridae*

2.3.1 Genus: Ampelovirus

2.3.1.1 *Little cherry virus* 1 and *Little cherry virus* 2 (LChV-1 and LChV-2)

Little cherry disease (LChD) is a complex and serious viral disease of cherry. It is distributed worldwide in ornamental and sweet cherry and has a great impact on fruit quality of infected trees. The first recorded outbreak of LChD was in the Kootenay valley in British Columbia, Canada, during the early 1930s.

The disease is associated with two different viruses, *Little cherry virus* -1 and -2 (LChV-1, LChV-2). Each virus induces distinct fruit symptoms such as dramatic reduction of size, color, and taste and a total loss of crops with a damaged vegetative growth. Other cultivars show only moderate fruit deformation or are tolerant. Mixed infections of LChV-1 and LChV-2 cause severe symptoms on sweet cherry cultivars (Hadidi & Barba, 2011).

LChV-1 particles are very long flexuous rods that range in length from 1786 to 1820 nm, whereas LChV-2 particles are filamentous with a length of 1667 nm and a width of 11.2 nm (Jelkmann & Eastwell, 2011).

Both viruses associated with LChD are readily graft transmissible. LChV-2 is also transmitted by the apple mealybug *Phenacoccus aceris* (Theilmann, Orban, & Rochon, 2004) and the grape mealybug (*Pseudococcus maritimus*) (Mekouria, Smith, Beers, & Eastwell, 2013) the most important vector of the disease in Canadian orchards (Raine, McMullen, & Forbes, 1986). There is no known vector associated with LChV-1.

2.4. Family: *Potyviridae*

2.4.1 Genus: Potyvirus

2.4.1.1 *Plum pox virus*

Sharka disease was detected in the early 1900s in Bulgaria and it was first described by Atanasoff (1932).

PPV induces sharka disease, the most devastating disease of stone fruit trees worldwide. Its presence in a country may create trade restrictions at the international, national, and local levels.

Most of the susceptible plum, apricot, peach, and cherry cultivars show leaf symptoms that appear as pale or yellowish green rings, spots, or leaf mottling. Some peach cultivars may also show discoloration on flowers. Affected plum fruits are deformed and show rings, irregular lines, and poxes on the surface. Colored rings and bands appear on the skin of apricot fruits and apricot stones show pale rings or spots. Peach fruit symptoms are mostly restricted to the skin where pale rings and diffuse bands appear before maturation (Barba, Hadidi, Candresse, & Cambra, 2011).

PPV consists of flexuous, filamentous virus particles, about 750 nm long and 15 nm wide. Nine strains or types have been identified so far, with different biological, serological, and molecular characteristics (Garcia, Glasa, Cambra, & Candresse, 2013; James, Varga, & Sanderson, 2013).

The virus is transmitted by grafting and other vegetative propagation techniques and by aphids such as *Aphis spiraecola* and *Myzus persicae* in a non-persistent manner. There is no evidence for either pollen or seed transmission (Glasa, Hricovsky, & Kudela, 1999; Myrta, Di Terlizzi, & Savino, 1998; Pasquini, Simeone, Conte, & Barba, 2000; Pasquini & Barba, 2006).

The costs associated with the disease involve not only direct losses in fruit production, commercialization, eradication, compensatory measures, and lost revenue but also indirect costs including those from preventive measures such as quarantine, surveys, inspections, control of nurseries, diagnostics, and the impact on foreign and domestic trade. The loss from sharka during the last few decades, at the global level, is estimated at about 18 million metric tons of apricots and about 45 million metric tons of European plums with an estimated value of 3600 and 5400 million Euros, respectively (Cambra, Capote, Myrta, & Llácer, 2006).

3. RELIABLE AND SENSITIVE DETECTION METHODS

Control measures depend on proper identification of diseases and of the causal agents. Therefore, diagnosis is the most important aspect of control temperate fruit plant viruses. Early detection of viruses in fruit trees or in the propagative material is prerequisite for the control of the diseases induced by these pathogens and to guarantee a sustainable agriculture.

Without the identification of the disease-causing agent, disease control measures can be waste of time and money and can lead to further plant losses.

Diagnostic techniques for viruses fall into two main categories: detection and identification based on biological properties and intrinsic properties of the virus.

Biological diagnosis is based on the interaction of the virus with its host and is characterized by the symptom inspection on the natural hosts or on susceptible indicator plants mechanically inoculated by sap inoculation (herbaceous indicator plants) or by grafting or chip budding (woody indicator plants).

All fruit tree viruses mentioned in this chapter are easily identified by characteristic symptoms induced on a specific host range of herbaceous and/or woody indicator plants.

Diagnosis based on intrinsic properties of the virus is based on the identification of the coat protein (CP) or of the nucleic acid.

Detection methods based on CP include all serological diagnostic techniques, in which the specific reaction antigen (Ag)–antibody (Ab) is displayed in different ways. The enzyme-linked immunosorbent assay (ELISA), in which the Ag–Ab reaction is visualized by a colorimetric reaction on a solid surface, is the most sensitive and suitable serological detection method for fruit tree viruses (Table 2).

ELISA, which was first applied for the detection of PPV (Clark & Adams, 1977), is widely used to detect the presence of fruit tree viruses in different matrices. Commercial ELISA kits are currently available through several companies, using polyclonal and monoclonal antibodies or antisera

Table 2 Serological detection methods reported for major viruses affecting temperate fruit trees

Virus	References
Family: *Betaflexiviridae* Genus: *Trichovirus*	
Apple chlorotic leaf spot virus	Flegg and Clark (1979), Llácer, Cambra, Laviña, and Aramburu (1985), Barba and Clark (1986), Poul and Dunez (1990), Cieslinska, Malinowski, and Zawadzka (1995), Malinowski, Cieślińska, Zawadzka, Interewicz, and Porebska (1997), Jelkmann (2001)
Cherry mottle leaf virus	Oldfield et al. (1995), James and Mukerji (1996), Gispert, Perring, and Creamer (1998), James and Howell (1998)
Family: *Betaflexiviridae* Genus: *Capillovirus*	
Apple stem grooving virus	Fuchs, Grüntzig, and Al Kai (1988)
Family: *Betaflexiviridae* Genus: *Foveavirus*	
Apple stem pitting virus	Jelkmann, Kunze, Vetten, and Lesemann (1992), Jelkmann and Keim-Konrad (1997), Paunovic, Maksimovic, Rankovic, and Radovic (1999), Gugerli and Ramel (2004)
Family: *Betaflexiviridae* Genus: Unassigned	
Cherry green ring mottle virus	Haufler and Ramsdell (1989), Zagula, Aref, and Ramsdell (1989), Zhang, Kirkpatrick, Smart, and Uyemoto (1998)
Family: *Bromoviridae* Genus: *Ilarvirus*	
Prunus necrotic ringspot virus	Torrance and Dolby (1984), Mink, Howell, Cole, and Regev (1987), Scott, Barnett, and Burrows (1989), Myrta et al. (2001), Mekuria et al. (2003)
Apple mosaic virus	Clark and Adams (1977), Barbara, Clark, Thresh, and Casper (1978), Torrance and Dolby (1984), Pasquini and Barba (1991)
Prune dwarf virus	Mink and Aichele (1984), Torrance and Dolby (1984), Abou-Jawdah et al. (2004)
Family: Closteroviridae Genus: *Ampelovirus* Genus: *Ilarvirus*	
Little cherry virus 1 and *Little cherry virus* 2	Jelkmann and Eastwell (2011), Theilmann et al. (2002)

Table 2 Serological detection methods reported for major viruses affecting temperate fruit trees—cont'd

Virus	References
Family: Potyviridae Genus: *Potyvirus*	
Plum pox virus	Clark and Adams (1977), Himmler et al. (1987), Cambra et al. (1994), Boscia et al. (1997), Candresse et al. (1998), Crescenzi et al. (1998), Myrta, Potere, et al. (1998), Myrta, Potere, Crescenzi, Nuzzaci, and Boscia (2000), Cambra, Boscia, et al. (2006), Glasa and Candresse (2005)

produced from recombinant proteins as the antiserum produced for LChV-2 (Jelkmann & Eastwell, 2011; Theilmann et al., 2002).

ELISA can be used also for quantitative virus analysis (Himmler et al., 1987) and has been used for the identification of different virus isolates or serotypes by using specific monoclonal antibodies, such as the 34 serogroups of PNRSV (Myrta et al., 2001) and the PPV serotypes or strains (Cambra, Boscia, Gil, Bertolini, & Olmos, 2011).

Detection methods based on the virus nucleic acid identification are more sensitive and specific and are very useful for rapid diagnosis of virus diseases in disease surveys, epidemiological studies, plant quarantine, and certification and breeding programs.

The most suitable molecular detection system for temperate fruit tree viruses is the polymerase chain reaction (PCR), an *in vitro* method for amplifying target nucleic acid sequences. Because known temperate fruit tree viruses are RNA viruses, PCR must be preceded by a retrotranscription (RT) of the RNA into cDNA. RT-PCR has been developed and routinely used for all mentioned fruit tree viruses (Hadidi et al., 2011).

During the last several years, diagnosis of fruit tree viruses has evolved significantly to achieve rapid and reliable detection of pathogens.

Molecular detection has been optimized through the use of nucleic acids automated extraction systems by columns or robotics. Multiplex PCRs have been produced for the simultaneous detection of many fruit tree viruses (Foissac et al., 2005; Hassan, Myrta, & Polak, 2006; Ito, Ieki, & Ozaki, 2002; James, 1999; Massart, Roussel, Kummert, Dutrecq, & Jijakli, 2008; Menzel, Zahn, & Maiss, 2003; Sanchez-Navarro et al., 2005). Quantitative real-time PCRs allow to monitor the titer of viruses in infected tissues (Capote et al., 2006, 2008; Marbot et al., 2003; Olmos, Bertolini, Gil, & Cambra, 2005; Roussel, Kummert, Salmon, Dutrecq, &

Jijakli, 2005; Schneider, Sherman, Stone, Damsteegt, & Fredeick, 2004; Varga & James, 2006).

The microarray technology allows to implement multipathogen detection systems and to obtain an easy genotyping and detection of the virus (Pasquini et al., 2008). Microarray detection of several temperate fruit viruses have been reported (for review see Barba & Hadidi, 2011). The loop-mediated isothermal amplification has been applied in the diagnostic field offering high sensitivity and low cost (Gao, Li, & Wang, 2011; Notomi et al., 2000). Novel sequencing technologies (next-generation sequencing) represent a great tool for the evaluation of the sanitary status of fruit tree species (Barba, Czosneck, & Hadidi, 2014; Hadidi & Barba, 2012; Olmos, Cambra, Marais, Faure, & Candresse, 2012; Yoshikawa, Yamagishi, Yaegashi, & Ito, 2012), using different strategies (see Barba et al., 2014 for details).

4. PRESENT CONTROL METHODS

Pome and stone fruit viruses are responsible for important economic losses. Direct or indirect damage associated with virus infections of fruit tree host plants (Bos, 1982; Waterworth & Hadidi, 1998) may include the following:

- Reduction in vegetative growth: growth reduction (including symptomless infection) and crop failure;
- Reduction in vigor such as increased sensitivity to frost and drought, and increased predisposition to other virus isolates elicit a disease of apricot and plum known as "pseudopox," characterized by depressions and protuberances that deform the fruits;
- Reduction in quality or market value: (a) defects of visual characteristics such as size, shape, and color, (b) reduced keeping quality, (c) reduced consumer appeal such as grading, taste, texture, and composition, and (d) reduced fitness for propagation;
- Cost of attempting to maintain crop health: (a) common cultural practice in orchards including vector control, (b) production of virus-free propagation material, (c) quarantine programs on commodities for export/import, (d) eradication programs, and (e) breeding for resistance to viruses and/or vectors.

The significance of diseases caused by these pathogens on tree health, fruit shape, and quality has resulted in the imposition of control measures, both nationally and internationally. In fact, global movement of plant material accelerates spread of systemic pathogens and increases the risk of inadvertent introductions of novel pathogens into new environments. These aspects need the creation of networks of scientists and regulatory personnel to be able to identify and respond to emerging disease situations in a timely manner. This includes the following:

- research on diagnosis of pathogens to facilitate their detection and identification in infected plant tissue, as a prerequisite for their efficient control;
- education to make field personnel aware of key signs of emerging threats.

Viral diseases are difficult to manage as they cannot be directly controlled by chemical application on infected plants.

The major control strategies applicable to all systemic pathogens that infect pome and stone fruit are the following:

1. Exclusion of the pathogen(s) by crop quarantine;
2. Exclusion of the pathogen(s) by crop certification;
3. Control of pathogen by eradication of infected cultivars and rootstocks;
4. Control of pathogen vectors;
5. Elimination of pathogen in planting material;
6. Selection of tolerant and/or resistant crop cultivars.

4.1. Exclusion of the pathogen(s) by crop quarantine

The establishment of trade routes provides a pathway whereby plants and their associated pests can be transported more rapidly. Quarantine regulations are designed primarily to exclude the movement of pathogens into geographic areas where they were not previously known to occur. Regulations have also been developed to restrict the movement of disease agents into areas where they may occur but distribution is limited. All these phytosanitary measures are overseen by governments based on agreements produced by international organizations.

National legislation is under the control of the national plant protection organization (NPPO) or may be at a higher international level. For example, the European Union (EU) Plant Health Directive is applicable to all EU member countries. Individual countries are required to implement this legislation into national regulations, and they can add additional "internal" requirements. In the USA, federal legislation is applicable to the whole country, with individual states having their own additional internal legislation if required (Reed & Foster, 2011).

International phytosanitary activities are governed by some organizations, the principal among which are the Agreement on the Application of Sanitary and Phytosanitary Measures under the World Trade Organization General Agreement on Tariffs and Trade (the WTO-SPS), the IPPC, administered by a Commission on Phytosanitary measures under the United Nations Food and Agriculture Organization (FAO).

The quarantine measures must be based on international standards if these exist, such as those produced by the IPPC, Regional Plant Protection Organizations (RPPOs), and the FAO. In Europe and surrounding countries, the applicable RPPO is the European and Mediterranean Plant Protection Organization (EPPO), which provides recommendations on pathogens which should be regulated by its 50 member countries. The North American Plant Protection Organization (NAPPO) has similar recommendations and provides guidelines for international movement of pome and stone fruit into NAPPO countries. The recommendations by RPPOs are not legally binding but only suggestions for regulations that national governments may implement based on their own circumstances and assessment of risk or importance of the host crop (Ebbels, 2003).

The import–export regulations are dependent by national or regional evaluation of the risk. For example, the EU requires freedom from specific fruit tree pathogens, for material from countries outside the EU, if they are not known to be present within the EU. Other fruit tree pathogens, such as PPV, are known to be established within the EU but not widely distributed, and they are also prohibited for their important economic impact on the crops.

4.2. Exclusion of the pathogen(s) by crop certification

The main way of systemic pathogens spread and dissemination is through the use of infected propagative materials such as rootstocks, buds, shoots, and grafted plantlets. Many strategies have been established to reduce this risk and to guarantee the production of high-quality planting material. Due to the upsurge of fruit tree germplasm exchange at national and international level during the last decades, several local and international (European in particular) frameworks have been established for evaluating technical and operative aspects of production, certification, and marketing of vegetatively propagated germplasm. Certification represents an effective way to guarantee the two requirements that propagation material must meet and maintain throughout the different steps of production: trueness to plant cultivar type and sanitary status. Certification is based on the collaboration between scientific and technical organizations and involves—at different levels and with different responsibilities—more than one component. Certification safeguards both the nurseryman, who sells vegetatively propagated plant material, and the grower, who buys these products. Several certification standards have been drafted and published by international organization such as NAPPO (Mink, 1998) and EPPO (Roy, 2011). When preparing a certification scheme, the following aspects are considered to: (a) define steps to be followed during certification, (b) describe the category of plant material produced in each step of the scheme, (c) identify the most damaging pathogens to be excluded from certified material, and (d) suggest the most reliable methods for their detection. For each steps of the certification schemes the plants multiplication history have to be known.

The main steps of a certification schemes are as follows (Barba, 1998; Mink, 1998):

- Selection of individual plants according to their pomological quality;
- Production and maintenance of nuclear stock. The plants belonging to the nuclear stock must show from official checks and tests varietal characteristics corresponding to those typical of that variety and must have been found free from the pathogens listed in the specific annexes. Each single plant has to be tested for virus and virus-like organisms using diagnostic methods listed also in specific annexes. They are generally maintained in a screen house or in an isolated area to prevent recontamination through viruliferous insects, grown in containers with sterile growth medium, isolated from the soil to ensure freedom from infection by nematode vectors of viruses;
- Production of propagation stock. Buds from nuclear stock plants are used to produce, by grafting on virus-free rootstocks, a suitable number of certified mother plants to maintain in open field. *In vitro* micropropagation can be used to speed the production of propagation stock. Plants must be grown in separate field away from any other noncertified plant material and maintained under continuous surveillance to control pests and diseases, and the soil must be free of virus transmitting nematodes;
- Production of certified plants. Buds from propagation stock plants are used to produce, by grafting on virus-free rootstocks, certified plants. They are kept in plots free of virus-transmitting nematodes and separated from other noncertified material plots. Certified plants are visually inspected for harmful organisms and disease that impair quality. The

official responsible authority issues the nurserymen the appropriate number of labels for individual labeling of certified plants.

4.3. Control of pathogens by eradication of infected cultivars and rootstocks

This principle aims at eliminating a pathogen after it is introduced into an area but before it has become well established or widely spread. It can be applied to individual plants, fields, or regions, but generally, it is not effective over large geographic areas.

Eradication involves surveying orchards and nurseries regularly, and immediately removing infected trees before the virus spreads. For example, many attempts were made to eradicate new foci of PPV in countries, regions or geographical areas, but it has been difficult to eliminate the pathogen. Eradication is effective only if it is done timely.

4.4. Controlling viral insect vectors

A small number of viruses infecting temperate fruit trees are transmissible by vectors. The most important example is represented by PPV, transmitted by different species of aphids in a non-persistent manner.

The control of insect vectors plays an important role in the management of systemic diseases, but it must be used together with other control measures such as eradication of infected plants and use of certified propagation material.

The use of insecticide treatments results only in reducing the populations of potential vectors, without preventing the transmission of nonpersistently transmitted viruses such as PPV. This is because these viruses are prevalently spread by transient species. However, oil application has been proposed to reduce the spread of PPV in nurseries.

In case of PPV, most aphid samples captured in orchards are *Prunus* noncolonizing species that make short flights in stone fruit orchards moving from one plant to another (Kring, 1972; Pickett, Wadhams, Woodcock, & Hardie, 1992). Noncolonizing species, collected in orchards, were viruliferous as shown by RT-PCR tests (Cambra et al., 2004), and more than 20 species of aphids were demonstrated to be potential vectors (Gildow et al., 2004; Labonne, Yvon, Quiot, Avinent, & Llacer, 1995). Chemical control specific for insect treatments cannot prevent spread of PPV in orchards but is useful for preventing aphid colonization of infected trees.

The efficacy of insecticide treatments can be enhanced by using additional control measures. For example, (a) elimination of infected trees before the major peak of aphid flights; (b) use of orchard cover crop, intercropping, or delimitation of isolation areas as a habitat for predators or as a method that allows to lose or decrease the virus that aphids acquired by probing PPV on neighboring nonhost trees (Carraro, Labonne, & Dallot, 2011); (c) the use of summer mineral oils may decrease the acquisition of the virus from infected trees and its transmission to healthy plants (Migliori et al., 1998).

4.5. Elimination of pathogen in planting material

Elimination of viruses from infected cultivars by thermotherapy, tissue culture, cryotherapy, or other methods is used to establish pathogen-tested foundation sources for such cultivars (Laimer & Barba, 2011). Currently, the most effective method for virus elimination from fruit tree plant material is a combination of thermotherapy and shoot-tip grafting or meristem shoot-tip culture.

Cryogenic techniques have proved to be helpful in the eradication of viruses from infected plant material. The first evidence was provided by Brison, de Boucaud, Pierronet, and Dosba (1997) who succeeded in eradicating PPV from *Prunus* interspecific rootstock by vitrification technique, obtaining high eradication rates (45–60%) irrespective of the size of the shoot tip used.

Encapsulation–dehydration was then applied for eradicating *Cucumber mosaic virus* and *Banana streak virus* from banana (Helliot et al., 2002) and *Grapevine Virus A* from grapevine (Wang et al., 2003). In the latter case, encapsulation–dehydration was also compared to vitrification and no difference in eradication rate was observed.

Cryotherapy combined with thermotherapy resulted in efficient elimination of *Raspberry bushy dwarf virus* (RDBV) from raspberry (Wang, Cuellar, Rajamaki, Hiraka, & Valkonen, 2008).

Cryogenic technique proves to be a suitable alternative to traditional methods of virus eradication (meristem culture, thermotherapy), and the only resource in those cases when the above mentioned techniques fail. For example, RDBV a pollen-transmitted virus is difficult to eliminate by meristem culture as it efficiently invades leaf primordia and all meristematic tissues except the least differentiated cells of the apical dome. It was also observed that elimination rate is independent of the size of shoot tips used.

This is another advantage over meristem culture, whose efficiency increases with very small shoot tips, which however decreases survival and regrowth.

4.6. Selection of tolerant and/or resistant crop cultivars

Apart from the control of the virus vector and the use of virus-free material, the development of virus-resistant varieties appears to be the most effective approach to achieve control of plant viruses, especially for perennial crops that can become infected during their long life span. The use of resistant or tolerant cultivars and/or rootstocks could be the most important aspect of virus disease management, especially in areas in which virus infections are endemic. The conventional breeding for virus-tolerant or -resistant fruit tree cultivars using available germplasm is a long-term strategy, and the development and production of these cultivars may take decades, if successful. In particular, the selection is slow and difficult due to the transfer of undesirable characteristics and other constraints typical of fruit trees such as long biological cycle with extended juvenile phase and high level of heterozygosity (Decroocq, Badenes, & Neumüller, 2011; Garcia et al., 2013).

Many studies have been performed on fruit tree cultivars for evaluation of their susceptibility to different viruses, based on field observations under natural infection pressure or involving models obtained by grafting or chip budding or by viruliferous vector transmission in the field or under greenhouse conditions. Obtained data in most cases were not comparable as they were influenced by virus isolates, virus inoculation methods, classification of plant response to viruses (Cooper & Jones, 1983), and types of diagnostic tests with different sensitivity used for evaluation of germplasm infection.

In these studies, only limited information is available on tolerance or resistance sources for the majority of these pathogens. With the exception of PPV in stone fruits (Decroocq et al., 2011) and possibly *Cherry leaf roll virus* in walnut, there are currently no conventional breeding efforts for tolerance or resistance to viruses of fruit trees.

Many efforts have been made for the evaluation of tolerance and resistance germplasm for PPV, the most detrimental for stone fruits cultivation, but with very few natural sources of resistance identified in *Prunus* species. Resistant apricot germplasm, originated from North America, has been used in breeding programs and used in crosses with local cultivars in several European countries (Audergon, 1997; Badenes & Llácer, 2006; Bassi, 2006; Karayiannis et al., 2006). PPV resistance seems to be controlled in apricot by at least two genes (Karayiannis, Thomidis, & Tsaftaris, 2008;

Martínez-Gómez & Dicenta, 2000; Moustafa, Badenes, Martínez-Calvo, & Llácer, 2001).

No peach germplasm was resistant to PPV, but resistance has been identified in wild species *Prunus davidiana*. This genotype was included in breeding programs for introducing PPV resistance in peach cultivars (Decroocq et al., 2005). Quantitative trait locus analysis performed on F1 progeny allowed the identification of several genomic regions involved in the resistance (Decroocq et al., 2005; Marandel, Salava, Abbott, Candresse, & Decroocq, 2009; Pilařová et al., 2010).

Also, interspecific peach rootstocks and almond cultivars seem to be a source of resistance to PPV (Pascal, Pfeiffer, & Kervella, 2002; Rubio, Martínez-Gómez, & Dicenta, 2003).

The hexaploid genome of plum has made it difficult to study PPV resistance in this species. No data on molecular markers are available for any trait in plum (Decroocq et al., 2011). The hypersensitive response (Kegler, Grüntzig, Fuchs, Rankovic, & Ehrig, 2001), characterized by a localized cell death, is an effective resistance mechanism to PPV under natural and artificial inoculation (Garcia et al., 2013) and has been used in plum breeding programs (Hartmann, 1998).

5. TRANSGENIC APPROACHES TO INDUCE VIRUS RESISTANCE IN TEMPERATE FRUIT TREES

Genetic engineering allows the introduction of specific DNA sequences for the desired resistance trait. However, transformation and regeneration of fruit trees are demanding and, despite constant efforts, genetic transformation of pome and stone fruits is still limited to few commercial genotypes (Gambino & Gribaudo, 2012; Petri & Burgos, 2005). Transgenic resistance strategies have been previously tested on herbaceous model plants (Ilardi & Di Nicola-Negri, 2011). Research has been mostly focused on PPV, the most important virus affecting fruit trees (Scholthof et al., 2011).

The first transgenic strategy developed to control virus infection was based on expression of virus-derived sequences in plants, the so-called pathogen-derived resistance (PDR) strategy (Sanford & Johnston, 1985). From the early 1990s, up to the discovery of post-transcriptional gene silencing (PTGS), the leading idea was to interfere with virus multiplication by expressing in plants wild-type (wt) or mutated virus-derived proteins. Almost all PPV gene sequences either wild type or mutagenized were used to confer virus resistance (reviewed in Ilardi & Di Nicola-Negri, 2011).

Regardless of the viral gene expressed, it soon became apparent that PPV resistance did not rely on the expression of the viral protein. In fact, (a) a very low percentage of the transgenic plants were resistant; (b) resistance did not correlate with the amount of transgenic protein and an inverse correlation between the amount of transgenic mRNA and/or protein and the resistance was observed; (c) in several PPV-resistant plants, transgenic proteins were undetectable; (d) some transgenic lines showed recovery from viral infection. This indicated that in the PPV-resistant plants, the resistance was based on an RNA-mediated mechanism (Simón-Mateo & García, 2011). The reason for only a minor fraction of transgenic plants showing PPV resistance is due to integration in the host genome of multiple and/or rearranged T-DNA copies which in turn can lead to transcription of aberrant or double-stranded RNAs (dsRNA), the inducers of RNA silencing mechanism (Baulcombe, 2004; Voinnet, 2008). The most important and exploited example of this finding is the plum clone C5, now renamed 'HoneySweet' (reviewed in Scorza et al., 2013). The plum cultivar Stanley was transformed with the sense PPV *CP* gene, and among the transgenic lines obtained, only C5 was resistant to PPV under greenhouse conditions. Molecular analysis showed that C5 possesses multiple and rearranged *CP* gene copies, expressing low levels of *CP* mRNA without detectable accumulation of CP (Scorza et al., 1994, 2001). Subsequent research on C5 showed that *CP* gene was methylated and that *CP*-specific small interfering RNA (siRNA) accumulated in this PPV-resistant plum clone (Scorza et al., 2001). All these findings indicated that C5 PPV resistance is the result of unpredictable PTGS activation against the *CP* sequence. The C5 clone was extensively tested in the field to validate its ability to resist PPV infection under natural environments. Plantings were made in Poland, Spain, Romania, and the Czech Republic at sites characterized by sharka presence (Malinowski et al., 2006; Polak et al., 2008). In the field, C5 plants were resistant to PPV when exposed to viruliferous aphids, while when inoculated with PPV by chip bud-, bark-, or rootstock accumulated a low level of PPV, mostly near the graft junction. To test the effect of heterologous virus infections on the efficacy and stability of C5 PPV resistance, plants were graft-inoculated with different combinations of PNRSV, ACLSV, PDV, and PPV-D (Zagrai et al., 2008) or PPV-Rec (Polak et al., 2008). These tests confirmed that a low level of PPV could only accumulate in C5 grafted plants and that heterologous virus infection sustained by PNRSV, ACLSV, or PDV did not suppress PPV resistance. Finally, the biosafety of C5 clone was evaluated in both field and laboratory tests culminating in the regulatory approval of 'HoneySweet' in the USA (Scorza et al., 2013).

Once understood that RNA silencing was the mechanism underlying most of the viral PDRs, it was possible to construct transgenes expressing in plants virus-derived dsRNAs leading to a preactivated host resistance. In the pioneering work of Smith et al. (2000), it was shown that virus-derived gene constructs encoding intron-spliced hairpin RNAs (ihpRNAs) can efficiently induce RNA silencing and virus resistance in most of the transgenic plants. In subsequent years, PPV constructs based on the ihpRNA technology were developed starting from virus isolates of the most economically important PPV strains (PPV-D and PPV-M) (Garcia et al., 2013; Ilardi & Di Nicola-Negri, 2011).

Pandolfini, Molesini, Avesani, Spena, and Polverari (2003) transformed *Nicotiana benthamiana* plants with PPV-D ihpRNA covering the polyprotein AUG translation initiation codon under the transcriptional control of the *rolC* promoter. As the *rolC* promoter drives transcription in phloem but not in epidermal and mesophyll cells, resistance was obtained only to systemic PPV infection, while local infection was unaffected.

Four ihpRNA constructs covering the 5′UTR, *P1*, *HC-Pro*, and *P3* sequences of PPV-M ISPaVe44 were introduced in *N. bethamiana* plants (Di Nicola-Negri, Brunetti, Tavazza, & Ilardi, 2005). All four constructs confer efficient resistance to the homologous ISPaVe44 isolate (Di Nicola-Negri et al., 2005) and also to seven PPV isolates of the D, M, and Rec strains (Di Nicola-Negri, Tavazza, Salandri, & Ilardi, 2010). Interestingly, lines expressing high levels of h-UTR/P1-derived siRNAs were also able to confer resistance to isolates of the distantly related PPV-EA and PPV-C strains although the overall nucleotide identity between the h-UTR/P1 transgene and the respective sequence of PPV-EA and PPV-C isolates is 77.8% and 71.2%, respectively (Di Nicola, Tavazza, Lucioli, Salandri, & Ilardi, 2014; Di Nicola-Negri et al., 2010). Importantly, h-UTR/P1-plants grown at low (15 °C) or high (30 °C) temperatures were also fully resistant to multiple PPV challenges, different PPV inoculum concentrations, and even to the most distantly related PPV-C isolates. Finally, infections of h-UTR/P1-plants with viruses belonging to *Cucumovirus*, *Potyvirus*, or *Tombusvirus*, all known to affect PTGS at different steps by their suppressors of gene silencing (Burgyán & Havelda, 2011), were not able to affect PPV resistance (Di Nicola et al., 2014). These analyses conducted on the model plant allowed to select, among the four constructs, the h-UTR/P1 as the best candidate for transforming *Prunus* plants. Transgenic plum clones were resistant to two PPV-D isolates in two different *in vitro* grafting virus inoculation procedures specifically set up in order to contain both genetically modified plants and the quarantine virus (García-Almodóvar et al., 2014; Monticelli, Nicola-Negri, Gentile, Damiano, & Ilardi, 2012).

Subsequently, ihpRNA constructs covering the *P1* and *HC-Pro* genes were produced and tested in *N. benthamiana* also by Zhang et al. (2006) confirming that these PPV sequences are good targets to induce PPV resistance.

In another set of experiments, ihpRNA constructs containing the full-length or the second half of the PPV *CP* gene under the transcriptional control of either CaMV 35S or peach *Cab* promoters were produced and introduced in *N. benthamiana* (Hily et al., 2007). The CaMV 35S full-length ihpRNA *CP* construct, which outperformed all the other constructs in conferring PPV resistance, was chosen for transforming plum. Transgenic plums were inoculated with PPV isolates of D, M, Rec, and EA strains over multiple cycles of vegetative growth. A few ihRNA-CP plants initially failed to prevent PPV replication in areas close to the grafting point, but following the fourth and final bud-break cycle the plants fully recovered (Ravelonandro, Scorza, Hily, & Briard, 2014).

Similarly, Wang et al. (2009) transformed plums with ihpRNA *P1* construct (containing 5′ portion of *P1* gene without the UTR sequences) that was shown to give the best resistance results in the model plant. One to six plants for each transgenic plum line were challenged by PPV-D-infected chip buds. Importantly, PPV was undetectable in 5 of the 10 T0 transgenic lines analyzed.

Liu, Scorza, Hily, Scott, and James (2007) reported the production of ihpRNA construct, pTRAP6i, created by fusion of gene fragments from six stone fruit-infecting viruses: *Tobacco ringspot virus*, PNRSV, PDV, PPV, *American plum line pattern virus*, and *Peach mosaic virus*. The idea behind this approach was to build a single construct that interferes with the most important stone fruit-infecting viruses. Unfortunately, the length composition of the construct (about 400–500 bp for each virus for a total of 2500 bp for each repeat) affected transformation ability in stone fruit plants as 1600 cherry leaf explants in eight experiments did not yield any transgenic shoots, whereas the same transformation procedure allowed the production of transgenic cherry resistant to PNRSV through a shorter ihpRNA PNRSV *CP* construct (Song et al., 2013).

The above examples indicate that RNA silencing and in particular ihpRNA technology can be successfully applied to control temperate fruit tree viruses. Although RNA silencing operates in a sequence-specific manner, the data obtained clearly show that it is possible to identify viral genomic regions that confer a broad spectrum of resistance to different known PPV strains

(Di Nicola-Negri et al., 2010; Polak et al., 2008; Ravelonandro et al., 2014; Zagrai et al., 2008). Moreover, two aspects of ihpRNA technology should be pointed out. First, ihpRNAs induce consistent PPV resistance in relatively high number of the transformed herbaceous and stone fruit plants (Di Nicola-Negri et al., 2005; García-Almodóvar et al., 2014; Hily et al., 2007; Song et al., 2013; Wang et al., 2009), an important aspect considering the low efficiency in obtaining transgenic stone fruits. Second, RNA silencing technology does not require the expression of transgenic proteins, thus avoiding possible risk of allergy and/or toxicity for the consumer, and of complementation and/or synergy on behalf of incoming viruses (Lemgo, Sabbadini, Pandolfini, & Mezzetti, 2013; Tepfer, 2002). In addition, the very low RNA steady-state level of transgenic transcripts as well as their double-stranded structure should reduce the likelihood, if any, of a hypothetical recombination between the ihpRNA and the RNA genome of an incoming virus.

Besides ihpRNA-mediated resistance, other technologies that make use of RNA silencing mechanism are developing. It has been shown that PPV chimeras harboring wt, but not mutated microRNA (miRNA) target sequences, have an impaired infectivity on wt *N. benthamiana* and *Arabidopsis thaliana* plants (Simón-Mateo & García, 2006) and that plant miRNA precursors can be engineered to express, in transgenic plants, artificial miRNAs (amiRNAs) complementary to viral genomes (Niu et al., 2006). *A. thaliana* plants engineered to express amiRNAs complementary to *P69* of *Turnip yellow mosaic virus* (TYMV) and *HC-Pro* of *Turnip mosaic virus* (TuMV) gene sequences were specifically resistant to TYMV and TuMV, respectively (Niu et al., 2006). It is important to note that as amiRNAs design requires only 20–24 nt complementary to the target, it should be possible to select sequences that efficiently target viral RNAs without having off-target effects (Jackson et al., 2003). Nevertheless, as viral genomes accumulate point mutation under selective pressure this aspect clearly assumes great importance for amiRNAs. In fact, it was shown that if viral replication is permitted, RNA viruses quickly evolve breaking down amiRNA-mediated resistance (Lin et al., 2009). Hence, for conferring long-lasting and broad-spectrum virus resistance, ihpRNAs appear, in principle, more suitable than amiRNAs. A not fully explored advantage of using amiRNAs respect of ihpRNAs is that amiRNA activity, at least in the model plant *A. thaliana*, is not affected (Niu et al., 2006) by temperatures (i.e., 15 °C) known to have detrimental effect on PTGS (Szittya et al., 2003). However, we have recently shown that PPV resistance is not broken in the 5′UTR/P1 ihpRNA transgenic *N. benthamiana* plants grown at 15 °C (Di Nicola et al., 2014), whereas the level of the artificial miR2b in agro-infiltrated

N. benthamiana leaves was reduced at 15 °C compared to that at 25 °C (Qu, Ye, & Fang, 2007). All together, these data indicate that a refined amiRNA-mediated approach together with the ihpRNA technology may result, as already established for the PPV-plum binomial, in a new generation of virus-resistant fruit trees.

The recent availability of several plant genome sequences together with the understanding of the molecular mechanisms underlying virus–plant interaction will allow the development of novel strategies to control viral infection. In particular, among host-encoded factors required for PPV infection, the eukaryotic translation initiation factor eIF(iso)4E has been shown to play an important role. Interestingly, disruption of the *eIF(iso)4E* gene by transposon tagging does not lead to any developmental modifications of *A. thaliana* plants, while it confers resistance to five PPV isolates of the D, M, Rec, EA, and C strains (Decroocq et al., 2006; Duprat et al., 2002). Importantly, when the *eIF(iso)4E* gene was silenced in plum by ihpRNA constructs containing the same plum gene, about 80% of the *eIF(iso)4E* silenced transgenic plants were resistant to chip-budding PPV inoculation even after two successive vegetative cycles (Wang et al., 2013). This work paves the way to a new class of biotechnological plants in which the consumer concerns would be minimized. The expression of host factors necessary for virus infection could be impaired by using (a) the targeting induced local lesions in genomes (Tilling) technique (McCallum, Comai, Greene, & Henikoff, 2000); (b) silencing technology, as in the case of Wang et al. (2013); and (c) target mutagenesis of the gene of interest through the use of the Cas9 RNA-guided endonuclease (Nekrasov, Staskawicz, Weigel, Jones, & Kamoun, 2013). Other future perspectives aimed to reduce biosafety concerns rely on the development of RNA silencing strategies that could make use of the transgrafting in which the transgenic rootstock confers virus resistance to the wild-type scion. This strategy would allow the production of nontransgenic pollen and fruits (Ali, Kobayashi, Yamaoka, Ishikawa, & Nishiguchi, 2013; Lemgo et al., 2013).

REFERENCES

Abou-Jawdah, Y., Sobh, H., Cordahi, N., Kawtharani, H., Nemer, G., Maxwell, D. P., et al. (2004). Immunodiagnosis of *Prune dwarf virus* using antiserum produced to its recombinant coat protein. *Journal of Virological Methods*, *121*, 31–38.

Ali, E. M., Kobayashi, K., Yamaoka, N., Ishikawa, M., & Nishiguchi, M. (2013). Graft transmission of RNA silencing to non-transgenic scions for conferring virus resistance in tobacco. *PLoS One*, *8*, e63257.

Amari, K., Burgos, L., Pallás, V., & Sánchez-Pina, M. A. (2009). Vertical transmission of *Prunus necrotic ringspot virus*: Hitch-hiking from gametes to seedling. *Journal of Virological Methods*, *90*, 1767–1774.

Aparicio, F., Myrta, A., Di Terlizzi, B., & Pallás, V. (1999). Molecular variability among isolates of *Prunus necrotic ringspot virus* from different *Prunus* spp. *Phytopathology*, *89*, 991–999.

Atanasoff, D. (1932). Jahrbuch Universität Sofia. *Agronomische Fakultät*, *11*, 49.

Audergon, J. M. (1997). Prospects for breeding apricot for resistance to diseases: Sharka, bacteria and apricot chlorotic leafroll. *Italus Hortus*, *4*, 24–28.

Badenes, M. L., & Llácer, G. (2006). Breeding for resistance in Spain. *OEPP/EPPO Bulletin*, *32*, 323–326.

Barba, M. (1998). Virus certification of fruit tree propagative material in Western Europe. In A. Hadidi, R. K. Khetarpal, & H. Koganezawa (Eds.), *Plant virus diseases control* (pp. 288–293). St. Paul, MN: APS Press.

Barba, M., & Clark, M. F. (1986). Detection of strains of *Apple chlorotic leaf spot virus* by $F(ab)_2$-based indirect ELISA. *Acta Horticulturae*, *193*, 297–304.

Barba, M., Czosneck, H., & Hadidi, A. (2014). Historical perspective, development and applications of next-generation sequencing in plant virology. *Viruses*, *6*, 106–136.

Barba, M., & Hadidi, A. (2011). DNA microarrays and other future trends in detection and typing of viruses, viroids, and phytoplasmas. In A. Hadidi, M. Barba, T. Candresse, & W. Jelkmann (Eds.), *Virus and virus-like diseases of pome and stone fruits* (pp. 361–372). St. Paul, MN: APS Press.

Barba, M., Hadidi, A., Candresse, T., & Cambra, M. (2011). Plum pox virus. In A. Hadidi, M. Barba, T. Candresse, & W. Jelkmann (Eds.), *Virus and virus-like diseases of pome and stone fruits* (pp. 185–198). St. Paul, MN: APS Press.

Barba, M., Pasquini, G., & Quacquarelli, A. (1986). Role of seeds in the epidemiology of two almond viruses. *Acta Horticulturae*, *193*, 127–130.

Barbara, D. J., Clark, M. F., Thresh, J. M., & Casper, R. (1978). Rapid detection and serotyping of *Prunus necrotic ringspot virus* in perennial crop by Enzyme-Linked Immunosorbent Assay. *The Annals of Applied Biology*, *90*, 395–399.

Barksdale, T. H. (1959). *Green ring mottle virus* as an entity distinct from the sour cherry ring spot and yellows viruses. *Phytopathology*, *49*, 777–784.

Bassi, D. (2006). Breeding for resistance in Italy. *OEPP/EPPO Bulletin*, *32*, 327–328.

Baulcombe, D. C. (2004). RNA silencing in plant. *Nature*, *431*, 356–363.

Bos, L. (1982). Crops losses caused by viruses. *Crop Protection*, *1*, 263–282.

Boscia, D., Zeramdini, H., Cambra, M., Potere, O., Gorris, M. T., Myrta, A., et al. (1997). Production and characterization of a monoclonal antibody specific to the M serotype of plum pox potyvirus. *European Journal of Plant Pathology*, *102*, 477–480.

Bradford, F. C., & Joly, L. (1933). Infectious variegation in the apple. *Journal of Agricultural Research*, *46*, 901–908.

Brison, M., de Boucaud, M. Y., Pierronet, A., & Dosba, F. (1997). Effect of cryopreservation on the sanitary state of a cv. *Prunus* rootstock experimentally contaminated with *plum pox potyvirus*. *Plant Science*, *123*, 189–196.

Burgyán, J., & Havelda, Z. (2011). Viral suppressors of RNA silencing. *Trends in Plant Science*, *16*(5), 265–272.

Caglayan, K., Ulubas Serce, C., Gazel, M., & Varveri, C. (2011). Prune dwarf virus. In A. Hadidi, M. Barba, T. Candresse, & W. Jelkmann (Eds.), *Virus and virus-like diseases of pome and stone fruits* (pp. 199–206). St. Paul, MN: APS Press.

Cambra, M., Asensio, M., Gorris, M. T., Perez, E., Camarasa, E., García, J. A., et al. (1994). Detection of plum pox potyvirus using monoclonal antibodies to structural and non structural proteins. *OEPP/EPPO Bulletin*, *24*, 569–577.

Cambra, M., Boscia, D., Gil, M., Bertolini, E., & Olmos, A. (2011). Immunology and immunological assays applied to the detection, diagnosis and control of fruit tree viruses.

In A. Hadidi, M. Barba, T. Candresse, & W. Jelkmann (Eds.), *Virus and virus-like diseases of pome and stone fruits* (pp. 303–314). St. Paul, MN: APS Press.
Cambra, M., Boscia, D., Myrta, A., Palkovics, L., Navrátil, M., Barba, M., et al. (2006). Serological detection and characterization of *Plum pox virus*. *OEPP/EPPO Bulletin, 36*, 254–261.
Cambra, M., Capote, N., Myrta, A., & Llácer, G. (2006). *Plum pox virus* and the estimated costs associated with sharka disease. *OEPP/EPPO Bulletin, 36*, 202–204.
Cambra, M., Gorris, M. T., Capote, N., Asensio, M., Martinez, M. C., Bertolini, E., et al. (2004). Epidemiology of *Plum pox virus* in Japanese plums in Spain. *Acta Horticulturae, 657*, 195–200.
Candresse, T., Cambra, M., Dallot, S., Lanneau, M., Asensio, M., Gorris, M. T., et al. (1998). Comparison of monoclonal antibodies and polymerase chain reaction assay for the typing of isolates belonging to the D and M serotypes of plum pox potyvirus. *Phytopathology, 88*, 198–204.
Capote, N., Gorris, M. T., Martínez, M. C., Asensio, M., Olmos, A., & Cambra, M. (2006). Interference between D and M types of *Plum pox virus* in Japanese plum assessed by specific monoclonal antibodies and quantitative real-time reverse transcription polymerase chain reaction. *Phytopathology, 96*, 320–325.
Capote, N., Pérez-Panadés, J., Monzó, C., Carbonell, E. A., Urbaneja, A., Scorza, R., et al. (2008). Assessment of the diversity and dynamics of *Plum pox virus* and aphid populations in transgenic European plums under Mediterranean conditions. *Transgenic Research, 17*, 367–677.
Carraro, L., Labonne, G., & Dallot, S. (2011). Vector control of fruit tree viruses and phytoplasmas. In A. Hadidi, M. Barba, T. Candresse, & W. Jelkmann (Eds.), *Virus and virus-like diseases of pome and stone fruits* (pp. 379–382). St. Paul, MN: APS Press.
Cieslinska, M., Malinowski, T., & Zawadzka, B. J. (1995). Studies on several strains of *Apple chlorotic leaf spot virus* (ACLSV) isolated from different fruit tree species. *Acta Horticulturae, 386*, 63–71.
Clark, M. F., & Adams, A. N. (1977). Characteristics of the microplate method of enzyme-linked immunosorbent assay for detection of the plant viruses. *Journal of General Virology, 34*, 475–483.
Cochran, L. C., & Hutchins, L. M. (1941). A severe ring-spot virosis on peach. *Phytopathology, 31B*, 860.
Cole, A., Mink, G. I., & Regev, S. (1982). Location of *Prunus necrotic ringspot virus* on pollen grains from infected almond and cherry trees. *Phytopathology, 72*, 1542–1545.
Cooper, J. I., & Jones, A. T. (1983). Responses of plants to viruses: Proposals for the use of terms. *Phytopathology, 73*, 127–128.
Crescenzi, A., d'Aquino, L., Nuzzaci, M., Comes, S., Piazzolla, P., Ostuni, A., et al. (1998). Synthetic peptides in diagnosis of plum pox potyvirus. *Acta Horticulturae, 472*, 491–494.
Decroocq, M., Badenes, L., & Neumüller, M. (2011). Breeding for resistance to *Plum pox virus*. In A. Hadidi, M. Barba, T. Candresse, & W. Jelkmann (Eds.), *Virus and virus-like diseases of pome and stone fruits* (pp. 403–408). St. Paul, MN: APS Press.
Decroocq, V., Foulongne, M., Lambert, P., Le Gall, O., Mantin, C., Pascal, T., et al. (2005). Analogues of virus resistance genes map to QTLs for resistance to sharka disease in *Prunus davidiana*. *Molecular Genetics and Genomics, 272*, 680–689.
Decroocq, V., Sicard, O., Alamillo, J. M., Lansac, M., Eyquard, J. P., Garcia, J. A., et al. (2006). Multiple resistance traits control *Plum pox virus* infection in *Arabidopsis thaliana*. *Molecular Plant-Microbe Interactions, 19*, 541–549.
Desvignes, J. C., Boye, R., Cornaggia, D., Grasseau, N., Hurtt, S., & Waterworth, H. (1999). *Virus diseases of fruit trees. (Diseases due to viroids, viruses, phytoplasmas and other undetermined infectious agents)* (p. 202). Paris; France: Centre Technique Interprofessionnel des Fruits et Legumes (CTIFL).

Dhingra, K. L. (1972). Transmission of apple mosaic by natural root grafting. *The Indian Journal of Horticulture, 29*, 348–350.
Digiaro, M., Savino, V., & Di Terlizzi, B. (1992). Ilarvirus in apricot and plum pollen. *Acta Horticulturae, 309*, 93–98.
Di Nicola, E., Tavazza, M., Lucioli, A., Salandri, L., & Ilardi, V. (2014). Robust RNA silencing-mediated resistance to Plum pox virus under variable abiotic and biotic conditions. *Molecular Plant Pathology, 15*(8), 841–847. http://dx.doi.org/10.1111/mpp.12132.
Di Nicola-Negri, E., Brunetti, A., Tavazza, M., & Ilardi, V. (2005). Hairpin RNA-mediated silencing of *Plum pox virus* P1 and HC-Pro genes for efficient and predictable resistance to the virus. *Transgenic Research, 14*, 989–994.
Di Nicola-Negri, E., Tavazza, M., Salandri, L., & Ilardi, V. (2010). Silencing of *Plum pox virus* 5′UTR/P1 sequence confers resistance to a wide range of PPV strains. *Plant Cell Reports, 29*, 1435–1444.
Diekmann, M., & Putter, C. A. J. (1996). *FAO/IPGRI technical guidelines for the safe movement of germplasm*. (pp. 1–110). Rome: No. 16. Stone Fruit. Food and Agriculture Organization of the United Nations, Rome/International Plant genetic Resources Institute.
Duprat, A., Caranta, C., Revers, F., Menand, B., Browning, K. S., & Robaglia, C. (2002). The Arabidopsis eukaryotic initiation factor (iso)4E is dispensable for plant growth but required for susceptibility to potyviruses. *Plant Journal, 32*, 927–934.
Ebbels, D. L. (2003). *Principles of plant health and quarantine*. (pp. 1–302). Oxfordshire, UK: CABI Publishing.
FAOSTAT, (2012). http://faostat.fao.org/site/339/default.aspx.
Flegg, C. L., & Clark, M. F. (1979). The detection of *Apple chlorotic leaf spot virus* by a modified procedure of enzyme-linked immunosorbent assay. *Annals of Applied Biology, 91*, 61–65.
Foissac, X., Svanella-Dumas, L., Gentit, P., Dulucq, M. J., Marais, A., & Candresse, T. (2005). Polyvalent degenerate oligonucleotides reverse transcription-polymerase chain reaction: A polyvalent detection and characterization tool for trichoviruses, capilloviruses, and foveaviruses. *Phytopathology, 95*, 617–625.
Fridlund, P. R. (1966). Transmission and lack of transmission of seven viruses through Prunus seed. *Plant Disease Reporter, 50*, 902–904.
Fuchs, E., Grüntzig, M., & Al Kai, B. (1988). Der serologische Nachweis mechanisch übertragbarer Viren des Kern- und Steinobstes. *Nachrichtenbl. Pflanzenschutz DDR, 10*, 208–211
Gambino, G., & Gribaudo, I. (2012). Genetic transformation of fruit trees: Current status and remaining challenges. *Transgenic Research, 21*, 1163–1181.
Gao, Y., Li, M., & Wang, J. (2011). Development of loop-mediated isothermal amplification (LAMP) assays for the detection of *Plum pox virus*. *Phytopathology, 101*, S58.
Garcia, J. A., Glasa, M., Cambra, M., & Candresse, T. (2013). *Plum pox virus* and Sharka: A model potyvirus and a major disease. *Molecular Plant Pathology, 15*, 226–241. http://dx.doi.org/10.1111/mpp.12083.
García Almodóvar, R. C., Clemente-Moreno, M. J., Díaz-Vivancos, P., Petri, C., Rubio, M., Padilla, I. M. G., et al. (2014). Greenhouse evaluation confirms in vitro sharka resistance of genetically engineered h-UTR/P1 plum plants. *Plant Cell, Tissue and Organ Culture*. http://dx.doi.org/10.1007/s11240-014-0629-7.
George, J., & Davidson, T. R. (1963). Pollen transmission of necrotic ringspot virus and sour cherry yellows viruses from tree to tree. *Canadian Journal of Plant Science, 43*, 276–288.
George, J. A., & Davidson, T. R. (1964). Further evidence of pollen transmission of necrotic ringspot and sour cherry yellows viruses in sour cherry. *Canadian Journal of Plant Science, 44*, 383–384.
Gildow, F., Damsteegt, V., Stone, A., Schneider, W., Luster, D., & Levy, L. (2004). Plum pox in North America: Identification of aphid vectors and a potential role for fruit in virus spread. *Phytopathology, 94*, 868–874.

Gilmer, R. M. (1965). Additional evidence of tree-to-tree transmission of sour cherry yellows virus by pollen. *Phytopathology, 55*, 482–483.

Gilmer, R., & Brase, K. (1962). Evidence against seed transmission of green ring mottle virus in mazzard cherry seedlings. *Plant Disease Reporter, 46*, 356.

Gispert, C., Perring, T. M., & Creamer, R. (1998). Purification and characterization of *Peach mosaic virus*. *Plant Disease, 82*, 905–908.

Glasa, M., & Candresse, T. (2005). *Plum pox virus*. N. 410 in CMI/AAB description of plant viruses. In A. T. Jones, D. J. Robinson, N. Boonham, & R. Mumford (Assoc. Eds.), *Applied biology*, Wellesbourne, UK.

Glasa, M., Hricovsky, I., & Kudela, O. (1999). Evidence for non-transmission of *Plum pox virus* by seed in infected plum and myrabolan. *Biologia Bratislava, 54*, 481–484.

Gugerli, P., & Ramel, M. (2004). Production of monoclonal antibodies for the serological identification and reliable detection of Apple stem pitting and Pear yellow vein viruses in apple and pear. *Acta Horticulturae, 657*, 59–69.

Hadidi, A., & Barba, M. (2011). Economic impact of pome and stone fruit viruses and viroids. In A. Hadidi, M. Barba, T. Candresse, & W. Jelkmann (Eds.), *Virus and virus-like diseases of pome and stone fruits* (pp. 1–8). St. Paul, MN: APS Press.

Hadidi, A., & Barba, M. (2012). Next generation sequencing: Historical perspective and current applications in plant virology. *Petria, 22*, 262–277.

Hadidi, A., Olmos, A., Pasquini, G., Barba, M., Martin, R. R., & Shamloul, A. M. (2011). Polymerase chain reaction for detection of systemic plant pathogens. In A. Hadidi, M. Barba, T. Candresse, & W. Jelkmann (Eds.), *Virus and virus-like diseases of pome and stone fruits* (pp. 343–362). St. Paul, MN: APS Press.

Hammond, R. W. (2011). Prunus necrotic ringspot virus. In A. Hadidi, M. Barba, T. Candresse, & W. Jelkmann (Eds.), *Virus and virus-like diseases of pome and stone fruits* (pp. 207–214). St. Paul, MN: APS Press.

Hartmann, W. (1998). Hypersensitivity—A possibility for breeding Sharka resistant plum hybrids. *Acta Horticulturae, 472*, 429–432.

Hassan, M., Myrta, A., & Polak, J. (2006). Simultaneous detection and identification of four pome fruit viruses by one-tube pentaplex RT-PCR. *Journal of Virological Methods, 133*, 124–129.

Haufler, K. Z., & Ramsdell, D. C. (1989). Detection of the virus causing green ring mottle disease in cherry by direct and indirect ELISA. *Acta Horticulturae, 235*, 169–176.

Helliot, B., Panis, B., Poumay, Y., Swenner, R., LePoivre, P., & Frison, E. (2002). Cryopreservation for the elimination of cucumber mosaic and banana streak viruses from banana (Musa spp.). *Plant Cell Reports, 20*, 1117–1122.

Hily, J. M., Ravelonandro, M., Damsteegt, V., Bassett, C., Petri, C., Liu, Z., et al. (2007). Plum pox virus coat protein gene intron hairpin RNA construct provides resistance to *Plum pox virus* in *Nicotiana benthamiana* and *Prunus domestica*. *Journal of the American Society for Horticultural Science, 132*, 850–858.

Himmler, G., Laimer, M., Stemkellner, H., Mattanovich, D., Griessler, B., & Katinger, H. (1987). Production of monoclonal antibodies against *Plum pox virus* for the diagnosis of sharka disease of stone fruit (in German). *Mitteilungen Klosterneuburg, 37*, 251–253.

Hunter, J. A., Chamberlain, E. E., & Atkinson, J. D. (1958). Note on the transmission of apple mosaic by natural root grafting. *New Zealand Journal of Agricultural Research, 1*, 80–82.

Ilardi, V., & Di Nicola-Negri, E. (2011). Genetically engineered resistance to *Plum pox virus* infection in herbaceous and stone fruit hosts. *GM Crops, 2*, 24–33.

Ito, T., Ieki, H., & Ozaki, K. (2002). Simultaneous detection of six citrus viroids and *Apple stem grooving virus* from Citrus plants by multiplex reverse transcription polymerase chain reaction. *Journal of Virological Methods, 106*, 235–239.

Jackson, A. L., Bartz, S. R., Schelter, J., Kobayashi, S. V., Burchard, J., Mao, M., et al. (2003). Expression profiling reveals off-target gene regulation by RNAi. *Nature Biotechnology, 21*, 635–637.

James, D. (1999). A simple and reliable protocol for the detection of *Apple stem grooving virus* by RT-PCR and in a multiplex PCR assay. *Journal of Virological Methods, 83*, 1–9.

James, D. (2011). Cherry mottle leaf virus. In A. Hadidi, M. Barba, T. Candresse, & W. Jelkmann (Eds.), *Virus and virus-like diseases of pome and stone fruits* (pp. 127–132). St. Paul, MN: APS Press.

James, D., & Howell, W. E. (1998). Isolation and partial characterization of a filamentous virus associated with peach mosaic disease. *Plant Disease, 82*, 909–913.

James, D., & Mukerji, S. (1996). Comparison of ELISA and immunoblotting techniques for the detection of *Cherry mottle leaf virus*. *Annals of Applied Biology, 129*, 13–23.

James, D., Varga, A., & Sanderson, D. (2013). Genetic diversity of *Plum pox virus*: Strains, disease and related challenges for control. *Canadian Journal of Plant Pathology, 35*, 431–441.

Jelkmann, W. (2001). Detection of viruses and virus-like disease of fruit trees-laboratory assays, bioassays and indicators. *Acta Horticulturae, 550*, 473–491.

Jelkmann, W., & Eastwell, K. C. (2011). Little cherry virus-1 and -2. In A. Hadidi, M. Barba, T. Candresse, & W. Jelkmann (Eds.), *Virus and virus-like diseases of pome and stone fruits* (pp. 153–160). St. Paul, MN: APS Press.

Jelkmann, W., & Keim-Konrad, R. (1997). An immuno-capture polymerase chain reaction and plate-trapped ELISA for the detection of *Apple stem pitting virus*. *Journal of Phytopathology, 145*, 499–504.

Jelkmann, W., Kunze, L., Vetten, H. J., & Lesemann, D. E. (1992). cDNA cloning of dsRNA associated with apple stem pitting disease and evidence for the relationship of the virus-like agents associated with apple stem pitting and pear vein yellows. *Acta Horticulturae, 309*, 55–62.

Jelkmann, W., & Paunovic, S. (2011). Apple stem pitting virus. In A. Hadidi, M. Barba, T. Candresse, & W. Jelkmann (Eds.), *Virus and virus-like diseases of pome and stone fruits* (pp. 35–40). St. Paul, MN: APS Press.

Jelkmann, W., Rott, M., & Uyemoto, J. K. (2011). Cherry green ring mottle virus. In A. Hadidi, M. Barba, T. Candresse, & W. Jelkmann (Eds.), *Virus and virus-like diseases of pome and stone fruits* (pp. 115–118). St. Paul, MN: APS Press.

Karayiannis, I., Malnou, A., Stylianides, D., Thomidis, T., Karayiannis, N., & Tsaftaris, A. (2006). Resistance to sharka disease new apricots of high quality selected in Greece. *Acta Horticulturae, 1*, 337–340.

Karayiannis, I., Thomidis, T., & Tsaftaris, A. (2008). Inheritance of resistance to *Plum pox virus* in apricot (*Prunus armeniaca* L.). *Tree Genetics & Genomes, 4*, 143–148. http://dx.doi.org/10.1007/s11295-007-0095-z.

Kegler, H., Grüntzig, M., Fuchs, E., Rankovic, M., & Ehrig, F. (2001). Hypersensitivity of plum genotypes to *Plum pox virus*. *Journal of Phytopathology, 149*, 213–218.

Kring, J. B. (1972). Flight behaviour of aphids. *Annual Review of Entomology, 17*, 461–492.

Labonne, G., Yvon, M., Quiot, J. B., Avinent, L., & Llacer, G. (1995). Aphids as potential vectors of *Plum pox virus*: Comparison of methods of testing and epidemiological consequences. *Acta Horticulturae, 386*, 207–218.

Laimer, M., & Barba, M. (2011). Elimination by thermotherapy, tissue culture or *in vitro* micrografting. In A. Hadidi, M. Barba, T. Candresse, & W. Jelkmann (Eds.), *Virus and virus-like diseases of pome and stone fruits* (pp. 391–396). St. Paul, MN: APS Press.

Lemgo, G. N. Y., Sabbadini, S., Pandolfini, T., & Mezzetti, B. (2013). Biosafety considerations of RNAi-mediated virus resistance in fruit-tree cultivars and in rootstock. *Transgenic Research, 22*, 1073–1088.

Lin, S., Wu, H., Elena, S., Chen, K., Niu, Q., Ye, S., et al. (2009). Molecular evolution of a viral Non-Coding sequence under the selective pressure of amiRNA-mediated silencing. *PLoS Pathogens*, *5*(2), e1000312.

Liu, Z., Scorza, R., Hily, J. M., Scott, S. W., & James, D. (2007). Engineering resistance to multiple Prunus fruit viruses through expression of chimeric hairpins. *Journal of the American Society for Horticultural Science*, *132*, 407–414.

Llácer, G., Cambra, M., Laviña, A., & Aramburu, J. (1985). Suitable conditions for detecting *Apple chlorotic leaf spot virus* in apricot trees by enzyme-linked immunosorbent assay (ELISA). *Agronomie*, *5*, 809–812.

Malinowski, T., Cambra, M., Capote, N., Zawadzka, B., Gorris, M. T., Scorza, R., et al. (2006). Field trials of plum clones transformed with the *Plum pox virus* coat protein (PPV-CP) gene. *Plant Disease*, *90*(8), 1012–1018.

Malinowski, T., Cieślińska, M., Zawadzka, B., Interewicz, B., & Porebska, A. (1997). Characterisation of monoclonal antibodies against *Apple chlorotic leaf spot virus* (ACLSV) and their application for detection of ACLSV and identification of its strains. *Phytopathologia Polonica*, *14*, 35–40.

Marandel, G., Salava, J., Abbott, A., Candresse, T., & Decroocq, V. (2009). Quantitative trait loci meta-analysis of *Plum pox virus* resistance in apricot (*Prunus armeniaca* L.): New insights on the organization and the identification of genomic resistance factors. *Molecular Plant Pathology*, *10*, 347–360.

Marbot, S., Salmon, M., Vendrame, M., Huwaert, A., Kummert, J., Dutrecq, O., et al. (2003). Development of real-time RT-PCR assay for detection of *Prunus necrotic ringspot virus* in fruit trees. *Plant Disease*, *87*, 1344–1348.

Martínez-Gómez, P., & Dicenta, F. (2000). Evaluation of resistance of apricot cultivars to a Spanish isolate of *Plum pox potyvirus*. *Plant Breeding*, *119*, 179–181.

Massart, S., Jijakli, M. H., & Kummert, J. (2011). Apple stem grooving virus. In A. Hadidi, M. Barba, T. Candresse, & W. Jelkmann (Eds.), *Virus and virus-like diseases of pome and stone fruits* (pp. 85–90). St. Paul, MN: APS Press.

Massart, S., Roussel, S., Kummert, J., Dutrecq, O., & Jijakli, M. H. (2008). Development of routine duplex RT-PCR tests for certification of fruit tree multiplication material. *Acta Horticulturae*, *781*, 107–112.

Mathioudakis, M. M., Maliogka, V. I., Dovas, C. I., Paunovic, S., & Katis, N. I. (2009). Reliable RT-PCR detection of Apple stem pitting virus in pome fruits and its association with quince fruit deformation disease. *Plant Pathology*, *58*, 228–236.

McCallum, C. M., Comai, L., Greene, E. A., & Henikoff, S. (2000). Targeted screening for induced mutations. *Nature Biotechnology*, *18*, 455–457.

McLarty, H. R. (1935). Cherry mottle leaf. (Abstr.). *Northwest Association of Horticulturists, Entomologists, and Plant Pathologists*, *1*, 5.

Mekouria, T. A., Smith, T. J., Beers, E., & Eastwell, K. C. (2013). Little cherry virus 2 is transmitted to sweet cherry by Pseudococcus maritimus (Ehrhorn), a new vector of this virus. *Plant Disease*, *97*, 851.

Mekuria, G., Ramesh, S. A., Alberts, E., Bertozzi, T., Wirthensohn, M., Collins, G., et al. (2003). Comparison of ELISA and RT-PCR for the detection of *Prunus necrotic ring spot virus* and *Prune dwarf virus* in almond (*Prunus dulcis*). *Journal of Virological Methods*, *114*, 65–69.

Menzel, W., Zahn, V., & Maiss, E. (2003). Multiplex RT-PCR-ELISA compared with bioassay for the detection of four apple viruses. *Journal of Virological Methods*, *110*, 153–157.

Migliori, A., Quiot, J. B., Labonne, G., Boudon, J. P., Lauriaut, F., Freydier, M., et al. (1998). Mineral oil, a means of preventive control against the agent of Sharka spread by aphids in nurseries. *Phytoma*, *51*, 32–35.

Milbrath, J. (1966). Severe fruit necrosis of sour cherry caused by strains of *Green ring mottle virus*. *Plant Disease Reporter*, *50*, 59–62.

Mink, G. I. (1992). Prunus necrotic ringspot virus. In J. Kumar, H. S. Chaube, U. S. Singh, & A. N. Mukhopadhyay (Eds.), *Plant diseases of international importance: 3. Diseases of fruit crops* (pp. 335–356). Englewood Cliffs, NJ: Prentice Hall.

Mink, G. I. (1998). Virus certification of deciduous fruit trees in the United States and Canada. In A. Hadidi, R. K. Khetarpal, & H. Koganczawa (Eds.), *Plant virus diseases control* (pp. 294–300). St. Paul, MN: APS Press.

Mink, G. I., & Aichele, M. D. (1984). Detection of Prunus necrotic ringspot and prune dwarf viruses in *Prunus* seed and seedlings by enzyme-linked immunosorbent assay. *Plant Disease, 68*, 378–381.

Mink, G. I., Howell, W. E., Cole, A., & Regev, S. (1987). Three serotypes of *Prunus necrotic ringspot virus* isolated from rugose mosaic-diseased sweet cherry trees in Washington. *Plant Disease, 71*, 91–93.

Mink, G. I., & Shay, J. R. (1959). Preliminary evaluation of some Russian apple varieties as indicators for apple viruses. *Plant Disease, 254*, 13–17.

Monticelli, S., Nicola-Negri, D., Gentile, A., Damiano, C., & Ilardi, V. (2012). Production and *in vitro* assessment of transgenic plums for resistance to *Plum pox virus*: A feasible, environmental risk-free, cost-effective approach. *Annals of Applied Biology, 161*, 293–301.

Moustafa, T. A., Badenes, M. L., Martínez-Calvo, J., & Llácer, G. (2001). Determination of resistance to sharka (*Plum pox*) virus in apricot. *Scientia Horticulturae, 91*, 57–70.

Myrta, A., Di Terlizzi, B., Boscia, D., Choueiri, E., Gatt, M., Gavriel, I., et al. (2001). Serological characterisation of Mediterranean *Prunus necrotic ringspot virus* isolates. *Journal of Plant Pathology, 83*, 45–49.

Myrta, A., Di Terlizzi, B., & Savino, V. (1998). Study on the transmission of *Plum pox potyvirus* through seeds. *Phytopathologia Mediterranea, 37*, 41–44.

Myrta, A., Matic, S., Malinowski, T., Pasquini, G., & Candresse, T. (2011). *Apple chlorotic leaf spot virus* in stone fruits. In A. Hadidi, M. Barba, T. Candresse, & W. Jelkmann (Eds.), *Virus and virus-like diseases of pome and stone fruits* (pp. 85–90). St. Paul, MN: APS Press.

Myrta, A., Potere, O., Boscia, D., Candresse, T., Cambra, M., & Savino, V. (1998). Production of monoclonal antibodies specific to El-Amar strain of *Plum pox virus*. *Acta Virologica, 42*, 248–250.

Myrta, A., Potere, O., Crescenzi, A., Nuzzaci, M., & Boscia, D. (2000). Production of two monoclonal antibodies specific to cherry strain of *Plum pox virus* (PPV-C). *Journal of Plant Pathology, 82*, 95–103.

Nekrasov, V., Staskawicz, B., Weigel, D., Jones, J. D., & Kamoun, S. (2013). Targeted mutagenesis in the model plant *Nicotiana benthamiana* using Cas9 RNA-guided endonuclease. *Nature Biotechnology, 31*, 691–693.

Nemeth, M. (1986). *Virus, mycoplasmas and rickettsia diseases of fruit trees*. Budapest: Akademiai Kiado.

Niu, Q. W., Lin, S. S., Reyes, J. L., Chen, K. C., Wu, H. W., Yeh, S. D., et al. (2006). Expression of artificial microRNAs in transgenic *Arabidopsis thaliana* confers virus resistance. *Nature Biotechnology, 24*, 1420–1428.

Nolasco, G., Neves, M. A., & Faria, E. A. (1991). Distribuição no Algarve de vírus do grupo Ilarvirus em amendoeira e suas consequências na produção: 1ª aproximação. *Revista de la Facultad de Ciencias Agrarias, 15*, 33–37.

Notomi, T., Okayama, H., Masubuchi, H., Yonekawa, T., Watanabe, K., Amino, N., et al. (2000). Loop-mediated isothermal amplification of DNA. *Nucleic Acids Research, 28*(12), E63.

Nyland, G., Gilmer, R. M., & Moore, J. D. (1976). Prunus ringspot virus group. In *U.S. Dept. of Agr. Handbook: 437. Virus diseases and noninfectious disorders of stone fruits in North America* (pp. 104–132). Washington, DC: U. S. Government Printing Office.

Oldfield, G. N. (1970). Mite transmission of plant viruses. *Annual Review of Entomology, 15*, 343–380.

Oldfield, G. N., Creamer, R., Gispert, C., Osorio, F., Rodriguez, R., & Perring, T. M. (1995). Incidence and distribution of peach mosaic and its vector, *Eriophyes insidiosus* (Acari: Eriophyidae) in Mexico. *Plant Disease*, *79*, 186–189.

Olmos, A., Bertolini, E., Gil, M., & Cambra, M. (2005). Real-time assay for quantitative detection of non-persistently transmitted *Plum pox virus* RNA targets in single aphids. *Journal of Virological Methods*, *128*, 151–155.

Olmos, A., Cambra, M., Marais, A., Faure, C., & Candresse, T. (2012). A first evaluation of siRNA next-generation sequencing for detection and characterization of *Prunus* viruses. *Petria*, *22*, 281.

Pandolfini, T., Molesini, B., Avesani, L., Spena, A., & Polverari, A. (2003). Expression of self-complementary hairpin RNA under the control of the rolC promoter confers systemic disease resistance to *Plum pox virus* without preventing local infection. *BMC Biotechnology*, *3*, 7.

Pascal, T., Pfeiffer, F., & Kervella, J. (2002). Preliminary observations on the resistance to sharka in peach and related species. *Acta Horticulturae*, *592*, 699–706.

Pasquini, G., & Barba, M. (1991). Production and application of monoclonal antibodies against *Apple mosaic virus*. *Petria*, *1*, 31–36.

Pasquini, G., & Barba, M. (2006). The question of seed transmissibility of Plum pox virus. *OEPP/EPPO Bulletin*, *36*, 287–292.

Pasquini, G., Barba, M., Hadidi, A., Faggioli, F., Negri, R., Sobol, I., et al. (2008). Microarray-based detection and genotyping of *Plum pox virus*. *Journal of Virological Methods*, *147*, 118–126.

Pasquini, G., Simeone, A. M., Conte, L., & Barba, M. (2000). RT-PCR evidence of the non-transmission through seed of *Plum pox virus* strains D and M. *Journal of Plant Pathology*, *82*, 221–226.

Paunovic, S., Maksimovic, V., Rankovic, M., & Radovic, S. (1999). Characterization of a virus associated with pear stony pit in cv. Wurttemberg. *Journal of Phytopathology*, *147*, 695–700.

Paunovic, S., Pasquini, G., & Barba, M. (2011). *Apple mosaic virus* in stone fruits. In A. Hadidi, M. Barba, T. Candresse, & W. Jelkmann (Eds.), *Virus and virus-like diseases of pome and stone fruits* (pp. 91–96). St. Paul, MN: APS Press.

Petri, C., & Burgos, L. (2005). Transformation of fruit trees. Useful breeding tool or continued future prospect? *Transgenic Research*, *14*, 15–26.

Petrzik, K., & Lenz, O., (2011). *Apple mosaic virus* in pome fruit. In A. Hadidi, M. Barba, T. Candresse, & W. Jelkmann (Eds.), *Virus and virus-like diseases of pome and stone fruits* (pp. 25–28). St. Paul, MN: APS Press.

Pickett, J. A., Wadhams, L. J., Woodcock, C. M., & Hardie, J. (1992). The chemical ecology of aphids. *Annual Review of Entomology*, *37*, 67–90.

Pilařová, P., Marandel, G., Decroocq, V., Salava, J., Krška, B., & Abbott, A. G. (2010). Quantitative trait analysis of resistance to *Plum pox virus* in the apricot F1 progeny "Harlayne" × "Vestar". *Tree Genetics & Genomes*, *6*(3), 467–475.

Polak, J., Pivalova, J., Kundu, J. K., Jokes, M., Scorza, R., & Ravelonandro, M. (2008). Behaviour of transgenic *Plum pox virus* resistant *Prunus domestica* L. clone C5 grown in the open field under a high and permanent infection pressure of the PPV-Rec strain. *Journal of Plant Pathology*, *90*, 33–36.

Posnette, A. F., & Ellenberger, E. C. (1957). The line pattern virus disease of plum. *The Annals of Applied Biology*, *45*, 74–80.

Poul, F., & Dunez, J. (1990). Use of monoclonal antibodies for the identification of different antigenic domains in *Apple chlorotic leaf spot virus*. *Archives of Virology*, *114*, 191–202.

Qu, J., Ye, J., & Fang, R. (2007). Artificial microRNA-mediated virus resistance in plants. *Journal of Virology*, *81*, 6690–6699.

Raine, J., McMullen, R. D., & Forbes, R. D. (1986). Transmission of the agent causing little cherry disease by the apple mealybug *Phenacoccus aceris* and the dodder *Cuscuta lupuliformis*. *Canadian Journal of Plant Pathology, 8*, 6–11.

Rampitsch, C., Eastwell, K. C., & Hall, J. (1995). Setting confidence limits for the detection of *Prune dwarf virus* in *Prunus avium* with a monoclonal antibody-based triple antibody-sandwich ELISA. *Annals of Applied Biology, 126*, 485–491.

Rasmussen, E. J., Berkeley, G. H., Cation, D., Hildebrand, E. M., Keitt, G. W., & Moore, J. D. (1951). Green ring mottle. In D. Cation, G. H. Berkeley, L. C. Cochran, F. P. Cullinan, & R. J. Haskell (Eds.), *Virus diseases and other disorders with virus-like symptoms of stone fruits in North America* (pp. 159–161). Washington, DC, USA: Department of Agriculture.

Ravelonandro, M., Scorza, R., Hily, J. M., & Briard, P. (2014). The efficiency of RNA interference for conferring stable resistance to Plum pox virus. *Plant Cell, Tissue and Organ Culture, 118*, 347–356.

Reed, P. J., & Foster, J. A. (2011). Exclusion of pome and stone fruit viruses, viroids and phytoplasmas by certification and quarantine. In A. Hadidi, M. Barba, T. Candresse, & W. Jelkmann (Eds.), *Virus and virus-like diseases of pome and stone fruits* (pp. 391–396). St. Paul, MN: APS Press.

Roussel, S., Kummert, J., Salmon, M., Dutrecq, O., & Jijakli, M. (2005). Development of RT-PCR assays using fluorogenic-3′ minor groove binder DNA probes for detection of fruit tree viruses. *Bulletin OEPP, 35*, 105–108.

Roy, A. S. (2011). Control measures of pome and stone fruit viruses, viroids and phytoplasmas: Role of international organizations. In A. Hadidi, M. Barba, T. Candresse, & W. Jelkmann (Eds.), *Virus and virus-like diseases of pome and stone fruits* (pp. 407–413). St. Paul, MN: APS Press.

Rubio, M., Martínez-Gómez, P., & Dicenta, F. (2003). Resistance of almond cultivars to *Plum pox virus* (sharka). *Plant Breeding, 122*, 462–464.

Sanchez-Navarro, A. J., Aparicio, F., Herranz, M. C., Minafra, A., Myrta, A., & Pallas, V. (2005). Simultaneous detection and identification of eight stone fruit viruses by one-step RT-PCR. *European Journal of Plant Pathology, 111*, 77–84.

Sanford, J. C., & Johnston, S. A. (1985). The concept of parasite-derived resistance. Deriving resistance genes from the parasite's own genome. *Journal of Theoretical Biology, 113*, 395–405.

Schneider, W. L., Sherman, D. J., Stone, A. L., Damsteegt, V. D., & Fredeick, R. D. (2004). Specific detection and quantification of *Plum pox virus* by real-time fluorescent reverse transcription-PCR. *Journal of Virological Methods, 120*, 97–105.

Scholthof, K. B., Adkins, S., Czosnek, H., Palukaitis, P., Jacquot, E., Hohn, T., et al. (2011). Top 10 plant viruses in molecular plant pathology. *Molecular Plant Pathology, 12*, 938–954.

Scorza, R., Callahan, A., Dardick, C., Ravelonandro, M., Polak, J., Malinowski, T., et al. (2013). Genetic engineering of *Plum pox virus* resistance: 'HoneySweet' plum—From concept to product. *Plant Cell, Tissue and Organ Culture, 115*, 1–12.

Scorza, R., Callahan, A., Levy, L., Damsteegt, V., Webb, K., & Ravelonandro, M. (2001). Post-transcriptional gene silencing in *Plum pox virus* resistant transgenic European plum containing the plum pox potyvirus coat protein gene. *Transgenic Research, 1054*, 1–9.

Scorza, R., Ravelonandro, M., Callahan, A. M., Cordts, J. M., Fuchs, M., Dunez, J., et al. (1994). Transgenic plum (*Prunus domestica* L.) express the *Plum pox virus* coat protein gene. *Plant Cell Reports, 14*, 18–22.

Scott, S. W., Barnett, O. W., & Burrows, P. (1989). *Prunus necrotic ringspot virus* in selected peach orchards in South Carolina. *Plant Disease, 73*, 913–916.

Simón-Mateo, C., & García, J. A. (2006). miRNA-guided processing impairs *Plum pox virus* replication but the virus readily evolve to escape this silencing mechanism. *Journal of Virology, 80*, 2429–2436.

Simón-Mateo, C., & García, J. A. (2011). Antiviral strategies in plants based on RNA silencing. *Biochimica et Biophysica Acta, 1809*, 722–731.
Smith, N. A., Singh, S. P., Wang, M. B., Stoutjesdijk, P. A., Green, A. G., & Waterhouse, P. M. (2000). Total silencing by intron-spliced hairpin RNAs. *Nature, 407*, 319–320.
Song, G. Q., Sink, K. C., Walworth, A. E., Cook, M. A., Allison, R. F., & Lang, G. A. (2013). Engineering cherry rootstocks with resistance to *Prunus necrotic ring spot virus* through RNAi-mediated silencing. *Plant Biotechnology Journal, 11*, 702–708.
Sweet, J. B. (1980). Fruit tree virus infections of woody exotic and indigenous plants in Britain. *Acta Phytopathologica, 15*, 231–238.
Szittya, G., Silhavy, D., Molnar, A., Havelda, Z., Lovas, A., Lakatos, L., et al. (2003). Low temperature inhibits RNA silencing-mediated defence by the control of siRNA generation. *EMBO Journal, 22*, 633–640.
Tepfer, M. (2002). Risk assessment of virus-resistant transgenic plants. *Annual Review of Phytopathology, 40*, 467–491.
Theilmann, J., Mozafari, J., Reade, R., Wu, Z., Xie, W., Jesperson, G., et al. (2002). Partial nucleotide sequence and genome organization of a Canadian isolate of *Little cherry virus* and development of an enzyme-linked immunosorbent assay-based diagnostic test. *Phytopathology, 92*, 87–98.
Theilmann, J., Orban, S., & Rochon, D. (2004). High sequence variability among *Little cherry virus* isolates occurring in British Columbia. *Plant Disease, 88*, 1092–1098.
Thomas, H. E., & Hildebrand, E. M. (1936). A virus disease of prune. *Phytopathology, 26*, 1145–1148.
Torrance, L., & Dolby, C. A. (1984). Sampling conditions for reliable routine detection by enzyme-linked immunosorbent assay of three ilarviruses in fruit trees. *The Annals of Applied Biology, 119*, 265–271.
Uyemoto, J. K., & Scott, S. W. (1992). Important diseases of *Prunus* caused by viruses and other graft-transmissible pathogens in California and South Carolina. *Plant Disease, 76*, 5–11.
Varga, A., & James, D. (2006). Detection and differentiation of *Plum pox virus* using real-time multiplex PCR with SYBR Green and melting curve analysis: A rapid method for strain typing. *Journal of Virological Methods, 123*, 213–220.
Voinnet, O. (2008). Post-transcriptional RNA silencing in plant–microbe interactions: A touch of robustness and versatility. *Current Opinion in Plant Biology, 11*, 464–470.
Wang, Q. C., Cuellar, W. J., Rajamaki, M. L., Hiraka, Y., & Valkonen, J. P. T. (2008). Combined thermotherapy and cryotherapy for virus eradication: Evaluation of virus distribution, subcellular changes, cell survival and viral RNA degradation in shoot tips to efficient production of virus-free plants. *Molecular Plant Pathology, 9*, 237–250.
Wang, X., Kohalmi, S. E., Svircev, A., Wang, A., Sanfacon, H., et al. (2013). Silencing of the host factor eIF(iso)4E gene confers *Plum Pox Virus* resistance in plum. *PLoS One, 8*(1), e50627. http://dx.doi.org/10.1371/journal.pone.0050627.
Wang, Q. C., Mawassi, M., Li, P., Gafny, R., Sela, I., & Tanne, E. (2003). Elimination of *Grapevine virus A* (GVA) by cryopreservation of *in vitro*-grown shoot tips of *Vitis vinifera* L. *Plant Science, 165*, 321–327.
Wang, A., Tian, L., Huang, T. S., Brown, D. C. W., Svircev, A. M., Stobbs, L. W., et al. (2009). The development of genetic resistance to *Plum pox virus* in transgenic *Nicotiana benthamiana* and *Prunus domestica*. *Acta Horticulturae, 839*, 665–672.
Waterworth, H. E., & Hadidi, A. (1998). Economic losses due to plant viruses. In A. Hadidi, R. K. Khetarpal, & H. Koganezawa (Eds.), *Plant virus disease control* (pp. 1–13). St. Paul, MN: The American Phytopathological Society Press.

Yaegashi, H., Yoshikawa, N., & Candresse, T. (2011). *Apple chlorotic leaf spot virus* in pome fruits. In A. Hadidi, M. Barba, T. Candresse, & W. Jelkmann (Eds.), *Virus and virus-like diseases of pome and stone fruits* (pp. 17–22). St. Paul, MN: APS Press.

Yanase, H. (1974). Studies on apple latent viruses in Japan. *Bulletin of the Fruit Tree Research Station C, 1*, 47–109.

Yanase, H. (1983). Back transmission of apple stem grooving virus to apple seedlings and induction of symptoms of apple topworking disease in Mitsuba Kaido (*Malus sieboldii*) and Kabano Zumi (*Malus sieboldii* var. *arborescens*) rootstocks. *Acta Horticulturae, 130*, 117–122.

Yoshikawa, N. (2001). *Apple chlorotic leaf spot* virus. *CMI/AAB Descriptions of Plant Viruses, 386* No. 30 revised.

Yoshikawa, N., Yamagishi, N., Yaegashi, H., & Ito, T. (2012). Deep sequence analysis of viral small RNAs from a green crinkle-diseased apple tree. *Petria, 22*, 292–297.

Zagrai, I., Zagrai, L., Ravelonandro, M., Gaboreanu, I., Pamfil, D., Ferencz, B., et al. (2008). Environmental impact assessment of transgenic plums on the diversity of *Plum pox virus* populations. *Acta Horticulturae, 781*, 309.

Zagula, K. R., Aref, N. M., & Ramsdell, D. C. (1989). Purification, serology and some properties of a mechanically transmissible virus associated with green ring mottle disease in peach and cherry. *Phytopathology, 79*, 451–456.

Zhang, Y. P., Kirkpatrick, B. C., Smart, C. D., & Uyemoto, J. K. (1998). cDNA cloning and molecular characterization of *Cherry green ring mottle virus*. *Journal of General Virology, 79*, 2275–2281.

Zhang, S. C., Tian, L. M., Svircev, A., Brown, D. C. W., Sibbald, S., Schneider, K. E., et al. (2006). Engineering resistance to *Plum pox virus* (PPV) through the expression of PPV-specific hairpin RNAs in transgenic plants. *Canadian Journal of Plant Pathology, 28*, 263–270.

CHAPTER FOUR

Cassava Virus Diseases: Biology, Epidemiology, and Management

James P. Legg*,[1], P. Lava Kumar[†], T. Makeshkumar[‡], Leena Tripathi[§], Morag Ferguson[§], Edward Kanju*, Pheneas Ntawuruhunga[¶], Wilmer Cuellar**

*International Institute of Tropical Agriculture (IITA), Dar es Salaam, Tanzania
[†]International Institute of Tropical Agriculture (IITA), Ibadan, Nigeria
[‡]Central Tuber Crops Research Institute (CTCRI), Thiruvananthapuram, India
[§]International Institute of Tropical Agriculture (IITA), Nairobi, Kenya
[¶]International Institute of Tropical Agriculture (IITA), Lusaka, Zambia
**Centro Internacional de Agricultura Tropical (CIAT), Cali, Colombia
[1]Corresponding author: e-mail address: j.legg@cgiar.org

Contents

Abstract

Cassava (*Manihot esculenta* Crantz.) is the most important vegetatively propagated food staple in Africa and a prominent industrial crop in Latin America and Asia. Its vegetative propagation through stem cuttings has many advantages, but deleteriously it means that pathogens are passed from one generation to the next and can easily accumulate,

Advances in Virus Research, Volume 91
ISSN 0065-3527
http://dx.doi.org/10.1016/bs.aivir.2014.10.001

threatening cassava production. Cassava-growing continents are characterized by specific suites of viruses that affect cassava and pose particular threats. Of major concern, causing large and increasing economic impact in Africa and Asia are the cassava mosaic geminiviruses that cause cassava mosaic disease in Africa and Asia and cassava brown streak viruses causing cassava brown streak disease in Africa. Latin America, the center of origin and domestication of the crop, hosts a diverse set of virus species, of which the most economically important give rise to cassava frog skin disease syndrome. Here, we review current knowledge on the biology, epidemiology, and control of the most economically important groups of viruses in relation to both farming and cultural practices. Components of virus control strategies examined include: diagnostics and surveillance, prevention and control of infection using phytosanitation, and control of disease through the breeding and promotion of varieties that inhibit virus replication and/or movement. We highlight areas that need further research attention and conclude by examining the likely future global outlook for virus disease management in cassava.

1. INTRODUCTION

1.1. Cassava: the plant, its cultivation and current economic importance

Cassava (*Manihot esculenta* subspecies *esculenta* Crantz) is a perennial shrub from the family Euphorbiaceae. Its geographical origins remain a topic of research debate, but the most recent evidence based on molecular markers suggests that the plant was domesticated within the southwestern rim of the Amazon basin (in modern day Brazil) and is derived from its closest wild relative, *M. esculenta* ssp. *flabellifolia* (Pohl) (Léotard et al., 2009; Olson & Schaal, 1999). Cassava plants typically reach 1–4 m in height at physiological maturity, and the tuberous roots produced may be harvested from 6 months to 4 years after planting. These roots, which typically have a dry matter content of 30–40%, provide an important source of starch, and in communities in South America that have cultivated the crop for many generations, a wide variety of processed products have been developed. The broad environmental adaptability of cassava and its tolerance of acid soils and sustained periods of drought were key factors in its widespread adoption throughout the tropical Americas. Although monoecious cassava plants produce fertile seeds, and these may be planted, the standard cultivation system makes use of stem cuttings for propagation and establishing a new crop. Vegetative propagation ensures uniformity of a crop variety from season to season and means that planting a new crop is relatively simple; however, this also has the negative consequence of sustaining pathogen populations from one cropping cycle to the next, a fact that is particularly significant in the epidemiology of viruses that infect the plant.

Cassava may have been domesticated as much as 10,000 years ago, but it did not spread beyond Latin America until the sixteenth century, when Portuguese traders introduced it to the western Atlantic shores of Africa, in the Gulf of Guinea (Carter, Fresco, Jones, & Fairbairn, 1997). Diffusion inland occurred slowly after that, but by the period of European "exploration" into the interior in the nineteenth century, cassava cultivation had become widespread throughout much of the tropical belt of West, Central, and East Africa. The eighteenth and nineteenth centuries also saw the introduction of cassava to much of south and southeast Asia (Onwueme, 2002). By the start of the twenty-first century, cassava was being widely cultivated throughout the tropics and had become a globally important crop, providing an essential source of carbohydrates to hundreds of millions of people and offering diverse commercial and industrial applications via transformation processes. Although Latin America remains a major producer, more than half of global production is currently in Africa (FAOSTAT, 2014). In spite of its preeminence in overall production of cassava, Africa has lower average yields (10.9 t/ha) than both South America (13.2 t/ha) and Asia (19.7 t/ha) (FAOSTAT, 2014). Much of the production in Asia is grown for either animal feed exports (Thailand and Vietnam) or starch (India). Europe provided the major market for these products in the twentieth century, but rapid growth in demand in recent years means that China is now the main importer of cassava products, importing more than 15 million tons in 2011 (FAOSTAT, 2014).

1.2. Threats to cassava production

Cassava is affected by a diverse set of constraints. Some of the most important of these are pests and diseases. Arguably, the greatest deleterious global impacts on cassava production have resulted from the inadvertent introduction of insect pests or disease-causing pathogens to regions in which they did not previously occur. The most important examples of this have been the introductions of the arthropod pests—cassava mealybug [*Phenacoccus manihoti* Mat.-Ferr.] (CM), cassava green mite [*Mononychellus tanajoa* (Bondar)] (CGM), and cassava bacterial blight (CBB) caused by *Xanthomonas axonopodis* pv. *manihotis*—to Africa in the 1970s, and the more recent introductions of CM and CBB to southeast Asia (CIAT, 2010). These alien invasive introductions have also been associated with some of the greatest successes in cassava pest and disease management in the form of classical biological control programs that have resulted in dramatic reductions in the incidence and damage caused by CGM in Africa and CM in both Africa and Asia (IITA, 2010a; Neuenschwander, 1994; Yaninek, Onzo, & Ojo, 1993). Relatively less success has been achieved, by contrast, in managing pests and diseases that are indigenous to their respective continents. Most notable among these are the virus diseases. A diverse set of virus species infect cassava in Latin America, although the most economically important of these are the species that give rise to the cassava frogskin disease (CFSD) syndrome. Although this has been recognized for many years (Pineda, Jayasinghe, & Lozano, 1983), there is still uncertainty about the precise etiology, and evidence has been presented for the involvement of phytoplasmas (Alvarez, Mejía, Llano, & Loke, 2009) as well as several virus species (Carvajal-Yepes et al., 2014). There are fewer virus groups affecting cassava in Africa and fewer still in Asia. However, the large and increasing economic impact of cassava mosaic geminiviruses (CMGs) that cause cassava mosaic disease (CMD) in Africa and Asia and cassava brown streak viruses (CBSVs) causing cassava brown streak disease (CBSD) in Africa is such that they are currently considered to be the greatest global threat to cassava production (Legg, Somado, et al., 2014). In this chapter, we examine the biology and epidemiology of the most important groups of viruses infecting cassava in the world, and review control tactics and disease management strategies. In so doing, we highlight areas that need further research attention and conclude by examining the likely future outlook for virus disease management in cassava.

2. BIOLOGY AND EPIDEMIOLOGY OF CASSAVA VIRUSES

2.1. Viruses of cassava

2.1.1 Africa and South Asia

2.1.1.1 Introduction

Following the introduction of cassava to Africa in the sixteenth century, it became infected by a unique set of viruses, none of which are recorded from the crop's center of origin in South America. About 15 virus species and several strains have been identified infecting cassava in Africa and its offshore islands (Table 1). Eleven of these are responsible for the two most devastating diseases, namely: CMD and CBSD. *Cassava green mottle virus*, Cassava virus C and Cassava Kumi virus A & B have been reported infecting cassava but they were not well characterized, and their significance is not known (Table 1).

Virus name	Genus/Family	References	Sequence(s) available[a]	Diagnostics[b]	Distribution
(a) Latin America					
Cassava common mosaic virus (CsCMV)	*Alphaflexiviridae/ Potexvirus*	Costa (1940), Silva, Kitajima, and Oliveira (1963), and Kitajima, Wetter, Oliveira, Silva, and Costa (1965)	NC_001658	ELISA/ RT-PCR	Colombia, Brazil (isolated cases from Africa, Asia)
Cassava vein mosaic virus (CsVMV)	*Caulimoviridae/ Cavemovirus*	Costa (1940), de Kochko et al. (1998)	NC_001648	PCR	Brazil
Cassava virus X (CsVX)	*Alphaflexiviridae/ Potexvirus*	Lennon, Aiton, and Harrison (1986)	NA[c]	ELISA/ RT-PCR	Colombia
Cassava new alphaflexivirus (CsNAV)	*Alphaflexiviridae/ Potexvirus*	Carvajal-Yepes et al. (2014)	KC505252	RT-PCR	Colombia
Cassava frogskin-associated virus (CsFSaV)	*Reoviridae/ Oryzavirus*	Calvert, Cuervo, Lozano, Villareal, and Arroyave (2008)	DQ139870	RT-PCR	Colombia, Brazil, Costa Rica, Argentina
Cassava polero-like virus (CsPLV)	*Luteoviridae/ Polerovirus*	Carvajal-Yepes et al. (2014)	KC505249	RT-PCR	Colombia, Costa Rica
Cassava torrado-like virus (CsTLV)	*Secoviridae/ Torradovirus*	Carvajal-Yepes et al. (2014)	KC505250, KC505151	RT-PCR	Colombia, Argentina
Cassava symptomless virus (CsSLV)	*Rhabdoviridae/ Nucleorhabdovirus*[c]	Kitajima and Costa (1979)	NA	NA	Brazil
Cassava Caribbean mosaic virus (CsCaMV)	*Alphaflexiviridae/ Potexvirus*[c]	Lennon et al. (1986)	NA	NA	Colombia
Cassava Colombian symptomless virus (CsCSLV)	*Alphaflexiviridae/ Potexvirus*[c]	Lennon et al. (1986)	NA	NA	Colombia
Cassava American latent virus (CsALV)	*Secoviridae/ Nepovirus*[c]	Walter, Ladeveze, Etienne, and Fuchs (1989)	NA	NA	Brazil, Guyana

Continued

Table 1 The viruses of cassava—cont'd

(b) Africa

Cassava mosaic disease					
African cassava mosaic virus (ACMV)	*Begomovirus/ Geminiviridae*	Morris, Coates, Lowe, Richardson, and Eddy (1990)	X17095, X17096	PCR and Real-time PCR	SSA[d]
African cassava mosaic Burkina Faso virus (ACMBFV)	*Begomovirus/ Geminivridae*	Tiendrébéogo et al. (2012)	HE616777, HE616778	PCR and Real-time PCR	Burkina Faso
Cassava mosaic Madagascar virus (CMMGV)	*Begomovirus/ Geminivridae*	Harimalala et al. (2012)	HE617299, HE617300	PCR and Real-time PCR	Madagascar
East African cassava mosaic Cameroon virus (EACMCV)	*Begomovirus/ Geminiviridae*	Fondong et al. (2000)	AF112354, AF112355	PCR and Real-time PCR	SSA and Comoros
East African cassava mosaic Kenya virus (EACMKV)	*Begomovirus/ Geminiviridae*	Bull et al. (2006)	AJ717580, AJ704965	PCR and Real-time PCR	East Africa, Madagascar, Seychelles, Comoros
East African cassava mosaic Malawi virus (EACMMV)	*Begomovirus/ Geminiviridae*	Zhou, Robinson, and Harrison (1998)	AJ006460, N/A	PCR and Real-time PCR	Malawi
East African cassava mosaic virus (EACMV)	*Begomovirus/ Geminiviridae*	Bull et al. (2006)	AJ717542, AJ704949	PCR and Real-time PCR	SSA
East African cassava mosaic virus-Ugandan Variant (EACMV-UG)	*Begomovirus/ Geminiviridae*	Pita et al. (2001)	AF126804-7	PCR and Real-time PCR	SSA
East African cassava mosaic Zanzibar virus (EACMZV)	*Begomovirus/ Geminiviridae*	Bull et al. (2006)	AJ717562, AJ704942	PCR and Real-time PCR	Zanzibar, Madagascar
South African cassava mosaic virus (SACMV)	*Begomovirus/ Geminiviridae*	Berrie, Rybicki, and Rey (2001)	AF155806, AF155807	PCR and Real-time PCR	South Africa, Madagascar, Zimbabwe

Cassava brown streak disease					
Cassava brown streak virus (CBSV)	*Ipomovirus/ Potyviridae*	Winter et al. (2010)	FN434436	RT-PCR, Real-Time RT-PCR, RT-LAMP, and ELISA	Kenya, Mozambique, Uganda, Tanzania, Malawi, Rwanda, Burundi and DR Congo.
Ugandan cassava brown streak virus (UCBSV)	*Ipomovirus/ Potyviridae*	Mbanzibwa, Tian, Mukasa, and Valkonen (2009)	FJ039520	RT-PCR, Real-Time RT-PCR, RT-LAMP, and ELISA	Kenya, Mozambique, Uganda, Tanzania, Malawi, Rwanda, Burundi and DR Congo.
(c) South Asia and minor viruses					
Cassava mosaic disease					
Indian cassava mosaic virus (ICMV)	*Begomovirus/ Geminiviridae*	Malathi, Nair, and Shantha (1985), Hong, Robinson, and Harrison (1993)	NC_001932, NC_001933	PCR	Southern India and Sri Lanka
Sri Lankan cassava mosaic virus (SLCMV)	*Begomovirus/ Geminiviridae*	Saunders et al. (2002)	AJ314737, AJ314738	PCR	Southern India and Sri Lanka
Cassava viruses not linked with any major disease[c]					
Cassava virus C (CsVC) (syn. Cassava Q virus)	*Ourmiavirus/* Unassigned	Calvert and Thresh (2002), Rastgou et al. (2009)	FJ157981-83	NA[e]	Ivory Coast
Cassava green mottle virus (CsGMV)	*Nepovirus/ Comoviridae*	Lennon, Aiton, and Harrison (1987)	NA	NA	Australasia and Pacific Islands, Solomon Islands
Cassava Ivorian bacilliform virus (CIBV)	*Anulavirus/ Bromoviridae*	Fargette, Roberts, and Harrison (1991), Scott, MacFarlane, McGavin, and Fargette (2014)	NA	NA	Ivory Coast
Cassava Kumi viruses A and B	Uncharacterized	Calvert and Thresh (2002)	NA	NA	Kumi district of Uganda

[a]GenBank accession numbers of reference isolates of viruses provided.
[b]Many methods are available. Most common and current methods) indicated.
[c]Limited knowledge on disease biology and causal virus, reliable diagnostic tools are yet to be developed.
[d]SSA, sub-Saharan Africa.
[e]NA, not available.

Cassava Ivorian bacilliform virus (CIBV) (genus *Anulavirus*; family *Bromoviridae*) has recently had its full genome characterized (Scott et al., 2014) but has only been recorded from Ivory Coast and has no known effect on cassava. This section will consider the two main groups of cassava viruses that cause CMD and CBSD.

2.1.1.2 Cassava mosaic geminiviruses

The causal agent of CMD was initially named as cassava latent virus (Bock, Guthrie, & Figueiredo, 1981) but was subsequently characterized and renamed as *African cassava mosaic virus* (ACMV) (Bock & Woods, 1983; Stanley & Gay, 1983) (genus *Begomovirus*; family *Geminiviridae*). Between 1983 and 2006, seven different Begomovirus species were identified in association with CMD in different regions of Africa (Table 1; Figs. 1 and 2) (Alabi, Kumar, & Naidu, 2011; Legg & Fauquet, 2004; Patil & Fauquet, 2009): ACMV, *East African cassava mosaic virus* (EACMV), *East African cassava mosaic Malawi virus* (EACMMV), *South African cassava mosaic virus* (SACMV), *East African cassava mosaic Cameroon virus* (EACMCV), *East African cassava mosaic Zanzibar virus* (EACMZV) and *East African cassava mosaic Kenya virus* (EACMKV). The most recent additions to this list are *Cassava mosaic Madagascar virus* (CMMGV, Harimalala et al., 2012) and *African cassava mosaic Burkina Faso virus* (ACMBFV, Tiendrébéogo et al., 2012). In addition, several strains of these viruses have been identified. The most notable of these is East African cassava mosaic virus-Uganda (EACMV-UG) also known as the "Uganda variant" (Zhou et al., 1997). EACMV-UG was the first recorded example of a begomovirus that has arisen through recombination between two distinct begomovirus species (EACMV and ACMV) (Zhou et al., 1997).

Indian cassava mosaic virus (ICMV) was the first CMG to be recorded from South Asia (Malathi et al., 1985), followed by *Sri Lankan cassava mosaic virus* (SLCMV) several years later (Saunders et al., 2002). Although SLCMV was initially reported from Sri Lanka, it was subsequently shown to occur also in southern India, together with ICMV (Anitha, Makeshkumar, & Edison, 2011; Patil, Rajasubramaniam, Bagchi, & Dasgupta, 2005).

Coinfection with more than one species or strain is a common feature in the etiology of CMD in Africa. Where mixtures are composed of ACMV and one of the EACMV-like CMG species, a synergistic interaction between the species occurs, resulting in an increased overall virus titer leading to more severe symptoms (Fondong et al., 2000; Harrison, Zhou, Otim-Nape, Liu, & Robinson, 1997; Ogbe, Thottappilly, Dixon, & Mignouna, 2003). Rapid regionwide spread of EACMV-UG and ACMV, frequently

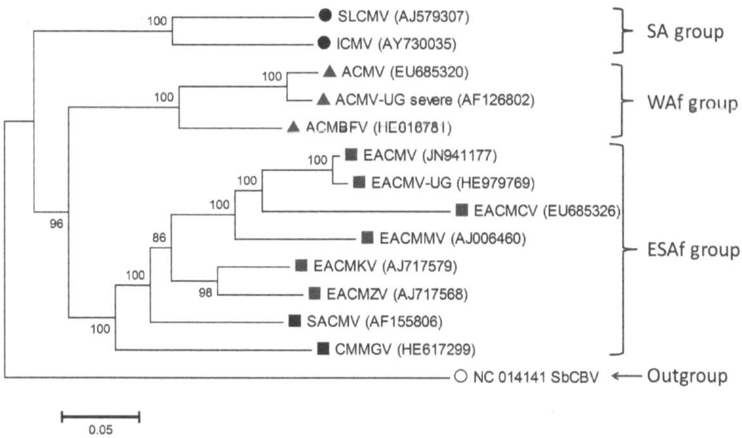

Figure 1 Phylogenetic relationships of the cassava mosaic geminiviruses. Phylogenetic relationship of 11 species of cassava mosaic geminiviruses based on the alignment of complete DNA-A using MEGA 5 software (Tamura et al., 2011). Sequences were aligned using the ClustalW algorithm, and the tree was constructed by the Neighbor-Joining method. The tree was rooted using *Soybean chlorotic blotch virus* (SbCBV) as an outgroup. The evolutionary distances were computed using the maximum composite likelihood method and are in the units of the number of base substitutions per site. The tree is drawn to scale, with branch lengths in the same units as those of the evolutionary distances used to infer the phylogenetic tree. The percentages of replicate trees in which the associated taxa clustered together in the bootstrap test (1000 replicates) are shown at the branch nodes. There were a total of 2627 positions in the final dataset. All positions containing gaps and missing data were eliminated from the dataset (complete deletion option). The NCBI GenBank accession numbers of 14 DNA-A sequences are indicated in parenthesis. ACMV: *African cassava mosaic virus*; ACMV-UG: ACMV-Uganda; ACMBFV: *African cassava mosaic Burkina Faso Virus*; CMMGV: *Cassava mosaic Madagascar virus*; EACMCV: *East African cassava mosaic Cameroon virus*; EACMKV: *East African cassava mosaic Kenya virus*; EACMMV: *East African cassava mosaic Malawi virus*; EACMZV: *East African cassava mosaic Zanzibar virus*; EACMV: *East African cassava mosaic virus*; EACMV-UG: EACMV-Uganda; ICMV: *Indian cassava mosaic virus*; SACMV: *South African cassava mosaic virus*; SLCMV: *Sri Lankan cassava mosaic virus*; SA: South Asian; WA: West African; ESA: Eastern and Southern African. (See the color plate.)

in mixed infections, was an important feature of the African severe CMD pandemic (Harrison et al., 1997; Legg, 1999; Otim-Nape et al., 1997).

The genome of CMGs comprises two circular single-stranded DNA molecules (DNA-A and DNA B) of about 2.8 kb each, encapsidated in 30 × 20-nm twinned icosahedral particles that replicate by rolling circle amplification through a dsDNA intermediary stage (Hanley-Bowdoin, Settlage, Orozco, Nagar, & Robertson, 1999). DNA-A carries six open reading frames (ORFs), with each encoding a specific protein: AC1, the replication-associated protein (Rep); AC2, the transcriptional activator

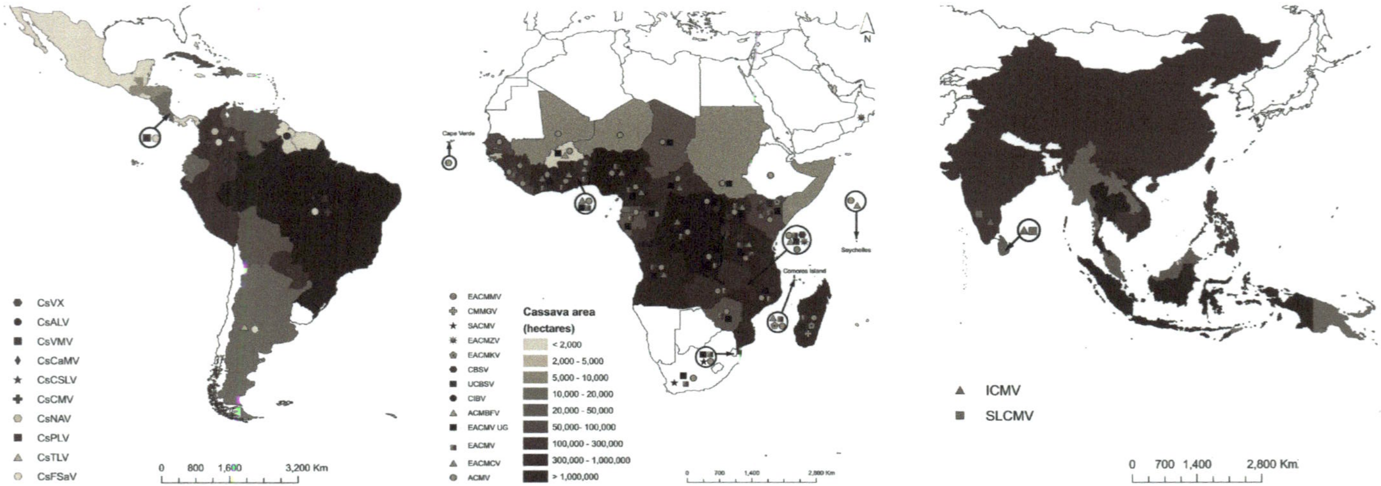

Figure 2 Global distribution of viruses affecting cassava. (See the color plate.)

protein (TrAP); AC3, the replication enhancer protein (REn); AC4, the RNA-silencing suppressor; AV1, the coat protein (CP); and AV2, the pre-coat protein. DNA-B has two ORFs: BV1 encodes the nuclear-shuttle protein and BC1 encodes the movement protein (MP).

2.1.1.3 Cassava brown streak viruses

CBSD was first reported in Tanzania by Storey (1936) and was considered to have a viral etiology from the outset. Although several efforts were made to identify the virus causing CBSD during the twentieth century (Bock, 1994a), it was not until early in the twenty-first century that its identity was confirmed (Monger, Seal, Isaac, & Foster, 2001) and the first sequence data provided. Following the characterization of whole genomes of several viruses isolated from CBSD-infected plants, it was shown that two species of CBSVs are involved in the etiology of CBSD: *Cassava brown streak virus* (CBSV) and *Ugandan cassava brown streak virus* (UCBSV) (genus *Ipomovirus*, family *Potyviridae*) (Mbanzibwa et al., 2009; Winter et al., 2010) (Table 1; Figs. 2 and 3).

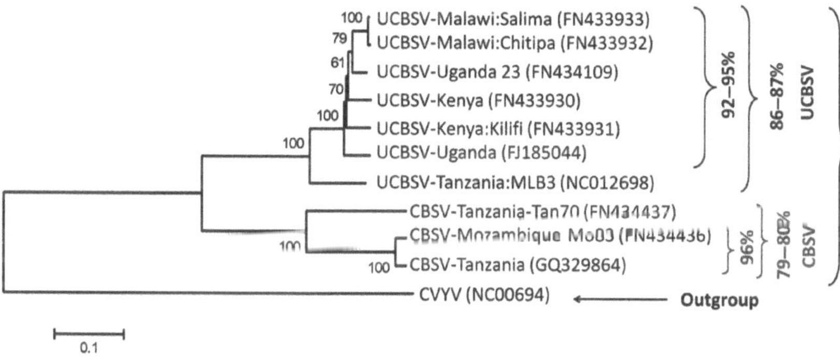

Figure 3 Phylogenetic relationships of the cassava brown streak viruses. Phylogenetic relationship of two cassava brown streak virus species based on the alignment of complete genome sequences using MEGA 5 software (Tamura et al., 2011). Sequences were aligned using the ClustalW algorithm, and the tree was constructed by the Neighbor-Joining method. The tree was rooted by using *Cucumber vein yellows virus* (CVYV) as an outgroup. The evolutionary distances were computed using the maximum composite likelihood method and are in the units of the number of base substitutions per site. The tree is drawn to scale, with branch lengths in the same units as those of the evolutionary distances used to infer the phylogenetic tree. The percentages of replicate trees in which the associated taxa clustered together in the bootstrap test (1000 replicates) are shown at the branch nodes. All positions containing gaps and missing data were eliminated from the dataset (complete deletion option). The NCBI GenBank accession numbers of 11 sequences are given in parentheses. Percentage sequence homologies within and between various clusters are indicated. CBSV: *Cassava brown streak virus*; UCBSV: *Ugandan cassava brown streak virus*. (See the color plate.)

The complete genomes of these two single-stranded RNA viruses (four CBSV isolates and eight UCBSV isolates, respectively) are 69.0–70.3% and 73.6–74.4% identical at the nucleotide and polyprotein amino acid sequence levels, respectively (Mbanzibwa et al., 2011). The viral genome is expressed as a polyprotein, which is subsequently cleaved by viral proteinases to produce the mature proteins. The genome structures of CBSV and UCBSV are similar, but they differ from other ipomoviruses (Mbanzibwa et al., 2009; Winter et al., 2010). CBSV and UCSBV encode a single P1 proteinase that functions as a suppressor of RNA silencing (Mbanzibwa et al., 2009). Neither of the CBSVs encodes a helper component proteinase (HC-Pro). The most unique feature of the CBSVs is the HAM1h protein, which is a putative nucleoside triphosphate pyrophosphatase, and is situated between the viral replicase (NIb) and the CP in the C-proximal part of the polyprotein. This has only previously been reported for *Euphorbia ringspot virus* (genus *Potyvirus*, family *Potyviridae*) (Mbanzibwa et al., 2009). Based on the analysis of CP genes of CBSV and UCBSV, the two viruses have been shown to be undergoing active but slightly different patterns of evolution (Mbanzibwa et al., 2011).

2.1.2 Latin America

Viruses reported infecting cassava in the "new world" are diverse and belong to virus families *Alphaflexiviridae*, *Reoviridae*, *Secoviridae*, and *Luteoviridae* among the RNA viruses, and to the family *Caulimoviridae* among the DNA viruses (Table 1). Unlike in Africa, no geminivirid or potyvirid sequences have been reported associated with diseases in cassava in the Americas (Calvert, Cuervo, & Lozano, 2012). Recent field surveys in Colombia have detected the common occurrence of *Cassava frogskin-associated virus* (CsFSaV) in fields affected by CFSD, the most economically important disease of cassava in Latin America. CsFSaV is a reovirus of isometric particles of ~70 nm containing a genome consisting of 10 dsRNA segments (Calvert et al., 2008). Such segments may differ in size distribution and distinct dsRNA patterns can be observed in samples collected in the Amazonas region when compared with samples collected in other regions of Colombia. This is confirmed by phylogenetic analyses of the replicase region from different isolates which indicate that the more diverse CsFSaV sequences are found in the Amazonas region (Cuervo, M. et al., unpublished results). Although CsFSaV is associated with the rot symptoms characteristic of CFSD, in single infections it does not induce leaf symptoms in the cassava indicator variety "Secundina" (Carvajal-Yepes et al., 2014; Fig. 4).

Figure 4 Virus symptoms observed in cassava plants of the landrace Secundina (COL2063) infected by viruses detected in Latin America. Symptoms shown develop around 3 weeks after grafting on noninfected Secundina rootstocks: (A) uninfected control, (B) infected with CsTLV, (C) infected with CsCMV, and (D) mixed infected with CsCMV, CsFSaV and CsTLV. Secundina plants single infected with CsFSaV rarely show symptoms in leaves. (See the color plate.)

The second-most common virus found in Colombia is Cassava torrado-like virus (CsTLV), a newly described virus species belonging to the *Torradovirus* genus. CsTLV has a bipartite genome of ~10,000 nt and isometric virion particles of ~25 nm (Carvajal-Yepes et al., 2014). Similar particles had been previously observed in preparations of CsFSaV from CFSD-affected plants (Calvert et al., 2008), and the virus has also been detected in frozen samples from symptomatic plants that were collected in the 1980s, suggesting that the virus has been present for longer than previously thought (Carvajal-Yepes et al., 2014). When analyzed at sequence level, a high degree of variability is detected, and even generic primers (Verbeek, Tang, & Ward, 2012) are not able to detect all isolates collected in the same region (Carvajal-Yepes et al., 2014).

Cassava common mosaic virus (CsCMV) (Costa, 1940) has been reported in several countries in South America in association with symptoms of mosaic and chlorosis in leaves (Calvert et al., 2012), although recent reexaminations of some of these samples have recorded the presence of additional newly described viruses that may have contributed to the symptoms initially attributed to CsCMV. CsCMV belongs to the *Potexvirus* genus with elongated, semiflexuous particles of 15 × 495 nm (Kitajima et al., 1965). To date, there are only two reported sequences of CsCMV isolates from Brazil, and because they only share ca. 80% nucleotide identity, it seems likely that other distinct strains of the virus exist.

Cassava vein mosaic virus (CsVMV), first reported from Brazil in 1940 (Costa, 1940), is the type species of the genus *Cavemovirus* (family *Caulimoviridae*). It is the only pararetrovirus of cassava, and has so far only been reported from Brazil. Virus particles are quasispherical, 45-50 nm in diameter, and encapsidate a single circular dsDNA genome of ~8159 bp (de Kochko et al., 1998). Virus symptoms in infected plants include chlorosis along the veins, mosaic, and leaf distortion. CsVMV spreads readily through vegetative propagation but transmission through seed or an insect vector has not been detected. There is scant information on the biology, epidemiology, or control of the disease caused. One study suggests that there are nonsignificant differences in yield between CsVMV-infected and uninfected plants (Santos et al., 1995). However, the 35S promoter sequence derived from the CsVMV genome is extensively used as a constitutive promoter in the genetic engineering of plants (Verdaguer et al. 1996).

A diverse range of mechanically transmitted potexviruses have been detected in symptomless cassava, including *Cassava virus X* (CsVX) and Cassava new alphaflexivirus (CsNAV) (Carvajal-Yepes et al., 2014; Harrison, Lennon, & Aiton, 1986). However, because these viruses do not cause symptoms, it is difficult to determine their distribution or to evaluate their importance. In fact, at least four other potexviruses infecting cassava have been reported since the 1960s (Table 1), but for these no sequences or even antisera are available, therefore it is not possible now to identify them to the species level.

2.2. Diseases caused by cassava viruses

2.2.1 Cassava mosaic disease

The first report of viruses affecting cassava was made from East Africa toward the end of the nineteenth century (Warburg, 1894). Here, the term "Kräuselkrankheit" was used to describe the mottled mosaic-like symptoms, leaf deformation, and stunted growth seen in affected plants. The first real evidence for a viral etiology was presented several decades later (Storey, 1938), although this was not to be definitively confirmed and causal viruses named until the development and application of early molecular techniques

(Bock & Woods, 1983). CMD in India has a more recent described history. The first published record of its presence was during the 1960s (Alagianagalingam & Ramakrishnan, 1966), although this source refers to an earlier report made a decade previously (Abraham, 1956). The earliest reports of CMD in India noted that the disease was restricted to the cassava-growing regions of southern India: primarily Kerala and Tamil Nadu, and to a lesser extent Karnataka and Andhra Pradesh (Malathi et al., 1985). More recently, CMD was also reported from Sri Lanka (Austin, 1986). Although there is variation in severity associated with the mix of species causing CMD in South Asia, in all cases the general symptoms are the same as those observed in Africa, hence the use of a single name to describe the disease, regardless of the CMG species causing the infection (Calvert & Thresh, 2002). CMD has not yet been recorded at any location in South America or Southeast Asia. It is assumed that this is a consequence of the absence of suitable insect vectors in these regions.

2.2.2 Cassava brown streak disease

CBSD was first reported during the early days of the Amani (northeastern Tanzania) cassava research program in the 1930s (Storey, 1936, 1938). At the time of this earliest record, it was already noted that CBSD occurred widely in the coastal region of East Africa. Three major symptom types were recognized: a blotchy yellow chlorosis of mature leaves often associated with minor veins; brown, round, or elongate streak-like lesions on the young green portion of stems; and dry, brown necrotic lesions in the tuberous roots (Storey, 1936). The viral etiology of CBSD was subsequently proved through the demonstration of Koch's postulates (Winter et al., 2010).

Prior to the twenty-first century, the distribution of the disease remained restricted to coastal East Africa—from northeastern Kenya in the north to Mozambique in the south, and inland to the shores of Lake Malawi (Hillocks & Jennings, 2003; Nichols, 1950). This situation changed abruptly in the early 2000s, however, as new outbreaks were reported from mid-altitude (>1000 m above sea level) areas of south-central Uganda (Alicai et al., 2007), western Kenya (H.M. Obiero, personal communication), and northwestern Tanzania (Jeremiah & Legg, 2008), precipitated by massive increases in populations of the whitefly, *Bemisia tabaci* (Legg et al., 2011). CBSD has subsequently been shown to be spreading as a pandemic through the major cassava-growing regions of East and Central Africa (Bigirimana, Barumbanze, Ndayihanzamaso, Shirima, & Legg, 2011; Legg et al., 2011) and threatens to spread further westwards into Central and West Africa (Legg, Somado, et al., 2014).

2.2.3 Cassava frogskin disease

Root symptoms associated with CFSD have been known in the Amazonas region for many years, but the first serious outbreak was reported in 1971 and occurred in the Cauca region of Colombia, across the Andes, which caused up to 90% yield losses. CFSD is characterized by the failure of the storage roots to accumulate starch and so affected roots develop a rough epidermis resembling the wart-like skin of toads or alligators, hence its name in Spanish (cuero de sapo) and in Portuguese (jacare). Only recently it was found that complex virus infections are associated with root symptoms of CFSD, that several novel viruses are part of that complex, and that neither CsFSaV nor phytoplasma can induce CFSD symptoms or the associated leaf symptoms in single infections (Alvarez et al., 2009; Carvajal-Yepes et al., 2014). More studies are needed to determine the specific role of each component.

As a consequence of this uncertainty, grafting is recommended for CFSD indexing in cassava. Interestingly, CsTLV is a member of the *Torradovirus* genus, is whitefly transmitted and has been associated with leaf symptoms in "Secundina" in single infections (Carvajal-Yepes et al., 2014).

2.3. Vectors of cassava viruses

2.3.1 Cassava mosaic geminiviruses

From the earliest period of research on CMD, it was suspected that the whitefly, *Bemisia tabaci* (Genn.), was the vector of the pathogens causing the disease, and these suspicions were provisionally confirmed by studies in both Central Africa (Kufferath & Ghesquière, 1932) and East Africa (Storey & Nichols, 1938). Dubern (1994) described the characteristics of transmission of ACMV and confirmed that *B. tabaci* was a relatively inefficient vector. Transtadial but not transovarial transmission was demonstrated. Adult whitefly transmission comprised a relatively long (minimum 3 h) acquisition access period (AAP) with a short (minimum 10 min) inoculation access period (IAP). A moderately long latent period (minimum 3 h) was recognized to be a consequence of the circulative transmission pathway.

The persistent mechanism of transmission, typical of begomoviruses, meant that ACMV particles were retained for an experimentally determined time of up to 9 days, a period that in nature might equate to the lifetime of the insect. This long-term association between virus and vector has important consequences for the pattern of spread of CMGs, and notably means that these viruses can be carried over long distances by their whitefly hosts. There have been no specific studies examining long distance flight of cassava-colonizing *B. tabaci*, but circumstantial evidence obtained from regional virus spread data suggests that *B. tabaci* populations can carry CMGs over distances of up to 38 km in a year (Legg, 2010). Research on *B. tabaci* elsewhere has reported individual flights of adult whiteflies of up to 7 km (Cohen, Kern, Harpaz, & Ben-Joseph, 1988).

CMGs from south Asia are transmitted in a similar way to their African relatives. ICMV was successfully transmitted from cassava to cassava by whiteflies reared on cassava, but not whiteflies reared on sweet potato (Antony et al., 2006), most likely since these were different *B. tabaci* genotypes. Virus-free cassava, generated by meristem-tip culture, has been used to study the transmission of viruses in cassava by *B. tabaci*. Using cassava-adapted whiteflies, symptoms appeared 25 days after inoculation and 85% of test plants became infected (Duraisamy et al., 2013).

2.3.2 Cassava brown streak viruses

Storey (1939) suggested that *Bemisia* whiteflies were the most likely vector of the viruses causing CBSD, but this was not definitively proved until many years later (Maruthi et al., 2005). Early experiments indicated transmission frequencies of about 22%, but more recent work has achieved higher efficiency levels and has indicated that transmission of these ipomoviruses is semipersistent (Jeremiah, 2012; Jeremiah, C. S. & Maruthi, M. N., unpublished data). Preliminary data from this work indicate a minimum AAP of 5 min, a minimum IAP of 30 min and a maximum retention time of 24 h. No significant differences in transmission have been observed for the two species of CBSVs (CBSV and UCBSV) (M.N. Maruthi, unpublished data). The shorter retention time of CBSVs by *B. tabaci* in comparison with CMGs, suggests that CBSVs are less likely to spread over long distances than CMGs.

2.3.3 Viruses associated with CFSD

There are no published studies of the vector transmission of Latin American viruses. Even the etiology of CFSD has yet to be fully described, as there is currently no confirmation of the vector(s) responsible for transmitting the viruses associated with the disease. Interestingly, it was observed that whiteflies (*Aleurotrachelus socialis* Bondar) were able to transmit agents responsible for leaf symptoms from CFSD-affected plants to healthy "Secundina" plants, but whether root symptoms could be elicited through similar "transmission" remained unclear (Angel, Nolt, & Pineda, 1987, Angel, Pineda, Nolt, & Velasco, 1989).

2.4. Epidemiology of cassava viruses

2.4.1 *Cassava mosaic geminiviruses*

During the early years of the Amani program in northeastern Tanzania, it was observed that the rate of spread of CMD was much greater at hot and moist low-altitude locations than it was at higher elevations where temperatures were cooler and the whiteflies less abundant (Storey, 1936, 1939). Similar epidemiological characteristics were noted from Ivory Coast in West Africa, during the intensive period of research of the ORSTOM/CIRAD program (Fauquet & Fargette, 1990; Fauquet, Fargette, & Thouvenel, 1988). Some of the key milestones achieved in developing knowledge of the epidemiology of CMD were as follows:

i. The most important source of new infection in initially CMD-free plantings was shown to be surrounding plantings of cassava (Fargette, Fauquet, Grenier, & Thresh, 1990; Fargette, Fauquet, & Thouvenel, 1985).

ii. Environmental spread gradients were demonstrated in which both whitefly vectors and new CMD infections were aggregated on upwind borders of fields (Fargette et al., 1985, 1990).

iii. Rates of spread were shown to vary greatly between seasons, and most of this variation could be attributed to changes in whitefly abundance and temperature (Fargette, Jeger, Fauquet, & Fishpool, 1993).

iv. A decline in susceptibility of cassava plants to new infection was demonstrated as plants matured (Fargette et al., 1993).

CMGs have been identified from several plants other than cassava (Alabi, Ogbe, et al., 2008; Robertson, 1985). However, the frequency of infection is typically very low, which coupled with the year-round presence of cassava means that cultivated cassava is considered to be the only epidemiologically significant host of CMGs.

As the focus of research switched from West to East Africa during the 1990s, great emphasis was placed on experimenting with CMD-resistant varieties, and large differences in the rate of spread into cassava varieties were demonstrated in Uganda (Otim-Nape, Thresh, & Shaw, 1998). Recovery (disappearance of symptoms during crop growth) and reversion (production of virus-free cuttings by infected parent plants) (Fargette, Thresh, & Otim-Nape, 1994) were shown to be important resistance mechanisms influencing epidemiology.

In Uganda, massive changes in the character of CMD spread were recorded from several fixed locations as an epidemic of unusually severe CMD expanded its geographical range to affect many parts of the country

(Gibson, Legg, & Otim-Nape, 1996; Legg & Ogwal, 1998; Otim-Nape et al., 1997). Some of the earliest molecular studies of CMGs helped to explain this phenomenon, by demonstrating that this unusually rapid spread of CMD was associated with the occurrence of mixed infections of two CMGs: ACMV and EACMV-UG (Harrison et al., 1997; Zhou et al., 1997). Synergism between the CMGs resulted in raised virus titers and increased virus spread between plants and fields. Arguably the most important "driver" of this rapid virus spread, however, was the massive increase in abundance of the *B. tabaci* whitefly vector that coincided with this outbreak in Uganda (Legg & Ogwal, 1998; Otim-Nape, Thresh, & Fargette, 1996). Molecular evidence based on cytochrome oxidase I sequences of mitochondrial DNA have suggested that genetically distinct populations of *B. tabaci* whiteflies are associated with the CMD pandemic in East and Central Africa (Legg, French, Rogan, Okao-Okuja, & Brown, 2002; Legg, Sseruwagi, et al., 2014).

Early studies of the epidemiology of CMD in India demonstrated primary spread through the use of diseased planting material (Shanta, 1978; Thankappan, 1978) and secondary spread through the *B. tabaci* whitefly vector (Nair, 1985). ICMV was reported to be transmitted by *B. tabaci* from cassava to cucumber (Mathew & Muniyappa, 1993; Menon & Raychaudhuri, 1970) as well as from cassava to cassava (Antony et al., 2006).

2.4.2 Cassava brown streak viruses

It was recognized from the earliest period of CBSD research in the 1930s that the disease spread more readily at low altitudes (Storey, 1936). The rapid spread that has occurred since 2004 in the Great Lakes region has demonstrated the importance of super-abundant populations of the whitefly vector, *B. tabaci* (Legg et al., 2011).

As is the case for the CMGs, the CBSVs appear to have no epidemiologically significant alternative host plants, although *Manihot glaziovii* has been shown to be infected by these viruses (Mbanzibwa et al., 2011). The question of the origins of the CBSVs remains an interesting one from an evolutionary perspective, as cassava arriving in East Africa in the eighteenth century must have been infected from a wild host somewhere in the region.

CBSVs are transmitted semipersistently, which means that they are not carried long distances by vectors, while the persistent transmission of CMGs means that they spread together with migrating populations of *B. tabaci*.

Field trials to determine the field-level epidemiological characteristics of CBSD have shown that gradients of spread are relatively steep from infected source plots to neighboring initially uninfected test plots (Jeremiah, 2012). The highest apparent rates of CBSD spread occurred in the cool, dry part of the year, suggesting that environmental factors (such as temperature, and/or soil moisture) may be important contributing factors to CBSD symptom expression (Jeremiah, 2012).

2.4.3 Latin American viruses

Almost nothing is known about the epidemiology of the viruses affecting cassava in South America (Calvert et al., 2012). It is apparent, however, that most of these viruses are almost entirely propagated through the use of infected planting material. If there were specific and more efficient vectors of these viruses, then they would certainly have become more widespread and spread more readily than they currently do. It is significant, at least for some of the viruses occurring in South America, that the *B. tabaci* genotype groups present in South America ("New World" and "MEAM1"; Dinsdale, Cook, Riginos, Buckley, & De Barro, 2010) do not colonize cassava (Carabali, Bellotti, Montoya-Lerma, & Cuellar, 2005). More importantly, the absence of cassava-colonizing populations of *B. tabaci* in South America means that the likelihood of the CMGs or CBSVs spreading there following inadvertent introduction from Africa is low.

3. MANAGEMENT OF CASSAVA VIRUSES

3.1. Management strategies for plant viruses in cassava

There are several general components that are essential for the effective management of plant virus diseases, and these are equally applicable to cassava. These components can be broadly described as: recognition and monitoring; prevention of infection; and control of infection. An important facet of the context for controlling viruses in cassava is that the majority of farmers growing this crop are subsistence producers who are either unable or choose not to allocate resources to the purchase of inputs for virus control.

3.1.1 Recognition and monitoring

Recognition of the occurrence of a disease is an essential prerequisite for its control. For cassava viruses, the ease with which this can be achieved varies greatly. Some viruses produce very obvious symptoms in the plant's foliage (CMGs, virus combinations producing CFSD), while others are much more cryptic, as symptoms may be confined to lower leaves or roots (CBSVs) or may not be visible at all in any plant part (CsSLV). For the major economically important diseases such as CMD in Africa and Asia, and CBSD in Africa, large-scale surveillance programs are required to record incidence levels and their changes over time and geography. These can be achieved effectively for CMD using visual symptoms and standardized survey protocols (Sseruwagi, Sserubombwe, Legg, Ndunguru, & Thresh, 2004), but virus testing may be required to confirm infection status for CBSD.

A vital aspect of recognition and monitoring systems is accurate diagnosis. Effective diagnosis relies on a combination of proper recognition of symptoms—where these are present—and the application of accurate, robust, and affordable laboratory-based diagnostics. Great progress has been made in the development and extension of diagnosis systems for the most important virus groups, such as the CMGs in Africa and Asia, and the CBSVs in Africa. Recent research in Latin America, however, using next-generation sequencing (NGS) approaches, has made it clear that current knowledge of the viruses affecting cassava is incomplete, and further work is required before fully comprehensive virus diagnosis systems can be developed. This will be particularly important for tissue culture-based virus indexing systems where it is necessary to provide assurance that tissue culture material is virus-free, for the purposes of local, regional, or international germplasm exchange.

3.1.2 Prevention of infection

Preventing infection can be an effective method of controlling disease, although in many situations this may be difficult to achieve. Many of the approaches of prevention might be considered as cultural control methods or phytosanitation. This starts with assuring the health of material to be planted, and then by selecting a site not close to other fields that might serve as sources of virus inoculum, or in a location known to be unfavorable for insect vectors. Unlike several other more commercial vegetatively propagated crops, cassava is almost never grown under protected conditions. The only exception to this general rule is when tissue culture plantlets are hardened off, usually following reception from a remote source during germplasm exchange programs. Infection in a newly planted crop can be prevented by assuring that the variety grown is resistant to vector-borne infection, and also by controlling the vectors themselves, although this is rarely practiced for cassava. Resistance may either be to the vector itself or be to infection by viruses carried by the vector. An important feature of varieties resistant to CMD is that infection is delayed relative to susceptible varieties. Delaying infection is an important contributing factor to the prevention of virus disease in cassava, notably since yield losses decline the later that cassava plants become infected by virus (Fauquet & Fargette, 1990).

3.1.3 Control of infection

Virus infection in cassava may be controlled by inhibiting virus replication or movement in the infected plant or by destroying infected plants. In the latter approach symptomatic or virus-positive asymptomatic plants are removed, typically through uprooting (=roguing). A wider range of techniques are used to inhibit virus function within infected plants. Meristem tip culture combined with thermotherapy excludes virus particles from meristem tissue, which is subsequently propagated through tissue culture. Both conventional and transgenic resistance approaches rely on modifications of the host plant's inherent defense mechanisms to restrict virus replication and/or movement. While these do not directly eliminate virus particles, they control infection by restricting the multiplication of the virus and thus its concentration to levels that do not result in economic damage or by restricting movement such that only a small proportion of plant parts are affected. Since virus-resistant varieties can be propagated for many years, the deployment of host plant resistance has been the most widely used control tactic in the control of cassava viruses. Although transgenic varieties offer the potential to make this resistance stronger, more efficient and more durable, their widespread use is likely to continue to be constrained by difficulties in resolving regulatory concerns.

In the following sections, we discuss each of the major cassava virus control approaches in greater detail.

3.2. Diagnostics and surveillance

Diagnostic methods used for the detection and identification of the viruses of cassava have been generally similar wherever the viruses occur. During the early years of research, from the 1930s to the 1970s, classical techniques of symptom description, grafting, mechanical inoculation, and the use of indicator plants were used. More recently, serological and nucleic acid-based techniques have facilitated more rapid, sensitive, and high volume diagnostic assessments. As nucleic acid sequencing has become cheaper and more widely accessible, whole genome sequencing has become a common "add-on" to diagnostic testing programs, and "fishing" for asymptomatic virus-like nucleic acids can now be achieved with NGS. This technique proved vital in the diagnosis and characterization of viruses associated with CFSD in Colombia (Carvajal-Yepes et al., 2014).

There are several groups of users of cassava virus diagnostics, and each has different requirements, as follows:

i. *Researchers*: diagnostics development; identification of new viruses; virus characterization; detection and identification of viruses in cassava research materials; virus indexing; surveillance; training. *Priorities*: novelty, sensitivity.
ii. *Plant protection/quarantine staff*: testing germplasm imports/exports; local quarantine; testing for seed certification. *Priorities*: accuracy, ease of use.
iii. *Commercial tissue culture laboratories*: virus indexing to assure quality of tissue culture material. *Priorities*: accuracy, ease of use, low cost.
iv. *Extension and NGO agricultural staff*: symptom recognition to provide advice to farmers. *Priorities*: ease of use in the field, robustness.
v. *Farmers*: symptom recognition to help in field management of cassava virus diseases. *Priorities*: ease of use in the field, robustness, and simplicity.

Although there has been great progress in the development and application of diagnostics for cassava viruses that can be used in laboratory situations by researchers and plant protection officers, there are currently no field-based testing options that can be used at "point-of-use" by extensionists or farmers themselves.

3.2.1 Cassava virus diagnostics in Africa

A number of diagnostic procedures utilizing symptoms, electron microscopy, serological methods and polymerase chain reaction (PCR)-based methods have been used for the diagnosis of CMD and CBSD (Alabi et al., 2011; Deng, McGrath, Robinson, & Harrison, 1994). Reliability, sensitivity, and operational convenience have led to the emergence of PCR or reverse transcription (RT)-PCR-based methods as the mainstay for routine diagnosis of CMGs and CBSVs, respectively (Abarshi et al., 2012, 2010; Alabi, Kumar, & Naidu, 2008; Aloyce, Tairo, Sseruwagi, Rey, & Ndunguru, 2013; Mbanzibwa et al., 2011; Monger, Seal, Cotton, & Foster, 2001; Were, Winter, & Maiss, 2004; Zhou et al., 1997). High-sequence diversity (intraspecific homology of ca. 85–99%; and ca. <75% nucleotide sequence homology between species) was such that early diagnostic methods for CBSVs encountered challenges such as "false negatives" due to mismatches in primer alignment. The enrichment of the public database with an increasing number of sequences for CBSV and UCBSV obtained from different locations in seven countries has enabled the development of oligonucleotide primers against the most conserved motifs, thereby improving the robustness of PCR-based diagnostic assays for CBSVs (e.g., Abarshi et al., 2012). The greatest challenge for the diagnosis of the CMGs is the relatively large number of species (9 in Africa), which means that current multiplex PCR technologies are unable to resolve all species in a single test. However, a multiplex PCR assay developed for the simultaneous detection of ACMV and EACMV-like viruses occurring in Africa (Alabi, Kumar, et al., 2008) was found to be useful for virus indexing due to its broad specificity. Another multiplex PCR method described can detect up to four CMG species in a single reaction (Aloyce et al., 2013), and protocols have also been described for the simultaneous detection of the two species of CBSVs and the two most widely distributed species of CMGs—ACMV and EACMV (Abarshi et al., 2012).

Real-time RT-PCR techniques have opened up the potential for increased throughput and the more sensitive detection of UCBSV and CBSV. Real-time Taqman assays developed for this purpose have been extensively used for the detection of CBSVs in cassava planting material produced for distribution within the countries of the Great Lakes region (Adams et al., 2013). Reverse transcription loop-mediated isothermal amplification (RT-LAMP)-based methods have also been developed for the detection of CBSVs (Tomlinson et al., 2013) and have the important practical advantage that they do not require thermal cycling equipment. These methods have been further modified through incorporating labeled primers for rapid detection of amplification products using lateral flow devices containing antibodies specific to the incorporated labels, avoiding the need for fluorescence detection or gel electrophoresis.

3.2.2 Cassava virus diagnostics in South Asia

A number of diagnostic methods utilizing symptoms, serology, and nucleic-acid-based techniques have been employed for the detection of CMGs in India. Several attempts have been made to detect ICMV using serological techniques (ELISA, DIBA, and TBIA) (Makeshkumar & Nair, 2001, Malathi et al., 1985; Malathi, Varma, & Nambisan, 1989), and histochemical and fluorochrome staining of infected tissues have also been used (Govindankutty, 2004). Different formats of PCR and multiplex PCR have been applied for the detection of ICMV and SLCMV (Anitha et al., 2011; Dutt, Briddon, & Dasgupta, 2005; Hegde, Jeeva, Makeshkumar, Misra, & Veena, 2010; Makeshkumar, Anoopsankar, Nair, & Edison, 2005). Nucleic acid spot hybridization techniques have been employed to detect CMGs in India for screening large numbers of samples (Makeshkumar et al., 2005). Most recently, real-time PCR methods have been developed and used for the detection of low titers of SLCMV in cassava plants showing symptom recovery, as well as for virus quantification (Deepthi, Makeshkumar, Unnikrishnan, & Winter, 2012).

3.2.3 Cassava virus diagnostics in Latin America

Viruses in Latin America for which sequence information is available are now being routinely detected by RT-PCR. Except for CsTLV isolates, the technique has proved to be efficient in detecting single infections and has significantly shortened the time for virus indexing when compared to the graft-indexing protocol using the cassava landrace indicator "Secundina" (Carvajal-Yepes et al., 2014). The great variability found among CsTLV isolates in Colombia alone means that different primer combinations are required to detect both RNA1 and/or RNA2 (Carvajal-Yepes et al., 2014; Verbeek et al., 2012). More sequences of diverse isolates need to be characterized in order to improve the effectiveness and reliability of CsTLV diagnostics.

For some cassava viruses reported in the 1980s (CsSLV, CsCaMV, CsCSLV, and CsALV), there are no sequences or antisera available (Table 1) and thus identification by indexing is impractical since it would require time-consuming and expensive virus purification and electron microscopy, and in some cases would not be possible at all. Moreover, some of these viruses could be related to (and possibly even the same as) the novel viruses recently identified. It is therefore suggested that they should be removed from the list of quarantine viruses that is used when controlling movements of cassava germplasm.

3.2.4 Cassava virus surveillance

Surveillance is an activity that has most commonly been implemented by research teams, and most frequently in Africa as a means to track the presence and national or regional spread of the CMGs and CBSVs. Some of the earliest countrywide surveys were implemented in Uganda at the time of the outbreak of the severe CMD associated with mixed ACMV+EACMV-UG infections (Otim-Nape et al., 1998). Similar approaches were also used in South Asia to determine the patterns of distribution and spread of ICMV and SLCMV (Patil et al., 2005). As the scale of the CMD pandemic in Africa became apparent, surveillance systems were increasingly organized at the regional (multicountry) level, within the framework of large CMD mitigation programs. This process led to an increasing standardization of protocols used which enabled datasets to be compared between countries. The first program to take this approach, focusing primarily on CMD, was a USAID-funded initiative that operated in eight countries of East and Central Africa between 1998 and 2008 (Legg, Kapinga, Teri, & Whyte, 1999; Legg, Owor, Sseruwagi, & Ndunguru, 2006). The most recent regional surveillance initiatives have combined assessments of both CMD and CBSD in countries of East, Central, and Southern Africa. Results from these were used to generate an online map database (IITA, 2012). Concerns about the speed with which results are made available, together with a demand for increased involvement of agricultural workers and farmers at the level of the communities affected, led to the piloting of alternative surveillance strategies. One example, referred to as the "Digital Early Warning Network," was employed for early warning of CBSD in northwestern Tanzania and Rwanda and combined farmer training with the use of a phone-based SMS reporting system (IITA, 2011a).

3.3. Quarantine systems

The frequency and volumes of cassava germplasm exchanged between and within counties are increasing dramatically, due to the increasing demands from crop diversification and crop improvement programs, and also to address insufficiency in planting material in food insecure parts of sub-Saharan Africa. This activity, however, involves an inevitable risk of accidentally introducing viruses along with the host plant material. Risk of pest and pathogen spread along with cassava germplasm is very high because:

i. The crop is clonally propagated.
ii. Cassava germplasm is predominantly exchanged as stem cuttings both within and even between countries, especially in sub-Saharan Africa.
iii. There is a high dependence on inter- and intracontinental exchange of cassava germplasm for research, crop diversification, and cassava improvement.
iv. A number of virus species associated with the cassava diseases in Africa, Asia, and Latin America are geographically restricted to those continents therefore representing threats to others (Table 1).
v. Vectors (e.g., *B. tabaci*) occur widely and are capable of spreading viruses from accidentally introduced sources.
vi. Knowledge of the distribution and diversity of several cassava viruses is limited and reliable diagnostics are not available (Table 1).

vii. Capacities of plant protection/quarantine organizations, as well as extension systems and the growers themselves, are currently insufficient for adequate monitoring of germplasm for pests and pathogens, and for the implementation of emergency control measures in case of accidental introductions.

Recent cases of the introduction of EACMV-like viruses into Indian Ocean islands (De Bruyn et al., 2012) and Oman (Khan, Akhtar, Al-Matrushi, Fauquet, & Briddon, 2013) are examples of virus spread that can be directly attributable to humans carrying infected planting material. Additionally, the emergence of the CBSD pandemic in the Great Lakes region of East/Central Africa demonstrates how the effects of natural spread by whiteflies can be significantly augmented through the propagation of infected planting material (Legg et al., 2011).

Technical guidelines for the safe movement of cassava germplasm have been established to minimize the risk of virus spread during germplasm exchange activities (Table 2) (Frison, 1994; Frison & Feliu, 1991). This requires the active engagement, however, of both national and regional plant protection bodies, working through established regulatory frameworks on germplasm exchange.

The measures listed in Table 2 have the potential to minimize the risk of spread of well-characterized viruses and also poorly characterized viruses,

Table 2 Basic guidelines for safe exchange of cassava germplasm

- The transfer of germplasm should be carefully planned in consultation with quarantine authorities and should be in amounts that allow adequate handling and examination. The material should be accompanied with the necessary documentation.
- Under no circumstances should germplasm be moved as rooted plant material except for *in vitro* plantlets.
- Cassava germplasm can be moved as seed, pathogen-tested *in vitro* material, or as cuttings from reestablished pathogen-tested *in vitro* material that has been grown under containment.
- All germplasm should be collected from healthy-looking plants and when possible from areas where quarantine pests are not present. Source plants should be tested for endemic viruses (e.g., CMGs and CBSVs for material sourced in East and Southern Africa)
- Only under special circumstances should the movement of untested, vegetative material be considered.
- Germplasm from areas where viruses of quarantine concern are known to occur should go through intermediate or postentry quarantine.

Source: Frison and Feliu (1991).

and even the exchange of planting material of unknown health status. In the absence of virus-free sources, planting material should be generated using thermotherapy and meristem culture techniques, which have been shown to eliminate viruses such as CMGs and CBSVs (IITA, 2010b; Kartha & Gamborg, 1975; Mohammed, Abarshi, Hillocks, & Maruthi, 2012; Zapata, Miller, & Smith, 1995) and regenerate virus-free cassava plants for international exchange. Recently, generic methods based on the high-throughput sequencing of viral RNAs have been established to overcome this bottleneck and discover viruses without any prior knowledge of their occurrence in host plants (Kreuze et al., 2009).

At local level, there have been emergency situations where it has been essential to speed the movement of virus-resistant germplasm from one country to neighbors. Such a situation occurred during the early stages of regional spread of the CMD pandemic from Uganda to neighboring countries. In this situation, open quarantine sites were established at locations in Tanzania and Kenya that were close to their respective borders with Uganda (Mohamed, 2002). These carefully managed sites received large quantities of CMD-resistant germplasm that was at the time only available in Uganda and provided a cheap and effective means of exchanging germplasm and safely multiplying resistant varieties in the target countries. Aspects of the exchange that were critical to its success were the fact that CMD symptoms can be readily seen in visual assessments thereby greatly minimizing the likelihood of inadvertent virus introductions, and that the pandemic-associated virus, EACMV-UG, had already spread into these border areas prior to the introduction of the new germplasm. Similar approaches have been proposed for the exchange of CBSD-resistant varieties between neighboring countries in East Africa. These have so far not been approved, however, in view of the more cryptic symptoms of CBSD and the greater likelihood, therefore, of inadvertently introducing CBSVs together with the exchanged germplasm.

3.4. Phytosanitation and clean seed

Phytosanitation typically refers to the set of measures that may be used to ensure the health of crop plants. With respect to cassava, there are three major components, which will be examined in turn.

3.4.1 Producing virus-free planting materials

3.4.1.1 Meristem tip culture and thermotherapy

Several methods have been used to produce virus-free planting material. High temperatures are known to inhibit replication in many virus species.

This phenomenon is exploited in cassava by growing young plants at a constant temperature of 35 °C for 1 month, and then excising 0.3-mm meristem tips and growing them on in tissue culture (Frison, 1994). These or similar procedures, coupled with virus indexing, are now used widely by laboratories working with tissue culture cassava where there is a need to ensure that the viruses are excluded from tissue culture material.

3.4.1.2 Field application of thermotherapy and hot water treatment

There have been various efforts to use this same principle to clean up cassava cuttings in field settings. Heat treatment and hot water treatment have been shown to eliminate certain CMGs in some cassava varieties (Gibson, 1994; Kaiser & Louie, 1982), but the "recovery" rates have been incomplete (i.e., a proportion remain infected after being returned to the normal temperature) and the approach has been shown to be ineffective in eliminating severe infections resulting from the cooccurrence of more than one CMG species. There has consequently been no widespread application of this technique for cassava virus control.

3.4.1.3 Field propagation of virus-free stocks of "clean seed"

Stocks of planting material (also referred to as "seed") may be "cleaned-up" in the field through regular monitoring and removal of infected plants. The effectiveness and speed of this "clean-up" process depends on several factors, such as the initial virus incidence in starting material, the abundance of whitefly vectors, the resistance of the variety to virus infection, and the distance of the field from neighboring (and potentially infected) cassava fields (Bock, 1994c; Thresh, Otim-Nape, & Fargette, 1998). By manipulating these factors, a conducive environment for the production of "clean seed" can be established, and within the space of two to three growing seasons, it can be possible to establish a "clean seed site" that is virtually virus-free. This approach is currently being implemented at a pilot level in Tanzania, for the production of "prebasic" "seed" of elite cassava varieties, and in the light of promising preliminary results, is being extended to other countries that have a similar goal of raising the quality standards of cassava planting material production.

3.4.2 *Managing the health of cassava crops in the field*

The relative merits of applying tactics for the management of the health of cassava crops in the field will vary depending on the nature of the viruses that affect those crops. CMD and CBSD provide a notable example of this.

Phytosanitation measures may be easy to apply for CMD, since clear foliar symptoms mean that it is straightforward to identify infected plants, but the benefits of phytosanitation may be limited, since crops may be readily infected by whiteflies arriving from distant inoculum sources. Conversely, phytosanitation measures may be less readily applied for CBSD, since symptoms are cryptic, but the benefits may be great, since cutting-borne infection is relatively more important than infection carried in by vectors from remote inoculum sources. This indicates that a thorough understanding of the epidemiology of cassava viruses is a key prerequisite to determining the most appropriate management strategy.

3.4.2.1 Roguing

Roguing is the removal of unwanted plants from within a planted crop. Plants are typically removed either because they are the wrong variety or because they are diseased. Roguing can be used to reduce levels of virus infection in cassava crops where the infecting virus(es) express clear foliar symptoms. The procedure is unpopular with farmers, and therefore hard to promote. Despite the fact that roguing is widely recommended by researchers and extension services trying to encourage cassava virus control, there is little evidence to suggest that it has real benefit for an individual farmer applying the tactic independently of his or her neighbors. Roguing in farmers' fields may provide little yield advantage for resistant varieties, but for susceptible varieties, it may have a significantly negative impact, because of the greatly reduced plant population that results from high virus incidence levels (Mallowa, Isutsa, Kamau, & Legg, 2011). The situation may be different; however, where the primary "product" of the field is planting material rather than tuberous root yield, since in such a situation the value of obtaining "healthy" planting material with little or no virus infection may outweigh the cost of the reduced plant population.

3.4.2.2 Selection

Selection usually refers to the choice by a grower or an agricultural worker of asymptomatic vegetative propagules—mature stems in the case of cassava. Like roguing, this practice was recommended for many years (Calvert & Thresh, 2002), although in contrast to roguing, field data have been presented to indicate that selection of healthy planting material results in significantly increased yields, even in relatively small plots (Mallowa et al., 2011). Studies on CMD in both Ivory Coast and Kenya indicated that the outcome of initiating cassava plantings with healthy cuttings was dependent on the inoculum pressure environment into which the crop was planted (Bock, 1994b; Fauquet, Fargette & Thouvenel, 1988). In India, virus-free stocks have been obtained by rigorous selection and through the use of meristem-tip therapy. These have been used in experiments and shown to remain largely free of CMD in areas where there is limited spread by whiteflies and substantial increases in yield have been achieved in this way (Nair, 1990; Nair & Thankappan, 1990). In Latin America, where the viruses affecting cassava are generally less aggressive and less readily spread by vectors, selection of healthy planting material has been widely applied for virus disease management (Calvert & Thresh, 2002). Vigorous selection has been shown by researchers to be effective in controlling CFSD, but many farmers are not inclined to inspect tuberous roots (where CFSD symptoms are expressed) before they collect and distribute the stem cuttings.

3.4.2.3 Crop management and disposition

Several crop management strategies can help to minimize infection of cassava crops in the field. Most act through reducing the likelihood of vector-borne infection. Where there are options to plant at different times of the year, dates can be selected which are least favorable for vector spread. For CBSVs, the relatively steep spread gradients mean that isolation of a new crop from neighboring fields can significantly reduce the degree of spread into the new crop (Jeremiah, 2012). Although the benefit of isolation appears to be less for CMGs, experimental trials have nevertheless demonstrated a strong reduction in infection within newly planted crops with increasing distance and reduced frequency of surrounding inoculum sources (Legg et al., 1997). In view of the obvious benefits of isolation, this has become an important requirement for the establishment of "clean seed" sites of cassava in recent years.

3.4.2.4 Intercropping

Intercropping is widely practiced by cassava growers in Africa, but is less frequent in the more commercially oriented production systems of Latin America and Asia. Several experimental studies have investigated the potential beneficial effect of intercropping on cassava virus control in Africa. Although some have suggested that there are significant benefits in intercropping either maize or cowpeas with cassava (Fondong, Thresh, & Zok, 2002), the benefits appear to be unpredictable and vary depending on the pattern of planting and mixture (Fargette & Fauquet, 1988). Perhaps in view of these mixed results, intercropping has not generally been advocated for the control of cassava viruses.

3.4.3 Implementing large-scale phytosanitation initiatives

Several organized and geographically extensive phytosanitation schemes have been implemented for the management of cassava viruses. These fall into two categories—certification and eradication.

3.4.3.1 Certification

Certification schemes are used for a wide range of crops as a means of assuring the quality of "seed" products. Interest in developing the commercial potential of cassava, combined with a recognition of the severe deleterious effects caused by viruses on planting material, have encouraged the pilot-scale development of programs promoting quality of cassava "seed." A set of guidelines designed to help growers produce cassava to defined quality standards, including virus disease tolerance levels, was first developed in the late 1990s in Uganda. This "Quality Management Protocol" (QMP) was first widely used within the framework of a regional cassava and banana disease mitigation project, the "Crop Crisis Control Project," which was implemented from 2006 to 2007 in six countries of East and Central Africa. The use of the QMP approach was extended during the follow-on Great Lakes Cassava Initiative (2007–2011) (IITA, 2011b), during which the system was broadened to include routine testing of CBSVs at higher level "seed" propagation sites. The application of this quality management system was effective in minimizing levels of virus disease at all stages of "seed" multiplication.

3.4.3.2 Eradication

All of the economically important cassava viruses (CMD, CBSD, and CFSD) are widely distributed and affect thousands, if not millions of cassava farms. In addition, some have significant reservoirs in noncassava plant hosts. Complete eradication of any of these is therefore an unrealistic goal for the foreseeable future. However, attempts have been made historically to achieve at least the regional-level eradication of some of these viruses. The best example of this is the large-scale replacement of heavily CMD-affected local cassava varieties by improved CMD-resistant material in northeastern Uganda in the 1950s. The removal of local materials was enforced by colonial authorities, with severe penalties imposed for failure to comply. Although the mode of implementation was authoritarian, the program greatly reduced the impact of CMD in that region of Uganda in subsequent years, and was considered to be a success.

It is clearly evident that the benefits of reducing virus disease inoculum levels accrue as the spatial scale of the application of these measures increases. Moreover, the benefits to be gained from area-wide phytosanitation for the control of CBSD are potentially greater than they would be for CMD, since the semipersistent transmission of the CBSVs means that they are less readily spread over long distances than CMGs. Based on this theoretical background, a pilot scheme to test community-wide phytosanitation was initiated in Tanzania in 2012. A key element of this ongoing program will be comprehensive community sensitization and collective planning, to ensure that all cassava growers in the four villages targeted in the first phase are aware of the stages of the process. The ambitious goal of the disease control work in each village will be to completely replace existing stocks of virus-diseased cassava plants with newly introduced virus-free planting material of the best virus-resistant varieties available. Since quantities of new material will initially be limiting, dissemination of the new material (coupled with removal of existing diseased material) will be achieved in a rolling process, in which new material initially received by the "Primary Recipient Group," will be shared in a second season with a larger "Secondary Recipient Group," who will in turn share planting material with any remaining cassava farmers in the village. If the approach proves to be successful, it may offer great potential for scaling up, and ultimately for regionwide reductions in the impact of cassava virus diseases.

3.5. Conventional breeding for resistance

3.5.1 Introduction

Efforts to improve cassava yield are generally not geared toward the highest possible yield under favorable conditions, but rather toward obtaining stable yields in marginal conditions where cassava is grown at present and is likely to expand in future. To achieve this, breeders address the key constraints of cassava production. Among the virus diseases, CMD, CBSD, and CFSD are currently the major constraints reducing cassava production, and these have been the target of resistance breeding work since the time that they were first recognized. Early successes that were achieved in this work provided encouragement for its further development, and this component of cassava virus disease management has consequently received most research attention and investment. Although vectors are a key component in the pathosystems of each of the major cassava virus groups, relatively less effort has been directed toward identifying sources of vector resistance and incorporating them into cassava germplasm.

3.5.2 Breeding for resistance to CMD in Africa

Breeding for CMD resistance was initiated in Tanzania (at Amani, near Tanga) by Storey in 1937 (Nichols, 1947) and was sustained through to 1958. The resistance of cultivated cassava to CMD was low in that part of Tanzania, as it was elsewhere in Africa. This encouraged researchers to exploit resistance sources from wild cassava relatives through interspecific hybridization. Moderately resistant clones with reasonable yield were selected from among the progeny of third backcrosses to cassava of ceara rubber (*M. glaziovii*) by cultivated cassava hybrids. Higher CMD resistance was subsequently obtained by intercrossing among resistant selections probably because it concentrated recessive genes (Jennings & Iglesias, 2002). These clones were consequently referred to as the "Amani hybrids."

Seeds of clone 5318/34 bred by the Amani group were introduced into Nigeria in 1956, and clone 58308 was subsequently selected as resistant to CMD in 1958 (Beck, 1960; Ekandem, 1970). Though breeding was discontinued in 1961, 58308 was maintained and it continued to show CMD resistance for nearly 30 years (Hahn & Theberge, 1985). However, this clone yielded poorly and produced low-quality tuberous roots (Hahn, Howland, & Terry, 1973).

The International Institute of Tropical of Agriculture (IITA)'s Roots and Tubers Improvement Program, established in 1971, gave priority to cassava, and its breeding strategy was to incorporate disease resistance into susceptible but well-adapted local cultivars. More crosses between clone 58308, local cultivars, and *M. glaziovii* were made. The improved families from these crosses showed superior resistance to CMD. After 10 years of research, the goal of developing high-yielding CMD-resistant material that increased yields of small-scale farmers by up to 40% was achieved (Nweke, 2009). Varieties TMS 30395, TMS 63397, TMS 30555, TMS 4(2)1425, and TMS 30572 were released in 1977 (Nweke, 2009), and TMS 30359 and TMS 30001 were reported to be almost immune to CMD (Hahn & Theberge, 1985).

During the 1990s, breeding work was extended to many of the major cassava-producing countries of Africa, through partnerships between IITA and national research programs. CMD gene pyramiding generated varieties with increasingly high levels of resistance, and many of these were selected for local environmental suitability in target countries and made available to farmers (Dixon et al., 2008, 2010; Ntawuruhunga et al., 2006).

The inheritance of CMD resistance was initially not well understood due to cassava's genetic heterogeneity and the limited amount of research dedicated to this topic. *M. glaziovii*-derived resistance was initially reported to be multigenic or polygenic (Doughty, 1958; Hahn, Howland, & Terry, 1977). Hahn et al. (1977) subsequently indicated that the resistance appeared to be additive with heritability of about 60% and that it was recessive (Hahn, Terry, & Leuschner, 1980). This polygenic, recessive resistance derived from *M. glaziovii* was subsequently designated CMD1, while a second source obtained from local Nigerian landraces was shown to be conferred by a major dominant gene and named CMD2 (Akano, Dixon, Mba, Barrera, & Fregene, 2002). Breeding lines expressing CMD1 are not immune to infection by CMGs but express mild or transient symptoms as a result of incomplete systemic infection leading to reversion of symptoms (Fargette et al., 1994). Several of these genotypes are moderately susceptible to CMD, especially when infected by more than one CMG species (Thresh & Cooter, 2005). West African landraces from which the CMD2 resistance source was derived (TME lines) show high levels of resistance to CMD, but the single gene character of this source of resistance means that it is more vulnerable to resistance breakdown than CMD1. By combining CMD1 and CMD2, breeders have been able to produce varieties with the agronomic and organoleptic qualities desired by cassava growers, while also possessing very high levels of durable CMD resistance (Dixon et al., 2003; Rabbi et al., 2014).

3.5.3 Breeding for resistance to CBSD in Africa

Breeding for resistance to CBSD was initiated together with the work on CMD during the Amani program in Tanzania (Nichols, 1947). A similar approach was used, in which resistance genes from wild relatives were introgressed and then backcrossed into cultivated cassava (Jennings, 2003; Nichols, 1947). One of the most resistant products of this program was the variety 46106/27 (Jennings, 1994; Nichols, 1947). Its resistance to CBSD has persisted up to the present in farmers' fields in coastal East Africa, where it is known locally as "*Kaleso*" in Kenya and "*Namikonga*" in Tanzania (Hillocks & Jennings, 2003).

CBSD resistance was recently reported to have both additive and nonadditive genetic effects. However, the additive effects were more important, implying that intrapopulation selection methods should be effective in accumulating favorable alleles in breeding materials (Moreno-Gonzalez & Cubero, 1993). In three different studies, Kaleso showed the highest general combining ability for resistance to CBSD (Kulembeka et al., 2012; Mtunda, 2009; Munga, 2008). This cultivar is now widely used by national breeding programs in the region to generate CBSD-resistant genotypes.

Following the outbreak of CBSD in the Great Lakes region in 2004, screening for CBSD resistance was initiated in Uganda using a large set of CMD-resistant families (Ntawuruhunga, Kiryowa, Okechukwu, Otim-Okello, & Kanju, 2012). CMD-resistant genotypes (MM and MH series) gave varied responses to CBSD, based on their foliar and root symptoms, and three categories of resistant (reduced incidence of symptom expression), tolerant (reduced severity of symptom expression) and susceptible genotypes were identified. Historically, much of the breeding work to combat CBSD has focused on tolerance, since resistance to infection is rare even in introgressed interspecific hybrids, and the expression of foliar symptoms has been considered as acceptable if root symptoms are absent, infrequent, or very mild. Yield loss experiments have indicated that the reduction in growth resulting from foliar symptoms can lead to larger reductions in yield than spoilage of roots due to CBSD-associated root damage (Hillocks, Raya, Mtunda, & Kiozia, 2001). This suggests that future breeding work needs to place greater emphasis on identifying sources of resistance to infection, which will prevent expression of symptoms in both foliage and roots. Recent studies have suggested an overlap in the resistance status of "tolerant" and "resistant" varieties, however, as concentrations of CBSVs in CBSD-tolerant varieties, expressing only mild foliar symptoms, have been shown to be significantly lower than concentrations of CBSVs in susceptible varieties (Maruthi, Bouvaine, Tufan, Mohammed, & Hillocks, 2014). This suggests that "tolerant" varieties possess molecular resistance mechanisms that impair the replication of CBSVs.

Several CBSD-resistant clones have been identified in Kenya (*Kaleso*, *Guzo*, *Gushe*, *Kibiriti Mweusi*, and *Ambari*), Mozambique (*Nikwaha*, *Chigoma Mafia*, *Nachinyaya*, *Xino Nn'goe*, *Likonde*, *Mulaleia*, and *Badge*), and Tanzania (*Namikonga*, *Kiroba*, *Nachinyaya*, *Kigoma Mafia*, *Kitumbua*, *Kalulu*, *Mfaransa*, *Muzege*, *Gezaulole*, and *Kibangameno*). These are now being used extensively in the respective countries' breeding programs as sources of resistance to generate new improved clones. Intercrossing among them concentrates resistance genes and allows recessive genes to be expressed (Hillocks & Jennings, 2003). Some of the derived resistant F1s have the ability to remain free of CBSD symptoms when exposed to infection and when infected the incidence and severity is low. Such F1s have been officially released in Kenya, Mozambique, Tanzania, and Uganda. Due to the current low levels of resistance available from *M. esculenta*, future efforts will also focus on evaluating and tapping natural resistance from wild relatives of cassava.

3.5.4 Breeding for resistance to CMD in South Asia

In India, cassava breeding is mainly carried out at the Central Tuber Crops Research Institute (CTCRI), Thiruvananthapuram, as well as in the State Agricultural Universities through the All India Coordinating Centers (Abraham, Nair, & Naskar, 2000; Unnikrishnan, EaswariAmma, Sreekumari, Sheela, & Mohan, 2002). During the last five decades of research at CTCRI, a large number of cassava varieties with varying reactions to ICMV and SLCMV have been released (Abraham et al., 2006; George, Kumar, & Unnikrishnan, 2012; Nair et al., 1998).

Mnga-1 is a breeding line from IITA (TMS 30001), which was received via CIAT in 1994 and which has been continuously evaluated for its response to Indian CMD under natural field conditions, showing high levels of resistance (Unnikrishnan et al., 2002). It was released for cultivation in Tamil Nadu as *Sree Padmanabha* and has also been used for developing more resistant clones through intervarietal hybridization. The evaluation of seedling populations of Mnga-1 for ICMV and SLCMV resistance resulted in 242 CMD-resistant lines with high yield and starch content. Two of these lines—CMR1 and CMR129—have proved to be popular among farmers (Unnikrishnan et al., 2011).

Fifty-six clones with resistance derived from West African landraces (CMD2) were introduced to India from the International Center for Tropical Agriculture (CIAT, Colombia) in 2005. These clones were crossed with inbred lines with the aim of developing heterotic hybrids. The CMD-resistant hybrids (CR21-10, CR43-11, and others) are now being evaluated in different parts of India under the All India Co-ordinated Research Project—Tuber Crops (Sheela, Abraham, & Moorthy, 2012).

In an interspecific breeding program, *M. glaziovii*, *Manihot caerulescens*, *Manihot tristis*, *Manihot flabellifolia*, *Manihot peruviana*, and *Manihot pseudoglaziovii* were used for developing improved CMD resistance. Among them, hybrids of *M. caerulescens* exhibited the highest levels of resistance and were used as donor parents for transferring resistance to elite Indian cultivars (Sheela, EaswariAmma, Unnikrishnan, & Nair, 2002; Sheela, Unnikrishnan, Edison, & EaswariAmma, 2004; Unnikrishnan et al., 2002). Among these crosses, one interspecific hybrid cassava with *M. caerulescens* (CMC-1) showed a very high level of resistance to CMD. Prebreeding lines developed at CTCRI also offer future potential for developing new CMD-resistant high-yielding hybrids.

3.5.5 Breeding for resistance to CFSD in Latin America

Field studies carried out by CIAT, Colombia show that there is a gradient of resistance to CFSD among cassava landraces. It also seems that different degrees of severity can be observed even between plants of a single cassava variety, which is almost certainly a consequence of the complex nature of the disease, comprising several types of mixed infections. In a 10-year assay carried out at CIAT's experimental station at Santander de Quilichao, Cauca, Colombia, more than 400 varieties were screened for resistance to CFSD. After a period of 5 years of evaluation, 70% of the landraces showed resistance to CFSD, including genotypes from Brazil, Peru, Colombia, and Paraguay (M. Cuervo and L. Calvert, unpublished data). In a further 3-year period of evaluation, these "resistant" landraces only showed very mild symptoms or no symptoms at all in the roots. To ensure high levels of disease pressure during this study, plants were inoculated by grafting using cuttings of a plant (code CM5460-10) showing severe CFSD root symptoms. When analyzing a sample of this inoculum using recently developed molecular techniques, mixed infections of at least four different viruses were identified, in addition to CsFSaV (Carvajal-Yepes et al., 2014). These preliminary results provide a strong indication that it is possible to control CFSD by the identification and use of varietal resistance. Molecular breeding through the identification of markers associated with resistance/susceptibility and better characterization of the causal agent(s) of the symptoms observed in CFSD-affected plants would improve the management of the disease.

3.5.6 The deployment of virus-resistant cassava varieties in Africa

Following the successes achieved in the development of CMD-resistant varieties in the 1970s, IITA worked with national program partners to distribute and evaluate these materials in many of the major cassava-producing countries of Africa (Manyong, Dixon, Makinde, Bokanga, & Whyte, 2000). Much of the country-based work involved various levels of selection, starting either from seed or clonal stages. Ultimately, on-farm trials with farmers completed evaluation procedures, and the best performing and most farmer-preferred varieties were formally released. As the capacities of national programs throughout Africa have grown stronger, an increasing level of independence of country-level breeding programs has been achieved, as crossing blocks have been established, and novel breeding strategies designed and implemented. As a consequence, the breeding process has become more sensitive to local environmental conditions, including differences in the presence of virus species and strains.

Between 1970 and 1998, it was estimated that 208 cassava varieties were released, many of which were selected primarily on the basis of resistance to CMD (Manyong et al., 2000). Effective partnerships were critical to the successful implementation of this geographically extensive program of germplasm exchange and evaluation (Ntawuruhunga et al., 2013; Nweke, 2009). Key partners included: national research systems, government and NGO extension agencies, subregional research organizations, and regional root crops networks such as the East Africa Root Crops Research Network (EARRNET) and the Southern Africa Root Crops Research Network (SARRNET).

After 1990, much of the large-scale variety deployment work in Africa was targeted toward the mitigation of virus disease pandemics, starting with the severe CMD pandemic in the early 1990s, and subsequently shifting emphasis to the CBSD pandemic from 2006 onward. Some of the most significant programs that were established for this purpose are summarized in Table 3. It is significant that over the period that these initiatives have been active, there has been a gradual increase in the scale of support and in the diversity of activities undertaken. There has been a significant impact in terms of the level of adoption of CMD-resistant varieties. Adoption was greatest in Uganda (where the National Agricultural Research Organization [NARO] first initiated large-scale dissemination of CMD-resistant varieties in 1993), where more than 50% of surveyed cassava farmers were growing CMD-resistant varieties by 2007 (Omongo et al., 2007). Similarly, while the frequency of CMD-resistant varieties grown as the predominant variety in eastern DR Congo was only 4.9% in 2007 (Tata-Hangy et al., 2007), by 2011 it was greater than 50% in six of the seven districts assessed (IITA, 2012). These changes have had a significant impact on the overall incidence of CMD in this part of East and Central Africa. Consequently, while average CMD incidence in Uganda was 83% in 1999 (Sseruwagi et al., 2004), by 2011 it had fallen to 12.9% (IITA, 2012).

Unfortunately, as farmers began to experience the yield benefits of effective CMD management provided by resistant varieties, the new outbreak of CBSD began to spread through the Great Lakes region. Since 2006, regional cassava virus disease management programs have therefore focused

Table 3 Regional cassava virus disease mitigation programs in sub-Saharan Africa: 1998–2017

Short title	Main themes	Executing institution[a]	Partner institutions[a]	Target countries[a]	Duration	Donor	Approx. budget ($)
Emergency mitigation of the CMD pandemic	Surveillance; varietal selection; germplasm exchange and multiplication; open quarantine; awareness raising	IITA	EARRNET, NARS, NPPOs, UA, CBOs, Extension	Bu, Ke, Rc, Rw, Tz, Ug	1998–2008	USAID	4,000,000 (400,000 per year)
Crop Crisis Control Project (C3P)	Surveillance; participatory varietal selection; germplasm multiplication; awareness raising; quality management	CRS	IITA, EARRNET, NARS, CBOs, Extension	Bu, Dc, Ke, Rw, Tz, Ug	2006–2007 (18 mths)	USAID	4,500,000 (Cassava part: 1,500,000 per year)
Great Lakes Cassava Initiative (GLCI)	Surveillance; participatory varietal selection; germplasm multiplication; awareness raising; quality management; CBSD focus; seed systems; virus diagnostics	CRS	IITA, NARS, NRI, FERA, KEPHIS, GTIL, CBOs	Bu, Dc, Ke, Rw, Tz, Ug	2007–2012	BMGF	22,000,000 (5,500,000 per year)
New Cassava Varieties and Clean Seed (5CP)	Germplasm exchange, germplasm evaluation; clean seed systems	IITA	NARS, NRI, KEPHIS, TOSCI, GTIL	Mw, Mz, Ke, Tz, Ug	2012–2016	BMGF	5,700,000 (1,425,000 per year)
Biotechnology applications to combat CBSD	QTL Mapping of CBSD resistance; RNASeq; identification of CBSD resistance genes; multilocation evaluation of resistance; marker-assisted breeding	IITA	ARI-Tanzania; NaCRRI	Ug, Tz	2009–2016	BMGF	3,900,000 (557,000 per year)
Building capacity in cassava virus diagnostics	CMD and CBSD virus diagnostics; physical and human capacity strengthening; surveillance; epidemiology; sustainable virus management	MARI	NARS, NCSU, NRI, RU	Ke, Mw, Mz, Rw, Tz, Ug, Zm	2009–2017	BMGF	11,000,000 (1,375,000 per year)

[a]Bu, Burundi; Dc, Democratic Republic of Congo; Ke, Kenya; Mw, Malawi; Mz, Mozambique; Rw, Rwanda; Tz, Tanzania; Ug, Uganda.
IITA, International Institute of Tropical Agriculture; CRS, Catholic Relief Services; MARI, Mikocheni Agricultural Research Institute.
EARRNET, East Africa Root Crops Research Network; NARS, National Agricultural Research Systems; NPPOs, National Plant Protection Organizations; UA, University of Arizona; CBOs, Community-based Organizations; NRI, Natural Resources Institute; FERA, Food and Environment Research Agency; KEPHIS, Kenya Plant Health Inspectorate Service; TOSCI, Tanzania Official Seed Certification Institute; GTIL, Genetic Technologies International Limited; ARI-Tanzania, Agricultural Research Institutes of Tanzania; NaCRRI, National Crop Resources Research Institute; NCSU, North Carolina State University; RU, Rutgers University.

increasingly on this. Since very little was known about the molecular character, biology and epidemiology of the viruses causing CBSD, considerable emphasis was placed on investing in critical research to strengthen knowledge about CBSD. The absence of good sources of resistance coupled with the greater potential for the application of phytosanitation have both favored an emphasis on developing seed systems that produce high-quality virus-free planting material. The longer term goal, however, will be to develop varieties that combine high levels of resistance to both CMD and CBSD. Conventional, gene editing, and transgenic approaches to breeding are all being explored with this goal in mind.

3.6. Molecular breeding using next-generation methods

Recent advances in NGS technologies have driven down sequencing costs and increased sequence capacity at an unprecedented rate (Varshney, Nayak, May, & Jackson, 2010). By harnessing these new technologies, molecular breeding can predict phenotypes from genotypes more efficiently and with greater accuracy than before. Some molecular breeding techniques rely on *a priori* knowledge of molecular markers associated with a trait (such as marker-assisted selection), and others, such as genomic selection (Goddard & Hayes, 2007; Heffner et al., 2009; Lorenz et al., 2011; Meuwissen, Hayes, & Goddard, 2001), use all markers across the genome to predict the performance of individuals (Jannink, Lorenz, & Iwata, 2010; Meuwissen et al., 2001). Genomic selection is being tested in cassava (Ly et al., 2013) with particular weighting for virus resistance in Uganda and Nigeria.

Efforts have been made to associate traits with markers, and early studies identified a microsatellite (SSRY28) and an RLFP locus (GY1) that flanked the single dominant CMD resistance gene known as CMD2 (Akano et al., 2002). These, together with additional markers (Fregene et al., 2006), were used to introgress CMD2 into Latin American germplasm for deployment in Africa (Okogbenin et al., 2007). Recently, Rabbi et al. (2014) using a high-density genotyping approach identified SNP locus S5214_780931 as being closest to the QTL peak for CMD2 and showed that all previous resistance linked markers cooccurred in the same chromosomal location indicating a single source of monogenic resistance. Other quantitative recessive resistance has been introgressed using purely conventional methods from wild *M. glaziovii* (Jennings, 1957, 1994; Nichols, 1947), although QTLs have not currently been identified for this source of resistance.

Efforts are underway to identify molecular markers associated with resistance and tolerance to CBSD in varieties including *Namikonga* (*Kaleso*), *Nachinyaya*, and *Kiroba* using a bi-parental QTL mapping approach. This approach is being supplemented by sequencing of the transcriptomes (RNASeq) of CBSV and UCBSV infected and control resistant and susceptible genotypes, with the aim of identifying candidate genes putatively involved in the resistance mechanism. Further functional genomics approaches are being implemented to validate these candidate genes.

Recent efforts are emphasizing the development of varieties with triple resistance to CMD, CBSD, and the whitefly vector. Accurate and robust phenotyping of germplasm for the multiple viral agents involved in CMD and CBSD is one of the greatest challenges. However, phenotyping procedures have been recently strengthened through the use of the *Agro*-inoculation of CMGs (Bi, Aileni, & Zhang, 2010) and a grafting method for CBSD (Wagaba et al., 2013), coupled with real-time PCR assays for quantification of CBSVs (Adams et al., 2013) and CMGs (Otti, Owati, Melaku, & Kumar, 2013). These novel techniques are greatly helping to improve our understanding of varietal response to virus and speeding the process of germplasm selection.

It is envisaged that as genomics tools are applied and the molecular basis of resistance/tolerance to virus diseases becomes better understood, greater progress will be made in deploying different combinations of resistance sources to enhance effectiveness and durability.

3.7. Transgenic approaches to strengthening host plant resistance

3.7.1 Introduction

The development of virus-resistant farmer-preferred cultivars is restricted by limitations inherent to traditional breeding (Ceballos, Iglesias, Perez, & Dixon, 2004). Transgenic technologies offer an alternative and novel way for introgressing beneficial traits into cassava. The concept of "pathogen-derived resistance" proposes that transforming plants with a pathogen's gene will disrupt the "normal" host–pathogen relationship and generate resistance against the pathogen (Sanford & Johnson, 1985). Powell-Abel et al. (1986) first demonstrated pathogen-derived resistance in transgenic tobacco plants conferring resistance against tobacco mosaic virus (TMV).

3.7.2 Transgenic approaches to developing CMD resistance

Moderate levels of resistance against several geminivirus species have been demonstrated in transgenic tobacco and tomato expressing the AC1 gene, which encodes the replication-associated protein (Rep) (Asad et al., 2003; Brunetti et al., 1997; Hong & Stanley, 1996; Noris et al., 1996; Sangare, Deng, Fauquet, & Beachy, 1999). Chellappan, Masona, Vanitharani, Taylor, and Fauquet (2004) reported that transgenic cassava expressing the ACMV AC1 gene showed resistance to both homologous and heterologous species of cassava-infecting CMGs. The high levels of resistance against CMGs were correlated with posttranscriptional gene silencing (PTGS) through the production of transgene-specific siRNAs. Zhang, Vanderschuren, Fütterer, and Gruissem (2005) demonstrated that transgenic cassava expressing antisense RNAs of ACMV Rep (AC1), TrAP (AC2), and REn (AC3) could resist ACMV infection via PTGS.

Vanderschuren et al. (2007) and Vanderschuren, Alder, Zhang, and Gruissem (2009) reported that the siRNAs, homologous to either the common region or AC1 in transgenic cassava, were able to suppress the replication of ACMV, leading to recovery after infection or immunity to infection by the virus. As the common region of ACMV does not share a high degree of sequence homology with other CMGs (e.g., EACMV and SACMV), the resistance was expected to be strain specific (Vanderschuren et al., 2007). In the last two decades, RNAi-based approaches have been tested in transgenic cassava and proved to confer robust CMD resistance in the model cassava cultivar TMS 60444. Constructs developed in this way need to be transferred to farmer-preferred cultivars, although the stability of the engineered CMD resistance remains to be demonstrated under field conditions and over multiple cycles of propagation.

3.7.3 Transgenic approaches to developing CBSD resistance

Several research groups have demonstrated that CBSD resistance can be engineered in cassava using RNAi-based approaches targeting the CP sequence of CBSVs (Ogwok et al., 2012; Vanderschuren, Moreno, Anjanappa, Zainuddin, & Gruissem, 2012; Yadav et al., 2011). Yadav et al. (2011) demonstrated that at least one of the two ipomoviruses, UCBSV, can be efficiently controlled using RNAi technology. Transgenic cassava plants constitutively expressing siRNA targeting the near full-length coat protein (FL-CP) of UCBSV showed 100% resistance to UCBSV in glasshouse experiments (Yadav et al., 2011) and confined field trials in Uganda (Ogwok et al., 2012). The transgenic plants expressing siRNA targeting the near FL-CP showed a 3-month delay in disease development, with 98% of clonal replicates remaining symptom free over the 11-month trial, whereas all nontransgenic control plants developed CBSD symptoms on aerial tissues by 6 months after planting. Highly effective suppression of UCBSV in transgenic plants under field conditions suggested that the coexpression of siRNAs from the CP sequences of both UCBSV and CBSV within the same plant holds promise for the integration of robust field resistance to CBSD into farmer-preferred cassava cultivars (Ogwok et al., 2012). The Donald Danforth Plant Science Center and IITA, in partnership with NARS in Uganda and Kenya, are developing and testing CBSD-resistant cassava using local farmer-preferred varieties under the Virus Resistant Cassava for Africa project (Taylor et al., 2012). In the most recent dataset, at least three transgenic lines showing near immunity to CBSV and UCBSV have been identified (Odipio et al., 2014).

Recently, Vanderschuren et al. (2012) demonstrated that sequences of the CP are highly conserved between CBSV and UCBSV and can therefore be used to engineer resistance against both viral species in the cassava model cultivar TMS 60444. This technology was transferred to the Nigerian cassava landrace TME7, which has natural resistance to CMD. This combination of natural and engineered virus resistance can be a promising approach to combat multiple cassava viral diseases in Africa (Vanderschuren et al., 2012). The RNAi approach is therefore a promising technology for engineering resistance to CBSD in farmer-preferred landraces. However, its value is greatly augmented where RNAi-derived transgenic resistance to CBSD is combined with conventionally bred CMD resistance. Many African countries have yet to finalize regulatory guidelines for the introduction, testing, and field planting of genetically modified (GM) crop plants, but the generally positive view that many African governments take of the practical benefits offered by GM mean that significant progress in resolving regulatory constraints is likely to be achieved in the near future.

3.8. Vector control

All of the economically important cassava viruses in Africa and South Asia are transmitted by the whitefly vector, *B. tabaci*. In spite of this fact, relatively little research attention has been directed toward developing strategies to control this insect. Host plant resistance, cultural methods, biological control, insecticides, and combinations of these tactics in integrated strategies are all used to counter the physical and virus-vectoring damage caused by whiteflies on other crop plants that are affected by whitefly transmitted viruses, such as cotton (Ellsworth & Martinez-Carrillo, 2001) or tomatoes (Lapidot & Friedmann, 2002). Expensive control methods, such as the application of insecticides, are almost never practiced by farmers in Africa since the cassava crop is primarily grown for subsistence purposes, rather than as a commercial crop. Some attention has also been given in India to the use of insecticides to control the whitefly vector in attempts to reduce the spread of CMD. However, the results have been unsatisfactory, and the routine use of insecticides is considered to be inappropriate on health and environmental grounds and for these reasons is not currently recommended (Malathi et al., 1985; Thankappan, Makeshkumar, & Edison, 1997).

4. CONCLUSIONS

Cassava has been grown for approximately 10,000 years, but it was only after its introduction to Africa in the sixteenth century, followed by Asia two centuries later, that the plant took on the role of a key component in global food systems. In the current environment of global warming—as one of the principal features of anthropogenic climate change—cassava is likely to become increasingly important (Jarvis, Ramirez-Villegas, Campo, & Navarro-Racines, 2012). Set against this, pests and diseases have become a major factor limiting production. For the pests, the greatest problems have arisen from inadvertent movements to places in which they did not previously occur. Outbreaks and epidemics of diseases have also arisen from external introductions (cassava bacterial blight), but more importantly from the movement of local viruses (CMGs and CBSVs) into the cassava crop. A key determining factor behind these scenarios is the adaptation of whitefly vectors to cassava, which has so far only happened in Africa and South Asia. The recent pandemics of severe CMD and CBSD in Africa seem to be associated with a second stage of adaptation of *B. tabaci* whiteflies to cassava, which has resulted in 100-fold increases in vector numbers in many areas and continues to affect new areas each year. Climate change threatens to exacerbate this situation still further, although studies are required to determine what the overall effects are likely to be on plant, vector, and viruses.

Set against this worrying future scenario is the rapid progress that has been made in both understanding the viruses of cassava and developing and delivering control solutions. This progress has been particularly significant in the last two decades, partly as a consequence of the increased severity of cassava virus problems. Some of the key requirements for future efforts to tackle cassava viruses are as follows:

i. *Awareness*. Highlight the continued importance of cassava viruses, their potential to recombine producing novel more virulent strains, and the risks posed by their inadvertent spread from country to country and continent to continent. Set up early warning networks to respond to transcontinental introductions and encourage preemptive measures to mitigate the westward spread of CBSD in Africa.

ii. *New technology*. Maximize the use of novel techniques both for virus/vector characterization, as well as for the development of resistant varieties through conventional, gene editing, and transgenic approaches. Combine resistance genes in order to ensure durability.

iii. *Surveillance*. Develop cheaper and more robust diagnostics to facilitate targeted but more frequent surveillance programs for monitoring new outbreaks and tracking progress in controlling them.

iv. "*Seed systems*." Invest in strengthening the quality and geographical coverage of "seed" systems for cassava to bring about widespread reductions in inoculum levels.

v. *Community action*. Work with communities in areas affected mainly by CBSD and CFSD to aim for local eradication through the implementation of area-wide phytosanitation programs.

vi. *Whitefly control*. Prioritize research on the integrated management of *B. tabaci* whitefly vectors—an area of work that has been neglected in recent decades.

vii. *Capacity development*. Train a cadre of researchers, plant protection officers, and agricultural development workers in order to equip them to drive the research and development activities necessary for the successful long-term management of cassava viruses. Training is required in all cassava-growing regions, but the needs are particularly acute in Africa.

Cassava is a highly adaptable crop that will continue to serve humanity into the future as a key source of food and industrial products. Viruses have posed a major challenge to this role. Research successes achieved over the last two decades, however, coupled with plans for future work such as those highlighted above, give us good reason to be hopeful about the future prospects, not just for virus management, but for the overall production of this important crop.

ACKNOWLEDGMENTS

This chapter was prepared with support from the Roots, Tubers, and Bananas (RTB) Program of the CGIAR. The technical assistance of Mr. Tunrayo Alabi in producing the virus distribution maps is gratefully acknowledged.

REFERENCES

Abarshi, M. M., Mohammed, I. U., Jeremiah, S. C., Legg, J. P., Kumar, L., Hillocks, R. J., et al. (2012). Multiplex RT-PCR assays for the simultaneous detection of both RNA and DNA viruses infecting cassava and the common occurrence of mixed infections by two cassava brown streak viruses in East Africa. *Journal of Virological Methods*, *179*, 176–184.

Abarshi, M. M., Mohammed, I. U., Wasswa, P., Hillocks, R. J., Holt, J., Legg, J. P., et al. (2010). Optimization of diagnostic RT-PCR protocols and sampling procedures for the reliable and cost-effective detection of cassava brown streak virus. *Journal of Virological Methods*, *163*, 353–359.

Abraham, A. (1956). *Tapioca cultivation in India: Farm Bulletin No. 17*. New Delhi, India: Indian Council of Agricultural Research, 20 pp.

Abraham, K., Edison, S., Unnikrishnan, M., Sheela, M. N., Vimala, B., Sreekumari, M. T., et al. (2006). *Tuber crops varieties released by CTCR: Technical Bulletin No.24*. CTCRI, 64 pp.

Abraham, K., Nair, S. G., & Naskar, S. K. (2000). Cassava breeding and varietal dissemination in India: Major achievements during the past 25–30 years of research. In *Sixth Asian cassava research workshop, February 21–25, 2000, Vietnam* (pp. 174–184).

Adams, I. P., Abidrabo, P., Miano, D. W., Alicai, T., Kinyua, Z. M., Clarke, J., et al. (2013). High throughput real-time RT-PCR assays for specific detection of cassava brown streak disease causal viruses, and their application to testing of planting material. *Plant Pathology*, *62*, 233–242.

Akano, A., Dixon, A., Mba, C., Barrera, E., & Fregene, M. (2002). Genetic mapping of a dominant gene conferring resistance to cassava mosaic disease. *Theoretical and Applied Genetics*, *105*, 521–525.

Alabi, O. J., Kumar, P. L., & Naidu, R. A. (2008). Multiplex PCR method for the detection of *African cassava mosaic virus* and *East African cassava mosaic Cameroon virus* in cassava. *Journal of Virological Methods*, *154*, 111–120.

Alabi, O. J., Kumar, P. L., & Naidu, R. A. (2011). Cassava mosaic disease: A curse to food security in sub-Saharan Africa. *APSnet Feature*. http://dx.doi.org/10.1094/APSnetFeature-2011-0701. http://www.apsnet.org/publications/apsnetfeatures/Pages/cassava.aspx.

Alabi, O. J., Ogbe, F. O., Bandyopadhyay, R., Kumar, P. L., Dixon, A. G., Hughes, J. d'A., et al. (2008). Alternative hosts of *African cassava mosaic virus* and *East African cassava mosaic Cameroon virus* in Nigeria. *Archives of Virology*, *153*, 1743–1747.

Alagianagalingam, M. N., & Ramakrishnan, K. (1966). Cassava mosaic in India. *South Indian Horticulture*, *14*, 71–72.

Alicai, T., Omongo, C. A., Maruthi, M. N., Hillocks, R. J., Baguma, Y., Kawuki, R., et al. (2007). Re-emergence of cassava brown streak disease in Uganda. *Plant Disease*, *91*, 24–29.

Aloyce, R. C., Tairo, F., Sseruwagi, P., Rey, M. E. C., & Ndunguru, J. (2013). A single-tube duplex and multiplex PCR for simultaneous detection of four cassava mosaic begomovirus species in cassava plants. *Journal of Virological Methods*, *189*, 148–156.

Alvarez, E., Mejía, J. F., Llano, G. A., & Loke, J. B. (2009). Characterization of a phytoplasma associated with frogskin disease in cassava. *Plant Disease*, *93*, 1139–1145.

Angel, S. J., Nolt, B. L., & Pineda, L. B. (1987). Estudios sobre transmission por moscas blancas (Homoptera: Aleyrodidae) de virus asociados con el cuero de sapo en yuca (*Manihot esculenta* Crantz). *Acta Agronomica*, *37*, 27–39, in Spanish.

Angel, S. J., Pineda, L. B., Nolt, B. L., & Velasco, A. C. (1989). Moscas blancas (Homoptera; Aleyrodidae) asociadas a transmisión de virus en yuca. *Fitopatologia Colombiana*, *13*, 65–71, in Spanish.

Anitha, J., Makeshkumar, T., & Edison, S. (2011). Survey of cassava mosaic disease in Kerala. *Journal of Root Crops*, *37*, 41–47.

Antony, B., Lisha, V. S., Palaniswami, M. S., Sugunan, V. S., Makeshkumar, T., & Henneberry, T. J. (2006). *Bemisia tabaci* and *Indian Cassava Mosaic Virus* Transmission. *International Journal of Tropical Insect Science, 26*, 176–182.

Asad, S., Harris, W. A., Bashir, A., Zafar, Y., Malik, K. A., Malik, N. N., et al. (2003). Transgenic tobacco expressing geminiviral RNAs are resistant to the serious viral pathogen causing cotton leaf curl disease. *Archives of Virology, 148*, 2341–2352.

Austin, M. D. N. (1986). Scientists identify cassava virus. *Asian Agribusiness, 3*, 10.

Beck, B. D. A. (1960). *Cassava,* Federal Department of Agricultural Research. Annual Report, 1958–1959, pp. 6–11.

Berrie, L. C., Rybicki, E. P., & Rey, M. E. C. (2001). Complete nucleotide sequence and host range of South African cassava mosaic virus: Further evidence for recombination amongst geminiviruses. *Journal of General Virology, 82*, 53–58.

Bi, H., Aileni, M., & Zhang, P. (2010). Evaluation of cassava varieties for cassava mosaic disease resistance jointly by agro-inoculation screening and molecular markers. *African Journal of Biotechnology, 4*, 330–338.

Bigirimana, S., Barumbanze, P., Ndayihanzamaso, P., Shirima, R., & Legg, J. P. (2011). First report of cassava brown streak disease and associated *Ugandan cassava brown streak virus* in Burundi. *New Disease Reports, 24*, 26.

Bock, K. R. (1994a). Studies on cassava brown streak virus disease in Kenya. *Tropical Science, 34*, 134–145.

Bock, K. R. (1994b). The spread of African cassava mosaic geminivirus in coastal and western Kenya. *Tropical Science, 34*, 92–101.

Bock, K. R. (1994c). Control of African cassava mosaic geminivirus by using virus-free planting material. *Tropical Science, 34*, 102–109.

Bock, K. R., Guthrie, E. J., & Figueiredo, G. (1981). A strain of cassava latent virus occurring in coastal districts of Kenya. *Annals of Applied Biology, 90*, 361–367.

Bock, K. R., & Woods, R. D. (1983). The etiology of African cassava mosaic disease. *Plant Disease, 67*, 994–995.

Brunetti, A., Tavazza, M., Noris, E., Tavazza, R., Caciagli, P., Ancora, G., et al. (1997). High expression of truncated viral rep protein confers resistance to tomato yellow leaf curl virus in transgenic tomato plants. *Molecular Plant-Microbe Interactions, 10*, 571–579.

Bull, S. E., Briddon, R. W., Sserubombwe, W. S., Ngugi, K., Markham, P. G., & Stanley, J. (2006). Genetic diversity and phylogeography of cassava mosaic viruses in Kenya. *Journal of General Virology, 87*, 3053–3065.

Calvert, L. A., Cuervo, M., & Lozano, I. (2012). Cassava viral disease in South America. In B. Ospina, & H. Ceballos (Eds.), *Cassava in the third millennium: Modern production, processing, use and marketing systems* (pp. 309–318). Cali, Colombia: CIAT.

Calvert, L. A., Cuervo, M., Lozano, I., Villareal, N., & Arroyave, J. (2008). Identification of three strains of a virus associated with cassava plants affected by frogskin disease. *Journal of Phytopathology, 156*, 647–653.

Calvert, L. A., & Thresh, J. M. (2002). The viruses and virus diseases of cassava. In A. C. Bellotti, R. J. Hillocks, & J. M. Thresh (Eds.), *Cassava: Biology, production and utilization* (pp. 237–260). Wallingford, UK: CABI.

Carabali, A., Bellotti, A. C., Montoya-Lerma, J., & Cuellar, M. E. (2005). Adaptation of *Bemisia tabaci* biotype B (Gennadius) to cassava, *Manihot esculenta* (Crantz). *Crop Protection, 24*, 643–649.

Carter, S. E., Fresco, L. O., Jones, P. G., & Fairbairn, J. N. (1997). *Introduction and diffusion of cassava in Africa.* IITA Research Guide 49, http://old.iita.org/cms/details/trn_mat/irg49/irg49.html.

Carvajal-Yepes, M., Olaya, C., Lozano, I., Cuervo, M., Castaño, M., & Cuellar, W. J. (2014). Unraveling complex viral infections in cassava (*Manihot esculenta* Crantz.) from Colombia. *Virus Research, 186*, 76–86.

Ceballos, H., Iglesias, C. A., Perez, J. C., & Dixon, A. G. O. (2004). Cassava breeding: Opportunities and challenges. *Plant Molecular Biology*, *56*, 503–516.

Chellappan, P., Masona, M. V., Vanitharani, R., Taylor, N. J., & Fauquet, C. M. (2004). Broad spectrum resistance to ssDNA viruses associated with transgene-induced gene silencing in cassava. *Plant Molecular Biology*, *56*, 601–611.

CIAT. (2010). *Cassava under threat: Pest and disease outbreaks put SE Asia on high alert*. *http://www.ciatnews.cgiar.org/2010/01/20/cassava-under-threat/*, Accessed March 8, 2014.

Cohen, S., Kern, J., Harpaz, I., & Ben-Joseph, R. (1988). Epidemiological studies of the tomato yellow leaf curl virus, in the Jordan Valley, Israel. *Phytoparasitica*, *16*, 259–270.

Costa, A. S. (1940). Observacóes sóbre o mosaic comun e o mosaic das nervuras da mandioca (Manihot utilissima Pohl. *Journal de Agronomia*, *3*, 239–248, in Portuguese.

De Bruyn, A., Julie Villemot, J., Lefeuvre, P., Villar, E., Hoareau, M., Harimalala, M., et al. (2012). East African cassava mosaic-like viruses from Africa to Indian Ocean islands: Molecular diversity, evolutionary history and geographical dissemination of a bipartite begomovirus. *BMC Evolutionary Biology*, *12*, 228. http://dx.doi.org/10.1186/1471-2148-12-228.

Deepthi, D. C., Makeshkumar, T., Unnikrishnan, M., & Winter, S. (2012). Quantitative analysis of geminiviruses in recovery phenotypes of cassava. In *Second scientific conference of the global cassava partnership for the 21st century (GCP-21-II), Kampala, Uganda, June 18–22, 2012*.

de Kochko, A., Verdaguer, B., Taylor, N., Carcamo, R., Beachy, R. N., & Fauquet, C. (1998). Cassava vein mosaic virus (CsVMV), type species for a new genus of plant double stranded DNA viruses? *Archives of Virology*, *143*, 945–962.

Deng, D., McGrath, P. F., Robinson, D. J., & Harrison, B. D. (1994). Detection and differentiation of whitefly-transmitted geminiviruses in plants and vector insects by the polymerase chain reaction with degenerate primers. *Annals of Applied Biology*, *125*, 327–336.

Dinsdale, A., Cook, L., Riginos, C., Buckley, Y. M., & De Barro, P. (2010). Refined global analysis of *Bemisia tabaci* (Hemiptera: Sternorrhyncha: Aleyrodoidea: Aleyrodidae) mitochondrial cytochrome oxidase I to identify species level genetic boundaries. *Annals of the Entomological Society of America*, *103*, 196–208.

Dixon, A. O. D., Akoroda, M. O., Okechukwu, R. U., Ogbe, F., Ilona, P., Sanni, L. O., et al. (2008). Fast track participatory approach to release of elite cassava genotypes for various uses in Nigeria's cassava economy. *Euphytica*, *160*, 1–13.

Dixon, A. G. O., Bandyopadhyay, R., Coyne, D., Ferguson, M., Ferris, R. S. B., Hanna, R., et al. (2003). Cassava: From a poor farmer's crop to a pacesetter of African rural development. *Chronica Horticulturae*, *43*, 8–14.

Dixon, A. G. O., Okechuku, R. U., Akoroda, M. O., Ilona, P., Ogbe, F., Egesi, C. N., et al. (2010). *Improved cassava varieties handbook*. Ibadan, Nigeria: IITA Integrated Cassava Project, 129 pp.

Doughty, L.R. (1958). Cassava breeding for cassava mosaic disease and brown streak virus. *Annual report of the East Africa Agricultural Forestry Research Organisation*, pp. 48–55

Dubern, J. (1994). Transmission of African cassava mosaic geminivirus by the whitefly (*Bemisia tabaci*). *Tropical Science*, *34*, 82–91.

Duraisamy, R., Natesan, S., Muthurajan, R., Gandhi, K., Lakshmanan, P., Karuppasamy, N., et al. (2013). Molecular studies on the transmission of Indian cassava mosaic virus (ICMV) and Sri Lankan cassava mosaic virus (SLCMV) in cassava by *Bemisia tabaci* and cloning ICMV and SLCMV replicase gene from cassava. *Molecular Biotechnology*, *53*, 150–158.

Dutt, N., Briddon, R. W., & Dasgupta, I. (2005). Identification of a second begomovirus, *Sri Lankan cassava mosaic virus*, causing cassava mosaic disease in India. *Archives of Virology*, *150*, 2101–2108.

Ekandem, M. J. (1970). *Cassava research in Nigeria before 1967: Memo No.103*. Ibadan, Nigeria: Federal Department of Agricultural Research.
Ellsworth, P. C., & Martinez-Carrillo, J. L. (2001). IPM for *Bemisia tabaci*: A case study from North America. *Crop Protection, 20*, 853–869.
FAOSTAT. (2014). *FAO database*. Rome, Italy: Food and Agriculture Organisation of the United Nations. *http://faostat.fao.org/site/567/default.aspx#anco*, Accessed February 16, 2014.
Fargette, D., & Fauquet, C. (1988). A preliminary study of intercropping maize and cassava on the spread of African cassava mosaic virus by whiteflies. *Aspects of Applied Biology, 17*, 195–202.
Fargette, D., Fauquet, C., Grenier, E., & Thresh, J. M. (1990). The spread of African cassava mosaic virus into and within cassava fields. *Journal of Phytopathology, 130*, 289–302.
Fargette, D., Fauquet, C., & Thouvenel, J. C. (1985). Field studies on the spread of African cassava mosaic. *Annals of Applied Biology, 106*, 285–294.
Fargette, D., Jeger, M., Fauquet, C., & Fishpool, L. D. C. (1993). Analysis of temporal disease progress of African cassava mosaic virus. *Phytopathology, 84*, 91–98.
Fargette, D., Roberts, I. M., & Harrison, B. D. (1991). Particle purification and properties of cassava Ivorian bacilliform virus. *Annals of Applied Biology, 119*, 303–312.
Fargette, D., Thresh, J. M., & Otim-Nape, G. W. (1994). The epidemiology of African cassava mosaic geminivirus: Reversion and the concept of equilibrium. *Tropical Science, 34*, 123–133.
Fauquet, C., & Fargette, D. (1990). African cassava mosaic virus: Etiology, epidemiology and control. *Plant Disease, 74*, 404–411.
Fauquet, C., Fargette, D., & Thouvenel, J. C. (1988). Some aspects of the epidemiology of African cassava mosaic virus in Ivory Coast. *Tropical Pest Management, 34*, 92–96.
Fondong, V., Pita, J. S., Rey, M. E. C., de Kochko, A., Beachy, R. N., & Fauquet, C. M. (2000). Evidence of synergism between African cassava mosaic virus and the new double recombinant geminivirus infecting cassava in Cameroon. *Journal of General Virology, 81*, 287–297.
Fondong, V., Thresh, J. M., & Zok, S. (2002). Spatial and temporal spread of cassava mosaic virus disease in cassava grown alone and when intercropped with maize and/or cowpea. *Journal of Phytopathology, 150*, 365–374.
Fregene, M. A., Morante, N., Sánchez, T., Marín, J., Ospina, C., Barrera, E., et al. (2006). Molecular markers for the introgression of useful traits from wild *Manihot* relatives of cassava; marker-assisted selection of disease and root quality traits. *Journal of Root Crops, 32*, 1–31.
Frison, E. (1994). Sanitation techniques for cassava. *Tropical Science, 34*, 146–153.
Frison, E. A., & Feliu, E. (1991). *FAO/IBPGR technical guidelines for the safe movement of cassava germplasm*. Rome, Italy: FAO-IBPGR, 48 pp.
George, J. Suresh, Kumar, P., & Unnikrishnan, M. (2012). *Descriptions of recommended/released varieties under AICRP on tuber crops 1975–2011: Technical Bulletin No. 51*. Trivandrum, India: CTCRI, 122 pp.
Gibson, R. W. (1994). Long-term absence of symptoms in heat-treated African cassava mosaic geminivirus-infected resistant cassava plants. *Tropical Science, 34*, 154–158.
Gibson, R. W., Legg, J. P., & Otim-Nape, G. W. (1996). Unusually severe symptoms are a characteristic of the current epidemic of mosaic virus disease of cassava in Uganda. *Annals of Applied Biology, 128*, 479–490.
Goddard, M., & Hayes, B. (2007). Genomic selection. *Journal of Animal Breeding and Genetics, 124*(6), 323–330.
Govindankutty, M. P. (2004). *Histopathological and fluorescence studies on cassava mosaic disease: CTCRI Annual Report 2003–2004*. Trivandrum, India: CTCRI, 58 pp.

Hahn, S. K., Howland, A. K., & Terry, E. R. (1973). Cassava breeding at IITA. In C. L. A. Leakey (Ed.), *Proceedings of the third symposium of the international society of root and tubers crops, Ibadan, Nigeria, 1973* (pp. 4–10). Ibadan, Nigeria: IITA.

Hahn, S. K., Howland, A. K., & Terry, E. R. (1977). Cassava breeding at IITA. In C. L. A. Leaky (Ed.), *Proceedings of the thirrd symposium, ISTRC-AB, IITA, Ibadan Nigeria, December 2–9, 1973* (pp. 4–10).

Hahn, S. K., Terry, E. R., & Leuschner, K. (1980). Breeding cassava for resistance to cassava mosaic disease. *Euphytica, 29*, 673–683.

Hahn, S. K., & Theberge, R. L. (1985). Techniques and advances in breeding cassava for disease resistance in Africa. In C. H. Hershey (Ed.), *Cassava breeding: A multi-disciplinary review. Proceedings of a workshop held in the Philippines, March 4–7, 1985.*

Hanley-Bowdoin, L., Settlage, S. B., Orozco, B. M., Nagar, S., & Robertson, D. (1999). Geminiviruses: Models for plant DNA replication, transcription, and cell cycle regulation. *CRC Critical Reviews of Plant Science, 18*, 71–106.

Harimalala, M., Lefeuvre, P., De Bruyn, A., Tiendrébéogo, F., Hoareau, M., Villemot, J., et al. (2012). A novel cassava-infecting begomovirus from Madagascar: Cassava mosaic Madagascar virus. *Archives of Virology, 157*, 2027–2030.

Harrison, B. D., Lennon, A. M., & Aiton, M. M. (1986). Properties of four previously uncharacterized cassava viruses. *Phytopathology, 76*, 1075.

Harrison, B. D., Zhou, X., Otim-Nape, G. W., Liu, Y., & Robinson, D. J. (1997). Role of a novel type of double infection in the geminivirus-induced epidemic of severe cassava mosaic in Uganda. *Annals of Applied Biology, 131*, 437–448.

Heffner, E. L., Sorrells, M. E., & Jannink, J.-L. (2009). Genomic selection for crop improvement. *Crop Sci, 49*, 1–12.

Hegde, V., Jeeva, M. L., Makeshkumar, T., Misra, R. S., & Veena, S. S. (2010). *Diagnostic techniques for diseases of tropical tuber crops: Technical Bulletin No. 50.* Trivandrum, India: CTCRI, 56 pp.

Hillocks, R. J., & Jennings, D. L. (2003). Cassava brown streak disease: A review of present knowledge and research needs. *International Journal of Pest Management, 49*, 225–234.

Hillocks, R. J., Raya, M., Mtunda, K., & Kiozia, H. (2001). Effects of brown streak virus disease on yield and quality of cassava in Tanzania. *Journal of Phytopathology, 149*, 1–6.

Hong, Y. G., Robinson, D. J., & Harrison, B. D. (1993). Nucleotide sequence evidence for the occurrence of three distinct whitefly-transmitted geminiviruses in cassava. *Journal of General Virology, 74*, 2437–2443.

Hong, Y., & Stanley, J. (1996). Virus resistance in *Nicotiana benthamiana* conferred by African cassava mosaic virus replication-associated protein (AC1) transgene. *Molecular Plant-Microbe Interactions, 9*, 219–225.

IITA. (2010a). *Thailand taps IITA's expertise as cassava mealybug invades Asia: IITA Bulletin No. 2005, 22–26 February, 2010.* http://www.iita.org/c/document_library/get_file?uuid=d9b886b5-374d-4478-8a1f-d9979a630720&groupId=25357, Accessed March 6, 2014.

IITA. (2010b). *Cassava* in vitro *processing and genebanking.* Ibadan, Nigeria: International Institute of Tropical Agriculture, 38 pp.

IITA. (2011a). *DEWN: A novel surveillance system.* http://r4dreview.org/2011/04/dewn-a-novel-surveillance-system/, Accessed, March 2, 2014.

IITA. (2011b). *Developing clean seed systems for cassava.* http://r4dreview.org/2011/04/developing-clean-seed-systems-for-cassava/, Accessed, March 2, 2014.

IITA. (2012). *Map database for GLCI project cassava pest and disease surveys conducted in six countries of East and Central Africa.* http://www.flickr.com/photos/iita-media-library/8485136117/in/photostream/, Accessed, March 2, 2014.

Jannink, J. L., Lorenz, A. J., & Iwata, H. (2010). Genomic selection in plant breeding: From theory to practice. *Briefings in Functional Genomics, 9*, 166–177.

Jarvis, A., Ramirez-Villegas, J., Campo, B. V. H., & Navarro-Racines, C. (2012). Is cassava the answer to African climate change adaptation? *Tropical Plant Biology*, *5*, 9–29.

Jennings, D. (1957). Further studies in breeding cassava for virus resistance. *East African Agricultural Journal*, *22*, 213–219.

Jennings, D. L. (1994). Breeding for resistance to African cassava mosaic virus in East Africa. *Tropical Science*, *34*, 110–122.

Jennings, D. L. (2003). Historical perspective on breeding for resistance to cassava brown streak virus disease. In J. P. Legg, & R. J. Hillocks (Eds.), *Cassava brown streak virus disease: Past, present, and future, Proceedings of an international workshop, Mombasa, Kenya, October 27–30* (pp. 55–57). Aylesford, UK: Natural Resources International Limited.

Jennings, D. L., & Iglesias, C. (2002). The viruses and virus diseases of cassava. In A. C. Bellotti, R. J. Hillocks, & J. M. Thresh (Eds.), *Cassava: Biology, production and utilization* (pp. 149–166). Wallingford, UK: CABI.

Jeremiah, S. (2012). *The role of whitefly (Bemisia tabaci) in the spread and transmission of cassava brown streak disease in the field* (Ph.D. thesis). Tanzania: University of Dar es Salaam, 229 pp.

Jeremiah, S. C., & Legg, J. P. (2008). *Cassava brown streak virus*. http://www.youtube.com/watch?v=nCJdws9CnUw, Uploaded, October 25, 2008.

Kaiser, W. J., & Louie, R. (1982). Heat therapy of cassava infected with African cassava mosaic disease. *Plant Disease*, *66*, 475–477.

Kartha, K. K., & Gamborg, O. L. (1975). Elimination of cassava mosaic disease by meristem culture. *Phytopathology*, *65*, 826–828.

Khan, A. J., Akhtar, S., Al-Matrushi, A. M., Fauquet, C. M., & Briddon, R. W. (2013). Introduction of *East African cassava mosaic Zanzibar virus* to Oman harks back to "Zanzibar, the capital of Oman" *Virus Genes*, *46*, 195–198.

Kitajima, E. W., & Costa, A. S. (1979). Rhabdovirus-like particles in tissues of five different plant species. *Fitopatologia Brasilera*, *4*, 55.

Kitajima, E. W., Wetter, C., Oliveira, A. R., Silva, D. M., & Costa, A. S. (1965). Morfologia do virus do mosaico comun da mandioca. *Bragantia*, *24*, 247–260, in Portuguese.

Kreuze, J. F., Perez, A., Untiveros, M., Quispe, D., Fuentes, S., Barker, I., et al. (2009). Complete viral genome sequence and discovery of novel viruses by deep sequencing of small RNAs: A generic method for diagnosis, discovery and sequencing of viruses. *Virology*, *388*, 1–7.

Kufferath, H., & Ghesquière, J. (1932). La mosaïque du manioc. *Compte-Rendu de la Société de Biologie Belge*, *109*, 1146–1148.

Kulembeka, H. P., Ferguson, M., Herselman, L., Kanju, E., Mkamilo, G., Masumba, E., et al. (2012). Diallel analysis of field resistance to brown streak disease in cassava (*Manihot esculenta* Crantz) landraces from Tanzania. *Euphytica*, *187*, 277–288.

Lapidot, M., & Friedmann, M. (2002). Breeding for resistance to whitefly transmitted geminiviruses. *Annals of Applied Biology*, *140*, 109–127.

Legg, J. P. (1999). Emergence, spread and strategies for controlling the pandemic of cassava mosaic virus disease in east and central Africa. *Crop Protection*, *18*, 627–637.

Legg, J. P. (2010). Epidemiology of a whitefly-transmitted cassava mosaic geminivirus pandemic in Africa. In P. A. Stansly, & S. E. Naranjo (Eds.), *Bemisia: Bionomics and management of a global pest* (pp. 233–257). Dordrecht-Heidelberg-London-New York: Springer.

Legg, J. P., & Fauquet, C. M. (2004). Cassava mosaic geminiviruses in Africa. *Plant Molecular Biology*, *56*, 585–599.

Legg, J. P., French, R., Rogan, D., Okao-Okuja, G., & Brown, J. K. (2002). A distinct, invasive *Bemisia tabaci* (Gennadius) (Hemiptera: Sternorrhyncha: Aleyrodidae) genotype cluster is associated with the epidemic of severe cassava mosaic virus disease in Uganda. *Molecular Ecology*, *11*, 1219–1229.

Legg, J., James, B., Cudjoe, A., Saizonou, S., Gbaguidi, B., Ogbe, F., et al. (1997). A regional collaborative approach to the study of ACMD epidemiology in sub-Saharan Africa.

In E. Adipala, J. S. Tenywa, & M. W. Ogenga-Latigo (Eds.), *Proceedings of the African crop science conference, Pretoria, January 13–17, 1997* (pp. 1021–1033).
Legg, J. P., Jeremiah, S. C., Obiero, H. M., Maruthi, M. N., Ndyetabula, I., Okao-Okuja, G., et al. (2011). Comparing the regional epidemiology of the cassava mosaic and cassava brown streak pandemics in Africa. *Virus Research, 159*, 161–170.
Legg, J. P., Kapinga, R., Teri, J., & Whyte, J. B. A. (1999). The pandemic of cassava mosaic virus disease in East Africa: Control strategies and regional partnerships. *Roots, 6*(1), 10–19.
Legg, J. P., & Ogwal, S. (1998). Changes in the incidence of African cassava mosaic geminivirus and the abundance of its whitefly vector along south-north transects in Uganda. *Journal of Applied Entomology, 122*, 169–178.
Legg, J. P., Owor, B., Sseruwagi, P., & Ndunguru, J. (2006). Cassava mosaic virus disease in East and Central Africa: Epidemiology and management of a regional pandemic. *Advances in Virus Research, 67*, 355–418.
Legg, J. P., Somado, E. A., Barker, I., Beach, L., Ceballos, H., Cuellar, W., et al. (2014). A global alliance declaring war on cassava viruses in Africa. *Food Security, 6*, 231–248.
Legg, J. P., Sseruwagi, P., Boniface, S., Okao-Okuja, G., Shirima, R., Bigirimana, S., et al. (2014). Spatio-temporal patterns of genetic change amongst populations of cassava *Bemisia tabaci* whiteflies driving virus pandemics in East and Central Africa. *Virus Research, 186*, 61–75.
Lennon, A. M., Aiton, M. M., & Harrison, B. D. (1986). *Cassava viruses from South America: Annual Report 1985*. Dundee, UK: James Hutton Institute, 167 pp.
Lennon, A. M., Aiton, M. M., & Harrison, B. D. (1987). Purification and properties of cassava green mottle, a previously undescribed virus from the Solomon Islands. *Annals of Applied Biology, 110*, 545–555.
Léotard, G., Duputié, A., Kjellberg, F., Douzery, E. J. P., Debain, C., de Granville, J. J., et al. (2009). Phylogeography and the origin of cassava: New insights from the northern rim of the Amazonian basin. *Molecular Phylogenetics and Evolution, 53*, 329–334.
Lorenz, A. J., Chao, S., Asoro, F. G., Heffner, E. L., Hayashi, T., Iwata, H., et al. (2011). Genomic selection in plant breeding: Knowledge and prospects. *Advances in Agronomy, 110*, 77–123.
Ly, D., Hamblin, M., Rabbi, I., Gedil, M., Bakare, M., Gauch, H. G., Jr., et al. (2013). Relatedness and genotype × environment interaction affect prediction accuracies in genomic selection: A study in cassava. *Crop Science, 53*, 1312–1325.
Makeshkumar, T., Anoopsankar, Nair, R. R., & Edison, S. (2005). Detection of Indian cassava mosaic virus through polymerase chain reaction and nucleic acid hybridisaton techniques. *Journal of Root Crops, 31*, 1–6.
Makeshkumar, T., & Nair, R. R. (2001). *Serological and nucleic acid based detection of ICMD. In Fifth international scientific meeting of cassava Biotechnology Network (CBN-V), St. Louis, USA, November 4–9, 2001*, http://www.danforthcenter.org/media/video/cbnv/session8/S8-12.html.
Malathi, V. G., Nair, N. G., & Shantha, P. (1985). *Cassava mosaic disease. Technical bulletin series: Vol. 5*. Trivandrum, India: Central Tuber Crops Research Institute.
Malathi, V. G., Varma, A., & Nambisan, B. (1989). Detection of Indian cassava mosaic virus by ELISA. *Current Science, 58*, 149–150.
Mallowa, S. O., Isutsa, D. K., Kamau, A. W., & Legg, J. P. (2011). Effectiveness of phytosanitation in cassava mosaic disease management in a post-epidemic area of western Kenya. *Journal of Agricultural and Biological Science, 6*(7), 8–15.
Manyong, V. M., Dixon, A. G. O., Makinde, K. O., Bokanga, M., & Whyte, J. (2000). *The contribution of IITA-improved germplasm to food security in sub Ssaharan Africa: An impact study* Ibadan, Nigeria: IITA.
Maruthi, M. N., Bouvaine, S., Tufan, H. A., Mohammed, I. U., & Hillocks, R. J. (2014). Transcriptional response of virus-infected cassava and identification of putative sources of

resistance for cassava brown streak disease. *PLoS One*, *9*(5), e96642. http://dx.doi.org/10.1371/journal.pone.0096642.

Maruthi, M. N., Hillocks, R. J., Mtunda, K., Raya, M. D., Muhanna, M., Kiozia, H., et al. (2005). Transmission of *Cassava brown streak virus* by *Bemisia tabaci* (Gennadius). *Journal of Phytopathology*, *153*, 307–312.

Mathew, A. V., & Muniyappa, V. (1993). Host range of Indian cassava mosaic virus. *Indian Phytopathology*, *46*, 16–23.

Mbanzibwa, D. R., Tian, Y., Mukasa, S. B., & Valkonen, J. P. (2009). Cassava brown streak virus (Potyviridae) encodes a putative Maf/HAM1 pyrophosphatase implicated in reduction of mutations and a P1 proteinase that suppresses RNA silencing but contains no HC-Pro. *Journal of Virology*, *83*, 6934–6940.

Mbanzibwa, D. R., Tian, Y. P., Tugume, A. K., Patil, B. L., Yadav, J. S., Bagewadi, B., et al. (2011). Evolution of cassava brown streak disease-associated viruses. *Journal of General Virology*, *92*, 974–987.

Menon, M. R., & Raychaudhuri, S. P. (1970). Cucumber, a herbaceous host of cassava mosaic virus. *Plant Disease Reporter*, *54*, 34–35.

Meuwissen, T. H., Hayes, B. J., & Goddard, M. E. (2001). Prediction of total genetic value using genome-wide dense marker maps. *Genetics*, *157*, 1819–1829.

Mohamed, R. (2002). Role of open quarantine in regional germplasm exchange. In J. P. Legg, & R. J. Hillocks (Eds.), *Cassava brown streak virus disease: Past, present, and future, Proceedings of an international workshop, Mombasa, Kenya, October 27–30* (pp. 63–65). Aylesford, UK: Natural Resources International Limited.

Mohammed, U. I., Abarshi, M. M., Hillocks, R. J., & Maruthi, M. N. (2012). Developing methods to eliminate cassava brown streak viruses from infected cassava varieties. In *Program and abstracts, second scientific conference of the global cassava partnership for the 21st century (GCP21-II), June 18–22, 2012, Kampala, Uganda*, SP05-05 (Abstract).

Monger, W. A., Seal, S., Cotton, S., & Foster, G. D. (2001). Identification of different isolates of *Cassava brown streak virus* and development of a diagnostic test. *Plant Pathology*, *50*, 768–775.

Monger, W. A., Seal, S., Isaac, A. M., & Foster, G. D. (2001). Molecular characterisation of *Cassava brown streak virus* coat protein. *Plant Pathology*, *50*, 527–534.

Moreno-Gonzalez, J., & Cubero, J. I. (1993). Selection strategies and choice of breeding methods. In M. D. Hayward, N. O. Bosemark, & I. Romagosa (Eds.), *Plant breeding: Principles and prospects*. London, UK: Chapman and Hall.

Morris, B., Coates, L., Lowe, S., Richardson, K., & Eddy, P. (1990). Nucleotide sequence of the infectious cloned DNA components of African cassava mosaic virus (Nigerian strain). *Nucleic Acids Research*, *18*, 197–198.

Mtunda, K. J. (2009). *Breeding, evaluation and selection of cassava for high starch content and yield in Tanzania* (Ph.D. thesis). Pietermaritzburg, South Africa: University of KwaZulu-Natal.

Munga, T. L. (2008). *Breeding for cassava brown streak disease resistance in coastal Kenya* (Ph.D. thesis). Pietermaritzburg, South Africa: University of KwaZulu-Natal.

Nair, N. G. (1985). Nature and extent of spread of cassava mosaic disease. In T. Ramanujam et al. (Eds.), *Tropical tuber crops: Proceedings of the national symposium 1985* (pp. 175–178), Trivandrum, India: CTCRI.

Nair, N. G. (1990). Performance of virus-free cassava (*Manihot esculenta* Crantz) developed through meristem tip culture. *Journal of Root Crops*, *16*, 23–131.

Nair, N. G., & Thankappan, M. (1990). Spread of Indian cassava mosaic disease under different agro-climatic locations. *Journal of Root Crops*, ISRC National Symposium Special Volume (pp. 216–219). Trivandrum, India: CTCRI.

Nair, S. G., Unnikrishnan, M., Sheela, M. N., Abraham, K., Vimala, B., Sreekumari, M. T., et al. (1998). *Tuber crops varieties released by CTCRI. Technical bulletin series: Vol. 24* (pp. 6–14). Trivandrum, India: CTCRI.

Neuenschwander, P. (1994). Control of the cassava mealybug in Africa: Lessons from a biological control project. *African Crop Science Journal*, *2*, 369–384.

Nichols, R. F. W. (1947). Breeding cassava for virus resistance. *East African Agricultural Journal, 12*, 184–194.

Nichols, R. F. J. (1950). The brown streak disease of cassava: Distribution, climatic effects and diagnostic symptoms. *East African Agricultural Journal, 15*, 154–160.

Noris, E., Accotto, G. P., Tavazza, R., Brunetti, A., Crespi, S., & Tavazza, M. (1996). Resistance to tomato yellow leaf curl geminivirus in *Nicotiana benthamiana* plants transformed with a truncated viral C1 gene. *Virology, 224*, 130–138.

Ntawuruhunga, P., Dixon, A. G. O., Kanju, E., Ssemakula, G., Okechukwu, R., Obiero, H., et al. (2013). Successful innovations and lessons learnt in cassava improvement and deployment by IITA in the Eastern African Region. *African Journal of Root and Tubers Crops, 10*, 41–54.

Ntawuruhunga, P., Kanobe, C., Ssemakula, G., Ojulong, H., Ragama, P., Whyte, J., et al. (2006). Evaluation of advanced cassava genotypes in Uganda. *African Crops Science Journal, 14*, 17–25.

Ntawuruhunga, P., Kiryowa, M., Okechukwu, R., Otim-Okello, F., & Kanju, E. (2012). Preliminary results of screening IITA improved germplasm for resistant to cassava brown streak disease (CBSD) in Uganda. In R. U. Okechukwu, & P. Ntawuruhunga (Eds.), *Proceedings of the 11th ISRTC-AB conference, Kinshasa, DR Congo, October 4–8, 2010* (pp. 398–405).

Nweke, F. (2009). *Controlling cassava mosaic virus and cassava mealybug in sub-Saharan Africa*, IFPRI Discussion Paper 00912.2020 Vision Initiative, Washington, DC: IFPRI.

Odipio, J., Ogwok, E., Taylor, N. J., Halsey, M., Bua, A., Fauquet, C., et al. (2014). RNAi-derived field resistance to cassava brown streak disease persists across the vegetative cropping cycle. *GM Crops and Food: Biotechnology in Agriculture and the Food Chain, 5*, 16–19. http://dx.doi.org/10.4161/gmcr.26408.

Ogbe, F. O., Thottappilly, G., Dixon, A. G. O., & Mignouna, H. D. (2003). Variants of East African cassava mosaic virus and its distribution in double infections with African cassava mosaic virus in Nigeria. *Plant Disease, 87*, 229–232.

Ogwok, E., Odipio, J., Halsey, M., Gaitan-Solis, E., Bua, A., Taylor, N., et al. (2012). Transgenic RNA interference (RNAi)-derived field resistance to cassava brown streak disease. *Molecular Plant Pathology, 13*, 1019–1031.

Okogbenin, E., Porto, M. C. M., Egesi, C., Mba, C., Espinosa, E., Santos, L. G., et al (2007). Marker-assisted introgression of resistance to cassava mosaic disease into Latin American germplasm for the genetic improvement of cassava in Africa. *Crop Science, 47*, 1895–1904.

Olson, K. M., & Schaal, B. A. (1999). Evidence on the origin of cassava: Phylogeography of *Manihot esculenta. Proceedings of the National Academy of Sciences, 96*, 5586–5591.

Omongo, C., Poro, S., Omara, J., Omara, T., Okao-Okuja, G., Obonyo, R., et al. (2007). *Monitoring and diagnostic survey of cassava mosaic virus disease (CMD) in Uganda*. http://www.iita.org/c/document_library/get_file?p_l_id=25420&folderId=178120&name=DLFE-1354.pdf, Accessed, March 3, 2014.

Onwueme, I. C. (2002). Cassava in Asia and the Pacific. In A. C. Bellotti, R. J. Hillocks, & J. M. Thresh (Eds.), *Cassava: Biology, production and utilization* (pp. 55–65). Wallingford, UK: CABI.

Otim-Nape, G. W., Bua, A., Thresh, J. M., Baguma, Y., Ogwal, S., Semakula, G. N., et al. (1997). *Cassava mosaic virus disease in Uganda: The current pandemic and approaches to control.* Chatham, UK: Natural Resources Institute, 65 pp.

Otim-Nape, G. W., Thresh, J. M., & Fargette, D. (1996). *Bemisia tabaci* and cassava mosaic virus disease in Africa. In D. Gerling, & R. T. Meyer (Eds.), *Bemisia 1995: Taxonomy, biology, damage, control and management* (pp. 319–350). Andover, UK: Intercept.

Otim-Nape, G. W., Thresh, J. M., & Shaw, M. W. (1998). The incidence and severity of cassava mosaic virus disease in Uganda: 1990–1992. *Tropical Science, 38*, 25–37.

Otti, G., Owati, A., Melaku, G., & Kumar, P. L. (2013). Quantitative detection of African cassava mosaic virus and East African cassava mosaic virus using TaqMan real time PCR. In *Program and book of abstracts of the 12th international plant virus epidemiology symposium, 28 January to 1 February 2013, Arusha, Tanzania* (p. 189), Ibadan, Nigeria: IITA.

Patil, B. L., & Fauquet, C. M. (2009). Cassava mosaic geminiviruses: Actual knowledge and perspectives. *Molecular Plant Pathology, 10*, 685–701.

Patil, B. L., Rajasubramaniam, S., Bagchi, C., & Dasgupta, I. (2005). Both Indian cassava mosaic virus and Sri Lankan cassava mosaic virus are found in India and exhibit high variability as assessed by PCR-RFLP. *Archives of Virology, 150*, 389–397.

Pineda, B., Jayasinghe, U., & Lozano, J. C. (1983). La enfermedad "cuero de sapo" en yuca (*Manihot esculenta* Crantz). *ASIAVA (Colombia), 4*, 10–12.

Pita, J. S., Fondong, V. N., Sangare, A., Otim-Nape, G. W., Ogwal, S., & Fauquet, C. M. (2001). Recombination, pseudorecombination and synergism of geminiviruses are determinant keys to the epidemic of severe cassava mosaic disease in Uganda. *Journal of General Virology, 82*, 655–665.

Powell-Abel, P., Nelson, R. S., Hoffmann, B. De N., Rodger, S. G., Fraley, R. T., & Beachy, R. N. (1986). Delay of disease development in transgenic plants that express the tobacco mosaic virus coat protein gene. *Science, 232*, 738–743.

Rabbi, I. Y., Hamblin, M. T., Kumar, P. L., Gedil, M. A., Ikpan, A. S., Jannink, J. L., et al. (2014). High-resolution mapping of resistance to cassava mosaic geminiviruses in cassava using genotyping-by-sequencing and its implications for breeding. *Virus Research, 186*, 87–96.

Rastgou, M., Habibi, M. K., Izadpanah, K., Masenga, V., Milne, R. G., Wolf, Y. I., et al. (2009). Molecular characterization of the plant virus genus *Ourmiavirus* and evidence of inter-kingdom reassortment of viral genome segments as its possible route of origin. *Journal of General Virology, 90*, 2525–2535.

Robertson, I. A. D. (1985). *Cassava/Whitefly Project. Part of Crop Virology Project R3177 at the Kenya Agricultural Research Institute. Final Report.* London, UK: Overseas Development Administration.

Sanford, J. C., & Johnson, S. A. (1985). The concept of parasite-derived resistance: Deriving resistance genes from the parasite's own genome. *Journal of Theoretical Biology, 115*, 395–405.

Sangare, A., Deng, D., Fauquet, C. M., & Beachy, R. N. (1999). Resistance to African cassava mosaic virus conferred by a mutant of the putative NTP-binding domain of the Rep gene (*AC1*) in *Nicotianabenthamiana. Molecular Biology Breeding, 5*, 95–102.

Santos, A. A., Gonçalvez, J. A., Quieroz, G. M., & Lima, R. N. (1995). Efeito do virus do mosaico das nervuras sobre os componentes produtivos da mandioca, cv Pretinha, em duas regioes fisiográficas do Ceára. *Fitopatologia Brasiliera, 20*, 506–508.

Saunders, K., Salim, N., Mali, V. R., Malathi, V. G., Briddon, S. R., Markham, P. G., et al. (2002). Characterisation of *Sri Lankan cassava mosaic virus* and *Indian cassava mosaic virus*: Evidence for acquisition of a DNA B component by a monopartite begomovirus. *Virology, 293*, 63–74.

Scott, S. W., MacFarlane, S. A., McGavin, W. J., & Fargette, D. (2014). Cassava Ivorian bacilliform virus is a member of the genus Anulavirus. *Archives of Virology, 159*, 2791–2793.

Shanta, P. (1978). Diseases of cassava. In N. Hrishi, & R. G. Nair (Eds.), *Cassava production technology* (pp. 62–65). Trivandrum, India: CTCRI.

Sheela, M. N., Abraham, K., & Moorthy, S. N. (2012). *Development of elite inbreds of cassava and their use in heterosis breeding: Annual Report 2011–2012.* Thiruvananthapuram, India: CTCRI, 25 pp.

Sheela, M. N., EaswariAmma, C. S., Unnikrishnan, M., & Nair, S. G. (2002). Expression of cassava mosaic tolerance in inter-specific backcross progeny of cassava (*Manihot esculenta* Crantz). In *International conference on vegetables, November 11–14, 2002, Bangalore, India*, Abstracts p. 58.

Sheela, M. N., Unnikrishnan, M., Edison, S., & EaswariAmma, C. S. (2004). *Manihot caerulescens*—A new source of resistance to cassava mosaic disease (ICMD). In *Sixth international scientific meeting of the Cassava Biotechnology Network, March 8–14, 2004*, Colombia: CIAT, Abstracts p. 48.

Silva, D. M., Kitajima, E. W., & Oliveira, A. R. (1963). Obtencao do virus do mosaico comúnda mandioca purificado. *Ciencia e cultura, 15*, 304, in Portuguese.

Sseruwagi, P., Sserubombwe, W. S., Legg, J. P., Ndunguru, J., & Thresh, J. M. (2004). Methods of surveying the incidence and severity of cassava mosaic disease and whitefly vector populations on cassava in Africa: A review. *Virus Research, 100*, 129–142.

Stanley, J., & Gay, M. R. (1983). Nucleotide sequence of cassava latent virus DNA. *Nature, 301*, 260–262.

Storey, H. H. (1936). Virus diseases on East African plants. VI. A progress report on the studies of the diseases of cassava. *East African Agricultural Journal, 2*, 34–39.

Storey, H. H. (1938). Virus diseases of East African plants. VII. A field experiment in the transmission of cassava mosaic. *East African Agricultural Journal, 3*, 446–449.

Storey, H. H. (1939). *Report of the plant pathologist,* Report of the East Africa Agricultural Research Station. p. 9.

Storey, H. H., & Nichols, R. F. W. (1938). Studies on the mosaic of cassava. *Annals of Applied Biology, 25*, 790–806.

Tamura, K., Peterson, D., Peterson, N., Stecher, G., Nei, M., & Kumar, S. (2011). MEGA5: Molecular evolutionary genetics analysis using maximum likelihood, evolutionary distance, and maximum parsimony methods. *Molecular Biology and Evolution, 28*, 2731–2739.

Tata-Hangy, K., Koffi-Tete, M., Okao-Okuja, G., Obonyo, R., Asiimwe, P., & Legg, J. P. (2007). *Monitoring and diagnostic survey of cassava mosaic virus disease (CMD) in Democratic Republic of Congo (part 2).* http://www.iita.org/c/document_library/get_file?uuid=5948dbc1-d31b-4eed-b1b0-2019af9ef78b&groupId=25357, Accessed, March 3, 2014.

Taylor, N. J., Halsey, M., Gaitán-Solís, E., Anderson, P., Gichuki, S., Miano, D., et al. (2012). The VIRCA project: Virus resistant cassava for Africa. *GM Crops, 3*, 93–103.

Thankappan, M. (1978). Diseases of Tapioca and their control. In N. Hrishi, & R. G. Nair (Eds.), *Cassava production technology* (pp. 69–75). Trivandrum, India: CTCRI.

Thankappan, M., Makeshkumar, T., & Edison, S. (1997). Management of virus diseases in tropical tuber crops. In *International conference of integrated plant disease management for sustainable agriculture, November 11–15, 1997, New Delhi, India.*

Thresh, J. M., & Cooter, R. J. (2005). Strategies for controlling cassava mosaic virus disease in Africa. *Plant Pathology, 54*, 587–614.

Thresh, J. M., Otim-Nape, G. W., & Fargette, D. (1998). The control of African cassava mosaic virus disease: Phytosanitation and/or resistance. In A. Hadidi, R. K. Khetarpal, & H. Koganezawa (Eds.), *Plant virus disease control* (pp. 670–677). St. Paul, Minnesota: A.P.S. Press.

Tiendrébéogo, F., Lefeuvre, P., Hoareau, M., Harimalala, M. A., De Bruyn, A., Villemot, J., et al. (2012). Evolution of *African cassava mosaic virus* by recombination between bipartite and monopartite begomoviruses. *Virology Journal, 9*, 67.

Tomlinson, J. A., Ostoja-Starzewska, S., Adams, I. P., Miano, D. W., Abidrabo, P., Kinyua, Z., et al. (2013). Loop-mediated isothermal amplification for rapid detection of the causal agents of cassava brown streak disease. *Journal of Virological Methods, 191*, 148–154.

Unnikrishnan, M., EaswariAmma, C. S., Sreekumari, M. T., Sheela, M. N., & Mohan, C. (2002). Cassava germplasm conservation and improvement in India. In *CIAT-VII Asian cassava research workshop, October 28–November 1, 2002*, Thailand Program & Abstract p. 4.

Unnikrishnan, M., Mohan, C., Santha, V., Ramanathan Pillai, S., Anantharaman, M., Makeshkumar, T., et al. (2011). *Varietal improvement in cassava for mosaic disease resistance, yield and quality: Annual Report 2010–2011* (pp. 26–29). Thiruvananthapuram, India: CTCRI.

Vanderschuren, H., Akbergenov, R., Pooggin, M. M., Hohn, T., Gruissem, W., & Zhang, P. (2007). Transgenic cassava resistance to African cassava mosaic virus is enhanced by viral DNA-A bidirectional promoter-derived siRNAs. *Plant Molecular Biology*, *64*, 549–557.

Vanderschuren, H., Alder, A., Zhang, P., & Gruissem, W. (2009). Dose dependent RNAi-mediated geminivirus resistance in the tropical root crop cassava. *Plant Molecular Biology*, *70*, 265–272.

Vanderschuren, H., Moreno, I., Anjanappa, R. B., Zainuddin, I. M., & Gruissem, W. (2012). Exploiting the combination of natural and genetically engineered resistance to cassava mosaic and cassava brown streak viruses impacting cassava production in Africa. *PLoS One*, 7, e4527.

Varshney, R. K., Nayak, S. N., May, G. D., & Jackson, S. A. (2010). Next-generation sequencing technologies and their implications for crop genetics and breeding. *Trends in Biotechnology*, *27*, 522–530.

Verbeek, M., Tang, J., & Ward, L. I. (2012). Two generic PCR primer sets for the detection of members of the genus *Torradovirus*. *Journal of Virological Methods*, *185*, 184–188.

Verdaguer, B., DeKochko, A., Beachy, R. N., & Fauquet, C. (1996). Isolation and expression in transgenic tobacco and rice plants of the Cassava vein mosaic virus (CVMV) promoter. *Plant Molecular Biology*, *31*, 1129–1139.

Wagaba, H., Beyene, G., Trembley, C., Alicai, T., Fauquet, C. M., & Taylor, N. J. (2013). Efficient transmission of cassava brown streak viral pathogens by chip bud grafting. *BMC Research Notes*, *6*, 516.

Walter, B., Ladeveze, I., Etienne, L., & Fuchs, M. (1989). Some properties of a previously undescribed virus from cassava: Cassava American Latent Virus. *Annals of Applied Biology*, *115*, 279–289.

Warburg, O. (1894). Die kulturpflanzen usambaras. *Mitteilungen aus den Deutschen Schutzgebieten*, 7, 131.

Were, H. K., Winter, S., & Maiss, E. (2004). Occurrence and distribution of cassava begomoviruses in Kenya. *Annals of Applied Biology*, *145*, 175–184.

Winter, S., Koerbler, M., Stein, B., Pietruszka, A., Paape, M., & Butgereitt, A. (2010). Analysis of cassava brown streak viruses reveals the presence of distinct virus species causing cassava brown streak disease in East Africa. *Journal of General Virology*, *91*, 1365–1372.

Yadav, J. S., Ogwok, E., Wagaba, H., Patil, B. L., Bagewadi, B., Alicai, T., et al. (2011). RNAi-mediated resistance to cassava brown streak Uganda virus in transgenic cassava. *Molecular Plant Pathology*, *12*, 677–687.

Yaninek, J. S., Onzo, A., & Ojo, J. B. (1993). Continental-wide releases of neotropical predators against the exotic cassava green mites in Africa. *Experimental and Applied Acarology*, *17*, 145–160.

Zapata, C., Miller, J. C., & Smith, R. H. (1995). An in vitro procedure to eradicate Potato viruses X, Y and S from russet norkotah and two of its strains. *In Vitro Cellular & Developmental Biology*, *31*, 153–159.

Zhang, P., Vanderschuren, H., Fűtterer, J., & Gruissem, W. (2005). Resistance to cassava mosaic disease in transgenic cassava expressing antisense RNAs targeting virus replication genes. *Plant Biotechnology Journal*, *3*, 385–397.

Zhou, X., Liu, Y., Calvert, L., Munoz, C., Otim-Nape, G. W., Robinson, D. J., et al. (1997). Evidence that DNA-A of a geminivirus associated with severe cassava mosaic disease in Uganda has arisen by interspecific recombination. *Journal of General Virology*, *78*, 2101–2111.

Zhou, X., Robinson, D. J., & Harrison, B. D. (1998). Types of variation in DNA-A among isolates of East African cassava mosaic virus from Kenya, Malawi and Tanzania. *Journal of General Virology*, *79*, 2835–2840.

CHAPTER FIVE

Control of Virus Diseases of Citrus

Richard F. Lee[1]
USDA ARS National Clonal Germplasm Repository for Citrus and Dates, Riverside, California, USA
[1]Corresponding author: e-mail address: Richard@theleeclan.us

Contents

Advances in Virus Research, Volume 91
ISSN 0065-3527
http://dx.doi.org/10.1016/bs.aivir.2014.10.002

Abstract

Citrus is thought to have originated in Southeast Asia and horticulturally desirable clonal selections have been clonally cultivated for hundreds of years. While some citrus species have nucellar embryony, most cultivation of citrus has been by clonal propagation to ensure that propagated plants have the same traits as the parent selection. Clonal propagation also avoids juvenility, and the propagated plants produce fruit sooner. Because of the clonal propagation of citrus, citrus has accumulated a large number of viruses; many of these viruses are asymptomatic until a susceptible rootstock and/or scion is encountered. The viruses reported to occur in citrus will be summarized in this review. Methods of therapy to clean selected clones from viruses will be reviewed; the use of quarantine, clean stock, and certification programs for control of citrus viruses and other strategies to control insect spread citrus viruses, such as mild strain cross-protection and the use of pest management areas will be discussed.

1. INTRODUCTION

Citrus and its relatives are native to Southeast Asia, New Caledonia, and Australia (Webber, Reuther, & Lawton, 1967). Citrus has been cultivated for hundreds of years and moved around the world, as seed and propagation materials, without regard to what viruses may be present until relatively recently. As a result, citrus probably has many more virus and virus-like diseases as compared to other commodities.

Citrus is clonally propagated (Swingle & Reese, 1967). There are several reasons for this. While some citrus species have nucellar embryony, and seedlings are genetically the same as the parent tree, seedling citrus plants express juvenility traits that are undesirable: excessive thorniness, upright growth, and lack of flower/fruit production for several years. Clonal propagation avoids juvenility and ensures that propagated plants will grow and produce fruit genetically like the original tree used as the source of buds. Citrus is normally grown on a rootstock. Research and experience has shown that rootstocks can influence the quality of fruit, tree size, and impart a level of tolerance to root rots, high water tables, and poor soil type (Castle, 1995).

Many citrus viruses occur as latent infections in some scions, or cultivars, but express symptoms when the cultivar is propagated on a susceptible rootstock or scion/rootstock combination. For example tristeza decline is not expressed when trees are grown on rough lemon rootstock, but when the trees are propagated on sour orange rootstock, decline and eventually death of the tree occurs. Many "old-line" bud sources are infected with multiple viruses and graft transmissible diseases, many which will be expressed only if the trees is propagated on a susceptible rootstock or reaches an advance age when the symptoms are then expressed. An example is psorosis disease which may not produce symptoms in the scion until the tree is from 5 to 12 years old (Roistacher, 1991).

In this review, the frequently encountered virus diseases of citrus will be briefly summarized with emphasis on their characteristics and properties which influence detection and their control. This is followed by a review of the control methods/strategies for citrus viruses.

2. COMMONLY ENCOUNTERED CITRUS VIRUSES AND GRAFT-TRANSMISSIBLE DISEASES

2.1. *Citrus tristeza virus*

Citrus tristeza virus (CTV) is probably the most economically important virus disease of citrus in the world, having killed more than 50 million trees on sour orange rootstock, beginning with the epidemic in Argentina and Brazil in the 1930s (Lee, Baker, & Rocha-Pena, 1994). CTV is a member of the closterovirus group, having a flexuous filament about 2000 by 12 nm in size, with a positive-sense RNA genome of about 20 kb, containing 12 open reading frames and two untranslated regions (Dawson, 2010). The virus is transmitted in a semi-persistent manner by a number of aphid species, with *Aphis gossypii* and *Toxoptera citricida* being the most important. *A. gossypii* is the most common vector in area where *T. citricida* is not yet present, such as Israel, the Mediterranean basin and the near East, and Texas, Arizona, and California (Rocha-Peña et al., 1995). In citrus producing areas where *T. citricida* has been introduced, the severity of CTV has increased (Halbert et al., 2004; Rocha-Peña et al., 1995).

CTV expresses a number of biological activities: decline on sour orange rootstock, seedling yellows in lemon and grapefruit, stem pitting on grapefruit and/or sweet orange. While the decline on sour orange symptom of CTV can be overcome by planting on CTV-tolerant rootstocks, the stem pitting activity is much more difficult to control. The stem pitting symptom of CTV is readily apparent in susceptible hosts, but the decline on sour strains and seedling yellows strains are latent in most varieties and cultivars, and become apparent only when propagated on susceptible rootstocks or in grapefruit or lemon varieties, respectively. Seedling yellows strains are almost always accompanied by decline on sour orange and/or stem pitting activities. Biological indexing is required to determine the biological activity of a specific CTV isolate. The biological indexing procedure has been standardized to include a battery of five indicator plants: Mexican lime as a sensitive universal indicator; sweet orange on sour orange rootstock to monitor for decline on sour orange; Duncan grapefruit seedlings to monitor for stem pitting on grapefruit; Madame vinous sweet orange seedlings to monitor for stem pitting on sweet orange; and sour orange as an indicator for seedling yellows (Garnsey, Barrett, & Hutchison, 1987; Garnsey, Gumpf, et al., 1987; Garnsey et al., 2005). Seedling yellows may also be detected on the Duncan grapefruit seedling indicators.

Quarantine, clean stock, and citrus certification programs are effective in preventing the introduction of new isolates of CTV into a citrus area, and in minimizing the spread of severe CTV isolates within a citrus area. In areas where stem pitting strains of CTV are not present, tristeza decline on sour orange may be controlled by the use of CTV tolerant rootstocks, such as *Poncirus trifoliata* and hybrids with *P. trifoliata* as one of the parents. When stem pitting strains of CTV are present, mild strain cross-protection (MSCP) has been shown to be a useful management strategy in Brazil, Australia, and Japan (Lee & Keremane, 2013; Roistacher, da Graça, & Müller, 2010). In California, large pest management areas for eradiation or suppression of tristeza have been established where citrus is surveyed on a regular basis, and when CTV is found, the infected trees are eradicated (Gottwald, Polek, & Riley, 2002). This has maintained CTV at very low levels in the Central Valley of California for the past five decades. *P. trifoliata* is immune to most strains of CTV: this immunity gene has been identified and may be useful in the future in developing horticulturally desirable citrus varieties which are immune to CTV (Garnsey, Barrett, et al., 1987; Garnsey, Gumpf, et al., 1987). However, initially in New Zealand and later in other citrus areas, isolates of CTV have been found which break the immunity shown by most isolates of CTV in *P. trifoliata* (Dawson & Mooney, 2000).

2.2. Citrus psorosis virus (including citrus ringspot)

Citrus psorosis was one of the first diseases realized to be graft transmissible and is characterized as a disease of citrus which induces bark scaling lesions on the trunk and limbs of sweet orange, mandarins, and grapefruit (Fawcett, 1934). The discovery that the disease could be graft transmitted to seedlings enabled the use of biological indexing to detect citrus psorosis virus (CPsV) in trees used for budwood. The ability to detect CPsV in trees led to the development of the first certification program in 1937 where budwood source trees could be declared to be psorosis-free (Hiltabrand, 1957).

The citrus ringspot disease was first reported in the late 1960s and was characterized by the development of localized ringspots when extracts from bark lesions from citrus were rubbed on the leaves of *Chenopodium quinoa*. Citrus ringspot disease was once considered a different disease from psorosis, but molecular characterization revealed that it is CPsV (Barthe, Ceccardi, Manjunath, & Derrick, 1998; Roistacher, 1993). Using citrus ringspot isolates because of their ability to cause local lesions when inoculated onto *C. quinoa*, it was realized that there were two centrifugal components of the associated virus when separated on sucrose density gradients, and both components were required for infectivity. Upon further characterization of the virus, three single stranded, negative-sense RNAs were found, with the positive strand of RNA1 encoding a p24 (unknown function) and the RNA-dependent RNA polymerase (280 kDa protein); RNA2 positive strand encoding a p54 of unknown function, and RNA3 positive strand encoding the coat protein (CP) (48 kDa) (Barthe et al., 1998; Derrick et al., 1988). The molecular characterization revealed that what formerly was referred to as psorosis A (no bark scaling) and psorosis B (with bark scaling) and citrus ringspot (bark scaling and causing localized ringspots on *C. quinoa*) are all isolates of CPsV, now the type member of the *Ophiovirus* genus (Milne, Garcia, & Grau, 2000).

CPsV has a means of natural spread, but the means of natural spread has not been identified (Roistacher, 1993). In Argentina, the natural spread of CPsV has made the virus hard to control, even when clean stock programs have been implemented. CPsV is present in most old-line bud sources and commonly found when testing imported germplasm for viruses. Symptoms of CPsV occur on sweet orange, mandarin, and grapefruit, although the bark scaling symptom is not displayed on trees in the field until the trees are 3–7 years of age, or even older. Sour orange and lemon will show symptoms on young leaves but usually do not show bark scaling symptoms.

Control of CPsV is by use of quarantine, clean stock and certification programs to prevent the introduction of new isolates or strains and to limit distribution of the disease within a growing region. Because natural spread can occur, although the means of spread is not known, foundation trees and mother trees should be re-indexed on a recurring basis (every 3–4 years) to ensure they are free of CPsV.

2.3. Concave gum disease

Concave gum disease (CGD) is a disease of unknown etiology, although it is graft transmitted. CGD was first reported in 1936, and is widespread in old-line budwood (Fawcett, 1936). The disease is characterized by concavities on the trunk and limbs, with gum-filled deposits in the wood beneath the concavities. The bark cracks occasionally and gum exudes from the cracks. Upon biological indexing, oak leaf patterns and vein clearing occur in sweet orange and Sweettangor indicator plants. As with CPsV, the control of the disease is by the use of quarantine, clean stock, and certification programs. In Florida, there appears to be a natural spread, especially in grapefruit (Powell, Pelosi, Sonoda, & Lee, 1998).

2.4. Impietratura disease

Impietratura disease is a graft transmissible disease presumably caused by an uncharacterized citrus virus (Caruso, Davino, & Terranova, 1993; Roistacher, 1991). The disease occurs in most countries in the Mediterranean basin. The leaves of affected trees show vein clearing and oak leaf patterns in the spring and some, but not all, of the fruit are small, malformed, and hard. The symptomatic fruit will display gum pockets in the albedo which appear as small bumps of the surface of the fruit. Most varieties will show symptoms, whereas citrons appear to be more tolerant. The disease results in fruit drop and many of the fruit are not marketable because they are malformed and hard. Upon biological indexing, the symptoms on sweet orange indicator plants are similar to CPsV (Roistacher, 1991).

2.5. Cristacortis disease

Cristacortis disease is a graft transmissible disease presumed to be caused by an uncharacterized citrus virus (Roistacher, 1991). It was first reported from Corsica (Vogel & Bové, 1968) and has been reported also in Italy, Spain, Northern Africa, Turkey, Mauritius, and San Paulo, Brazil. The disease is characterized by vertical depressions or pockets in the wood, with gumming occurring at the bottom of the depression when the bark is cut away. The symptoms appear similar to stem pitting caused by some strains of CTV, but the pits can occur on sour orange rootstocks, also on mandarins which tend to be more tolerant of CTV induced stem pitting. Leaf symptoms are vein flecking and oak leaf patterns, similar to those produced by CPsV. Natural spread is not reported. Quarantine, clean stock, and certification programs effectively control this disease.

2.6. Citrus vein enation

Citrus vein enation (CVEV), also known as woody gall, was first described in California as a graft-transmissible disease which caused enations in the leaf veins of sour orange (Wallace & Drake, 1953). CVEV also caused woody galls on the trunks or rootstocks of acid lime, rough lemon, Rangpur lime, and *Citrus volkameriana*. The disease is transmitted in a persistent manner by several aphid species including *T. citricida*, *Myzus persicae*, and *A. gossypii* (Roistacher, 1991). The disease has been reported from Australia, California, India, Japan, Peru, and Spain. Using electron microscopy, spherical virus particles about 28 nm in diameter have been visualized, leading to the suggestion that the virus associated with citrus vein enation was a luteovirus (Maharaj & da Graça, 1988). Recently, an Enamovirus has been associated with CVEV and sequenced by deep sequencing of the small RNAs, confirming that the virus associated with CVE is a luteovirus (Vives et al., 2013). Control of the disease involves using propagation material that is free of CVEV and to protect budwood source trees from aphids to keep the budwood source trees from becoming infected via aphids.

2.7. Citrus blight disease

Citrus blight is a disease of unknown etiology, but the disease is reproducibly graft transmitted using inoculum that comes from the rootstock (Tucker, Lee, Timmer, Albrigo, & Brlansky, 1984). Different rootstocks vary in susceptibility to citrus blight as indicated by the age where blight symptoms become evident (Derrick & Timmer, 2000). Rough lemon, Rangpur lime, *P. trifoliata*, and Carrizo citrange are especially susceptible, whereas Cleopatra mandarin, sour orange, and sweet orange are more tolerant. Once trees develop the disease, the decline is irreversible, and the trees become unproductive. Early symptoms of citrus blight are zinc deficiency symptoms in the leaves, zinc accumulation in the phloem, and eventually high zinc levels occur in the xylem. Blockage of xylem tissues with amorphous plugs occurs, resulting in reduced water uptake (Lee, Marais, Timmer, & Graham, 1984). The canopy of affected trees thins and upper shoots develop small leaves, pointing upright, the foliage has a dull green color. Overall tree decline continues with twig dieback and root loss. Other physiological changes follow such as off season flowering and production of water sprouts inside the canopy (Tucker et al., 1984). The first incidence of blight occurs at random in a citrus grove, but the subsequent incidences of blight tend to be within one to two trees of trees which are already symptomatic, suggesting a natural means of spread that is yet unknown (Castle & Gottwald, 2005). A serological assay has been developed to test for citrus blight (Derrick et al., 1990). Management of citrus blight includes use of blight tolerant rootstocks and testing of budwood source trees for lack of reactivity with the blight-associated protein (Derrick et al., 1990). Blight losses currently are estimated at about 4–7% of the tree population in Florida (Wang & Brlansky, 2013).

2.8. Citrus viroids

Citrus viroids are a group of small, low molecular weight (371–330 nt), circular, infectious RNA molecules which cause various symptoms in citrus and citrus relatives (Table 1) (Brlansky & Timmer, 2014; Duran-Vila, Roistacher, Rivera-Bustamante, & Semancik, 1988). Viroids are not encapsidated in a protein. The discovery that Etrog citron was a good indicator plant for viroids (Calavan, Frolich, Carpenter, Roistacher, & Christiansen, 1964) and the development of the sequential polyacrylamide gel electrophoresis (sPAGE) for diagnosis of the different viroids (Flores, Duran-Vila, Pallas, & Semancik, 1985) has led to the discovery of additional viroid pathogens present in citrus (Ito et al., 2001, 2007). Viroids are easily transmitted mechanically, and care must be made to sterilize clippers, saws, and other tools used to cut or prune trees to prevent spread (Roistacher, 1991). There are no reported vectors. Quarantine, clean stock, and certification programs are effective in control of viroids.

Citrus exocortis viroid (CEVd) was the first citrus viroid described (Roistacher, 1991). CEVd is associated with bark shelling on trees grown on *P. trifoliata* or *P. trifoliata* hybrids (citranges and citrumelos), and can cause dwarfing of trees grafted on these rootstocks. CEVd is the largest citrus viroid at 371 nt, and produces the most pronounced symptoms of leaf epinasty, stunting, and necrosis of the leaf midvein. CEVd is a species of the genus *Pospiviroid*.

Cachexia, previously also called xyloporosis and synonymous with *Hop stunt viroid*, (Childs, 1950), was first demonstrated to be viroid-like in 1980 when Roistacher, Nauer, and Wagner (1980) demonstrated the ability to mechanically transmit the disease and found that thermotherapy would not eliminate the pathogen. The viroid was first characterized by Semancik, Roistacher, Rivera-Bustamante, and Duran-Vila (1988). Cachexia causes dwarfing and poor growth of tangelo, mandarin and *Citrus macrophylla* as scions or as rootstocks. On susceptible trees, cachexia causes depressions in the wood of the tree with corresponding pegs in the bark. Cachexia (299 nt) is a species in the *Hostuviroid* genus.

Citrus bark cracking viroid, synonymous with Citrus viroid IV, causes leaf drooping on Etrog citron indicator plants, and general necrosis of the leaf midribs. *Citrus bark cracking viroid* (284 nt) is a species in the *Cocadviroid* genus.

Table 1 Properties of viroids reported from citrus

Viroid	Size (bases)[a]	Symptoms in citron	Parson's special mandarin reactivity	Symptoms on *Poncirus trifoliata*
Citrus exocortis viroid (CEV)	394–369	Stunting, epinasty, midvein necrosis, petiole necrosis, leaf tip browning	None	Bark scaling, stunting
Citrus viroid 1	CV-1a: 340 CV-1b: 330	Stunting, leaf bending, point necrosis of midvein	None	Unknown
Citrus viroid II-a, b	305–295	Slight petiole necrosis and leaf tip browning	None (CVd IIa)	Bark cracking
Citrus viroid IIb		Asymptomatic in citron	Strong symptoms	Unknown
Citrus viroid IIc		Asymptomatic in citron	Strong symptoms	Unknown
Citrus viroid III-a to d	291–296	Stunting, leaf dropping, necrosis of midvein, petiole necrosis	None	Grooving possible
Citrus viroid IV	275	Stunting, leaf dropping, necrosis of midvein, petiole necrosis	None	Unknown
Citrus viroid V[b]	293–294	Mild stunting, mild bark cracking and gum exudates	None	Unknown
Citrus viroid VI[c]	328–331	Stunting, mild leaf bending, petiole necrosis	None	Unknown

[a]From https://ebi.ac.uk/genomes/viroid.html; Genome Pages—Viroid.
[b]Serra, Barbosa, Daròs, Flores, and Duran-Vila (2007).
[c]Ito, Icki, Ozaki, and Ito (2001).

Citrus bent leaf viroid, synonymous with Citrus viroid I, causes leaf bending on Etrog citron and point necrosis on the leaf midribs. *Citrus bent leaf viroid* (318 nt) is a species in the *Apscaviroid* genus.

Citrus dwarfing viroids, synonymous with Citrus viroid III, cause dwarfing on *P. trifoliata*, citrange, and citrumelo rootstocks. Citrus viroid IIIa (297 nt) and Citrus viroid IIIb (294 nt) are species in the *Apscaviroid* genus.

Citrus viroid V produces mild symptoms in Etrog citron. Citrus viroid V (294 nt) is a species in the *Apscaviroid* genus (Serra et al., 2007).

Citrus viroid VI produces mild symptoms in Etrog citron. Citrus viroid VI (330 nt) is a species in the *Apscaviroid* genus.

2.9. Citrus tatterleaf virus

Citrus tatterleaf virus (CiTLV), synonyms *Citrange stunt virus*, *Apple stem grooving virus*, was first reported in 1962 from Meyer lemon which had been imported from China in 1908 (Wallace & Drake, 1962). CiTLV causes stunting or dwarfing, necrosis at the bud union, and virus-induced bud union incompatibility on scions grafted onto *P. trifoliata*, citrange, or citrumelo rootstocks (Iwanami, Kano, & Koizumi, 1991). The virus is carried asymptomatically in most citrus varieties, and symptoms often are not apparent until the tree is 3–7 years of age (Roistacher, 1991). The major method of transmission is by use of infected propagation materials. CiTLV is widespread in China, Japan, and Korea and has been reported in South Africa, Australia, and in the United States via the importation of Meyer lemon from China. Control of CiTLV is by use of quarantine, clean stock, and certification programs.

2.10. Citrus leaf blotch

Citrus leafblotch (CLBV) was first reported from Nagami kumquat on Troyer citrange rootstock plants that had a bud union crease (Vives, Galipienso, Navarro, Moreno, & Guerri, 2002). The CLBV is a member of the family *Flexiviridae.* The virus has filamentous virions about 960 × 14 nm in size with a single-stranded, positive-sense genomic RNA of 8747 nt. CLBV is seed transmitted and there are no known vectors. A similar if not identical virus was intercepted in California while indexing a Cleopatra mandarin variety introduced from Florida in 1968 and called *Dweet mottle virus* (DMV) (Krueger, Bash, & Lee, 2005). Sequencing of DMV has confirmed a 96% homology with CLBV, but the reported host range differs slightly (Hajeri, Ramadugu, Keremane, Vidalakis, & Lee, 2010). CLBV has been associated with bud union crease with *P. trifoliata* or trifoliate hybrid rootstocks with Pera sweet orange in Brazil, Marsh grapefruit and Roble sweet orange in Florida, and Nules Clementine plants in Italy, and Nules, Clementine, Navelina, and Navelate sweet orange in Spain (Vives et al., 2002).

2.11. Measles disease

Measles is a disease which causes pale yellow spots on leaves as they mature. The affected trees generally appear smaller and branches have a droopy appearance. Usually only a few trees in a grove will show symptoms and will continue to show symptoms year after year. There appears to be a natural spread but it is very slow with only a few additional trees developing symptoms once the disorder appears in a grove. While initially thought to be a genetic disorder, the measles symptoms has been graft transmitted in Florida (Lee, Derrick, Futch, & Tucker, 1993) and in California (Lee, Yokomi, & Vidalakis, 2008), but the causal agent has not been identified. Sour orange seedlings in a cool (20–27 °C) greenhouse are good indicator plants for measles disease.

2.12. Yellow vein

Yellow vein is a graft transmissible disease reported from California in the 1950s by Weathers (1960). The disease is characterized by yellow veins in the leaves. The fruit also may develop irregular yellow blotches in the rind and are smaller and flatter than normal fruit. Many varieties and types of citrus show symptoms of yellow vein when graft inoculated, but many of these plants recovery and do not show yellow veins on the leaves of new flushes. Limes and lemons appear to be the most symptomatic. There is no evidence of natural spread. A similar disease has been reported in Pakistan (Bové, 1995). The molecular characterization of *Citrus yellow vein clearing virus* has been recently reported (Loconsole et al., 2012).

2.13. Citrus leprosis

Citrus leprosis is one of the most important citrus viral diseases and has spread northward from South America to Central America and Mexico in recent years. The leprosis disease is vectored by the *Brevipalpus* species mites (commonly called flat mites) in a persistent manner (Rodrigues, Kitajima, Chilers, & Chagas, 2003). Once thought to be cause by a Rhabdovirus, research has revealed that while there are no difference in symptomology, there are two types of citrus leprosis viruses: cytoplasmic and nuclear (Guerra-Moreno, Manjunath, Brlanksy, & Lee, 2005). The cytoplasmic-type leprosis (CiLV-C) is limited to the cytoplasm. CiLV-C is the type member of a new genus *Celivirus*, has membrane-bound bacilliform virions 50–60 × 110–120 nm. The genome is a bipartitie, single-stranded, positive-sense RNA with RNA1 having 8729 nts and RNA2 having 4969 nts (Locali-Fabris et al., 2006). Recently, a variant isolate of CiLV-C was reported from Colombia which is different enough from the previous characterized isolates of CiLV-C from Panama and Brazil that serological and RT-PCR tests designed for the detection of these previous characterized isolates did not detect the Colombian variant, referred to as CiLV-C2 (Roy et al., 2013). CiLV-C2 has a similar genome to CiLV-C with the RNA1 having 8717 nts and RNA2 with 4989 nts. Both RNA1 and RNA2 had only 58% and 50% homology with RNA1 and RNA2 of CiLV-C. Symptoms produced by CiLV-C and CiLV-C2 are similar.

The genome sequence of the nuclear-type leprosis virus (CiLV-N) obtained from Mexico was recently reported (Roy et al., 2013). RNA1 had 6268 nts and RNA2 had 5847 nts. The genome organization closely resembles that reported for *Orchid fleck virus* with 90–91% homology with OFV at the nucleotide level and 93–98% homology with amino acid sequence identities.

Quarantine, clean stock, and certification programs are a key to prevent leprosis from becoming established in new areas and slowing spread within a citrus area. Mite control slows the spread of the disease once it is established in an area. For isolated trees in a grove which become infected with leprosis, probably because the tree did not receive the miticide application, severe pruning to eliminate all symptoms of leprosis, followed by miticide applications are successful in recovering the trees.

2.14. *Citrus yellow mosaic*

Citrus yellow mosaic (CYMV) (synonym *Citrus mosaic badnavirus*) is an important disease in India where infection rates are as high as 70% (Ahlawat et al., 1996). The disease infects most citrus cultivars and relatives, oranges, grapefruit, and mandarins usually show strong symptoms, while Mexican lime is symptomless. The mature leaves show bright yellow mottling for vein flecking, fruit production is reduced and on chronically infected trees, symptoms are present on the fruit. The disease is spread by the citrus mealy bug, *Planococcus citri* (Huang & Hartung, 2001). Spread by the mealy bug vector appears to be limited, widespread spread is by the use of infected propagating source materials.

CYMV is a badnavirus with nonenveloped bacilliform particles 150×30 nm in size having a DNA genome of 7559 bp in length and six putative open reading frames, all on the plus-strand of the genome (Huang & Hartung, 2001). Control is by the use of quarantine, clean stock, and certification programs, and control of the mealy bug vector.

2.15. Satsuma dwarf

Satsuma dwarf is a virus disease originally described in Japan where it causes serious problems (Changyong, Xueyuan, Yuanhui, & Xinhua, 1993; Cui,

Gu, & Roistacher, 1991). Trees infected with *Satsuma dwarf virus* (SDV) have spoon shaped leaves, enations, multiple flushing, stunting or dwarfing, fewer leaves, and small fruit having a thick peel (Tanaka, 1972). There are variants of SDV, with one variant being called *Citrus mosaic virus* (not to be confused with CYMV virus which is a badnavirus) (Ahlawat, Chenulu, Viswanath, Pandey, & Bhagabati, 1985; Dakshinamurti & Reddy, 1975). The SDV virion is an isometric virus particle about 26 nm in diameter. Two RNAs are present, RNA1 with about 7000 nt and RNA2 with about 5400 nt (Karasev, Han, & Iwanami, 2001). SDV has been reported from Japan, China, Iran, Korea, and Turkey, while the citrus mosaic variant of SDV has been reported only from Japan.

SDV is readily transmitted by grafting, and soil transmission has been reported but not confirmed (Koizumi, Kano, Ieki, & Mae, 1988). No vector has been reported.

3. OTHER INSECT-SPREAD DISEASES CAUSED BY PROKARYOTES WHICH NEED TO BE CONSIDERED IN CONTROL/MANAGEMENT OF CITRUS VIRUSES AND VIRUS-LIKE DISEASES

3.1. Huanglongbing

Huanglongbing (HLB), also known as citrus greening disease, is perhaps the most important citrus disease in areas where the disease and its vector are both present (Bové, 2006). The symptoms of include an asymmetrical mottling of the leaves; frequently the midribs of the leaves are yellowed. Sectors of the canopy decline and dieback, followed by the decline and dieback of the entire canopy. Once the tree is nearly totally infected, the tree will produce yellow shoots. The yellow shoots are characteristic of HLB, and HLB in Chinese translates to "yellow shoot." Symptomatic fruit are lobsided, usually contain aborted seed, and have an off flavor. Fruit production is reduced, symptomatic fruit is small, and fruit often drops prematurely. Over 2–3 years the tree declines and dies.

Three forms of HLB have been identified: the Asiatic form which is associated with *Candidatus* Liberibacter asiaticus (Las); the African form which is associated with *Candidatus* L. africanus (Laf), and the American form which is associated with *Candidatus* L. americanus (Lam) (Jagoueix, Bové, & Garnier, 1994; Teixeira et al., 2005). The Asiatic form of HLB expresses symptoms at a warmer temperature whereas the African from which expresses strongest symptoms under cooler temperatures. The American form expresses symptoms under cooler temperatures and extended periods of warmer temperature eliminate the disease from infected plants (Lopes et al., 2009). In Brazil where Lam was first reported, the Las has out competed Lam and Lam is rarely encountered now under field conditions (Lopes et al., 2009).The host range of HLB is limited to citrus and citrus relatives.

There are two psyllid vectors of HLB: *Diaphorina citri* and *Trioza erytreae* (Bové, 2006). Both species of the vector can transmit either the Asiatic form or the African form of HLB, and *D. citri* has been shown to transmit the American form in Brazil (Bové, 2006; Lopes et al., 2010). *T. erytreae* occurs mostly in Africa, while the *D. citri* occurs in northern Africa, Asia, and the Americas. *D. citri* is the more efficient vector (Bové, 2006).

Control of HLB requires quarantine, clean stock, and certification programs in order to produce healthy plants and prevent movement of infected nursery stock. The psyllid vectors must be controlled. Florida and Texas have begun using area-wide sprays to reduce the psyllid population during the winter months when the psyllids are not as active and only the overwintering adults are present (Bassanezi et al., 2013). In areas where HLB is not already established, a three pronged approach to control is effective: regular surveys to identify early symptoms on trees which are then removed; control of the psyllid vector by survey and pesticide application; and use of clean plant material for replanting (Bové, 2006). Detection of the bacterium associated with HLB is by PCR or real-time PCR. Testing psyllids for the presence of the bacterium associated with HLB by real-time PCR has proven to provide an earlier warning of the presence of the disease in an area where HLB is not already established (Manjunath, Halbert, Ramadugu, Webb, & Lee, 2008).

3.2. Stubborn disease

Stubborn was described as a disease in 1944 in California. It is now known to be caused by a helical mollicute *Spiroplasma citri* and Koch's postulates have been fulfilled (Bové et al., 2002; Saglio et al., 1973). The *S. citri* genome has been sequenced (Ye et al., 1992). Stubborn causes stunting, hence the name stubborn as the infected plants do not grow. This is especially apparent in when young trees become infected (Roistacher, 1991). Fruit on affected trees are often lopsided, small in size, and have aborted seed. Color inversion is often seen with the stylar end remaining green and the peduncular end showing color. The symptoms on the tree canopy are often localized into sectors, especially when larger trees become infected. Leaves are small, cupped, and often have an upright appearance. Mottling can occur on the leaves. Diagnosis of stubborn is by *in vitro* culture, or by use of PCR and/or real-time PCR (Yokomi, Mello, Fletcher, & Saponari, 2010). For biological indexing, Madame Vinous sweet orange seedling maintained in warm conditions (37 °C) day temperature and 27 °C night temperature are inoculated with bud chips or side shoot grafts (Roistacher, 1991). Symptoms expressed under these conditions are slow growth, small cupped leaves, short internodes between leaves, and leaf mottling. Most citrus varieties and cultivars are susceptible to stubborn.

Stubborn has several planthoppers as vectors: *Circulifer tenellus* and *Scaphytopius nitidus* common in California and Arizona, and *Circulifer haematocepts* (syn. *Neoaliturus haematocepts*) in the Mediterranean region (Calavan & Bové, 1989). The *Spiroplasma* multiplies in the vector but there is no evidence of transovarial passage. Most spread of stubborn in citrus is primary spread: there is little evidence of citrus to citrus transmission (Yokomi et al., 2010). The leafhopper vectors have a wide host range and citrus is a temporary host if more preferred hosts are not available. Stubborn has not been reported in tropical or subtropical regions.

Control is by the use of quarantine, clean stock, and certification programs to ensure that healthy plants are planted in the field. The movement of the planthopper vectors is seasonal, and planting of new citrus when the vectors are likely to move to the young actively growing citrus should be avoided. Young plants which become infected with stubborn should be removed and replaced with a healthy plant. The use of trap plants which are more attractive to the vectors than citrus has been reported (Schwarz, 1965).

3.3. Citrus variegated chlorosis

Citrus variegated chlorosis (CVC) first appeared in Brazil in 1987 and rapidly became established throughout the Brazil citrus industry (Lee, Derrick, Beretta, Chagas, & Rosetti, 1991). CVC has spread to countries surrounding Brazil and most recently was reported in Costa Rica (Aguilar et al., 2005). The disease was named because of the chlorotic appearance of the tree, resembling a zinc deficiency, once the tree becomes infected. Newly affected trees show sectoring of symptoms, while chronically infected trees have a yellow appearance. As the leave mature, small, light brown gummy lesions form on the underside of the leaves, corresponding to the yellow chlorotic areas on the upper side of the leaf. As the leaf ages, the lesions on the underneath of the leaf may become dark brown and even become necrotic. Fruit size on CVC affected trees is greatly reduced and the fruit has a hard rind. Normal fruit thinning after blossom set does not occur on a CVC affected tree, resulting in the tree carrying more fruit that is smaller. CVC affected fruit ripens earlier than fruit on a healthy tree and the sugar content is higher. Fruit tends to occur in clusters, rather than one or two fruit as normal, especially on Pera sweet orange trees. CVC affected trees become stunted, the branches dieback, and the canopy thins, but the tree rarely dies. The disease has been shown to be caused by a pathovar of *Xylella fastidiosa* (Hartung, Beretta, Brlansky, Spisso, & Lee, 1994).

CVC is vectored by numerous sharpshooters; several species have been identified as vectors (Hopkins & Purcell, 2002). Control of CVC is by quarantine, clean stock, and certification programs so that clean propagation materials are used and plants going to the field are healthy. Vector control is important, and sharpshooter populations are controlled by insecticide applications.

4. METHODS OF CONTROL OF GRAFT-TRANSMISSIBLE PATHOGENS OF CITRUS

Control of citrus viruses begins though three programs: (1) quarantine (safe introduction of germplasm into an area or region), (2) clean stock program (to test, therapy if necessary, and produce/provide sources of pathogen-tested propagating stock for an industry), and (3) certification to ensure the maintenance and use of pathogen-tested propagating material for commercial use (Lee, 2000; Lee, Lehman, & Navarro, 1999; Navarro, 1993). These three programs will be discussed in greater detail, Figs. 1 and 2 illustrate the relationship among these three programs.

4.1. Quarantine programs

Quarantine programs allow for the safe introduction of germplasm into an area or region. In a viable citrus industry, there is a constant demand for new varieties for possible commercial production or for use in citrus breeding programs. A quarantine program provides a mechanism for introduction of new varieties or cultivars into a citrus area without introducing citrus viruses or other graft-transmissible pathogens. Quarantine programs require regulatory authority and are operated under the guidance of the Ministry of Agriculture or the state Department of Agriculture.

Desired germplasm is requested using an import permit so that everyone concerned knows the material is being imported, and once the material

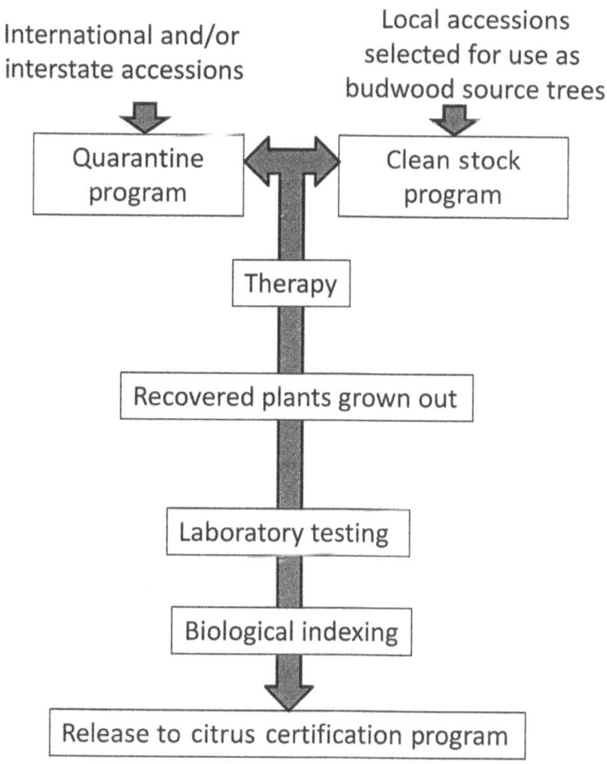

Figure 1 Diagram showing the relationship between a quarantine program and a clean stock program and the general scheme of therapy and testing before incoming accessions may be released into the citrus certification program or made available to the citrus industry. (See the color plate.)

arrives, it is usually inspected to make sure insects, fungi, and other pests are not present. The imported germplasm is then established in a quarantine greenhouse or in a region removed from the main citrus production area, or by the use of *in vitro* quarantine by placing the surface sterilized budstick in media so that emerging buds may be used for shoot tip grafting (Lee, 2000; Navarro, 1993). The imported germplasm must be therapied to eliminate graft-transmissible pathogens.

The traditional methods of therapy have been thermotherapy (Grant, 1957) or shoot tip grafting (Navarro, Roistacher, & Mirashige, 1975). More recently, cryotherapy has been applied to eliminate graft-transmissible pathogens (Wang, Panis, Engelmann, Lambardi, & Valkonen, 2009).

Thermotherapy involves subjecting the buds, grafted onto a rootstock, to an extended heat cycle (Roistacher, 1991). Thermotherapy of citrus

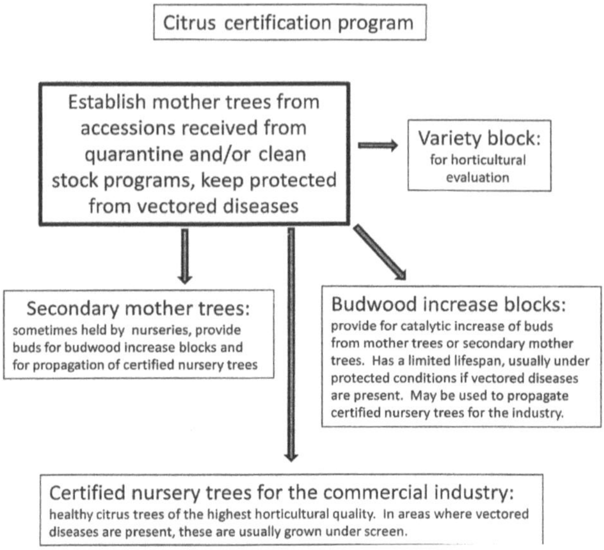

Figure 2 Diagram of the general structure of a citrus certification program. The clean material received from the quarantine or clean stock program is used to establish the mother trees and maintained under conditions to protect them from exposure to vector-borne pathogens. Buds cut from the mother trees may be used to establish secondary mother trees, usually maintained under protected conditions in a commercial nursery, for budwood increase blocks to rapidly increase the number of buds available for propagation, or for direct propagation of certified nursery trees. A variety block is established from buds obtained from the mother trees to evaluate the horticultural trueness to type of the budline and to obtain production data. (See the color plate.)

requires grafting a bud onto a rootstock, usually a citrange, keeping the bud wrapped with grafting tape so that it does not grow or push, and placing the grafted rootstock in a heat treatment chamber having a 16-h light cycle at 40 °C and 8-h dark cycle at 30 °C for 12–16 weeks. Once the heat treated plants are removed, the grafting tape is removed, and the buds are allowed to push. Once the buds have grown for 12–15 weeks, testing is performed to determine if the pathogens have been eliminated. This method is effective at removing most citrus pathogens which are detected by biological indexing under cool conditions (Table 2), but it is not effective at eliminating citrus viroids or CiTLV. The disadvantage of this approach is the time required for thermotherapy, 12–16 weeks.

Therapy by use of shoot tip grafting has been the approach of choice for removal of citrus viruses (Navarro et al., 1975). Using this procedure, the meristem tips, about 0.1 mm thick, are cut from emerging buds and grafted onto etiolated rootstocks grown *in vitro*. The tips are usually free of graft-

Table 2 A summary of the required tests for citrus viruses and virus-like diseases for release from quarantine status in California[a]

Virus/disease	Indicator plant	Symptoms	Index condition[b]	Laboratory diagnostic[c]
Citrus tristeza virus/tristeza	Mexican lime	Leaf vein clearing, leaf cupping, stunting, stem pitting	Cool	ELISA
Citrus psorosis virus/psorosis including citrus ringspot and pathogens which test similar (impietratura, concave gum, cristicortis)	Sweet orange, Dweet tangor	Leaf vein flecking, oak-leaf patterns on young leaves, bark scaling may occur with some isolates; psorosis and ringspot will cause shock symptoms	Cool	
Infectious variegation virus, *Citrus leaf rugose virus*, *Citrus crinkly leaf virus*	Sour orange, may be supplemented with Etrog citron, Eureka lemon	Leaf variegation and/or crinkle on sour orange; chlorotic spotting on lemon	Cool	
Apple stem grooving virus syn. *Citrus tatter leaf virus*/citrange stunt	*Citrus excelsa*, rusk citrange on rough lemon	*C. excelsa*: deformed leaves with chlorotic zones; rusk citrange: deformed leaves with chlorotic leaf blotch, stem with zigzag growth	Cool	
Citrus leaf blotch virus/*Dweet mottle virus*	Dweet tangor	Mottle pattern on leaves	Cool	RT-PCR
Citrus viroids including exocortis,	Arizona 861-SI Etrog citron	Leaf epinasty and/or stunting, midvein necrosis,	Warm	sPAGE from inoculated citrons

Continued

Table 2 A summary of the required tests for citrus viruses and virus-like diseases for release from quarantine status in California—cont'd

Virus/disease	Indicator plant	Symptoms	Index condition	Laboratory diagnostic
cachexia, and five other viroid species	on rough lemon	petiole necrosis, leaf tip browning		grown under warm conditions
Spiroplasma citri/stubborn	Sweet orange	Stunting, small leaves, leaf cupping	Warm	Culture from source tree held under warm conditions
Citrus vein enation virus/vein enation	Sour orange/ Mexican lime	Vein enations; galls may form on Mexican lime where wounds occur	Cool	RT-PCR
Huanglongbing associated with *Candidatus* Liberibacter species/citrus greening disease	Source plant observation	Asymmetrical mottle on leaves, yellowing (zinc deficiency symptoms)	Warm (Asiatic form); cool (African and American forms)	qPCR, observation of source plant
Xylella fastidiosa/citrus variegated chlorosis	Source plant observation	Zinc deficiency symptoms, gummy lesion on underside of leaves	Warm	PCR
Miscellaneous and unknown	Source plant			dsRNA

ELISA, enzyme-linked immunosorbent assay; RT-PCR, reverse transcriptase polymerase chain reaction; sPAGE, sequential polyacrylamide gel electrophoresis; qPCR, quantitative real-time polymerase chain reaction; PCR, polymerase chain reaction; dsRNA, double-stranded RNA.
[a]When performing biological indexing, positive controls *in planta* must be available in order to perform complete and definite testing. Exotic pathogens (such as Huanglongbing and *Xylella fastidiosa* may be used as DNA extracts for positive control for the laboratory testing).
[b]Cool indexing conditions is 24–27 °C, while warm indexing conditions is 28–40 °C.
[c]All laboratory testing is performed on the source plant unless otherwise described.

transmissible pathogens, and the resultant recovered plant is usually free of pathogens and does not undergo juvenility. This approach has been shown to be effective at elimination of all citrus viruses (Navarro, Civerolo, Juarez, & Garnsey, 1989). A modification of the procedure is referred to as *in vitro* quarantine (Lee et al., 1999). Here the budstick, once received under permit, is surface sterilized, placed in media in a glass tube and placed under lights at 25–26 °C. When the buds emerge, shoot tips are cut and grafted as in the normal shoot tip grafting protocol. Depending on the skill of the technician performing the shoot tip grafting procedure, it is most likely the resultant recovered plant will be free of pathogens.

An alternative method of therapy, referred to a cryotherapy, is where vegetative buds are excised, treated in a buffer to remove excess water within the cells, frozen in liquid nitrogen, then recovered under controlled conditions and grafted onto an etiolated rootstock as done with shoot tip grafting (Wang et al., 2009). The protocol we are using for citrus is very similar to that used for cryopreservation of citrus (Volk, Bonnart, Krueger, & Lee, 2012), except that the freezing step is limited to 1–2 h. The advantage of this procedure is that vegetative buds are used, not the more tender emerging buds as in shoot tip grafting. The buds are 1–1.5 mm when cut. After the freezing step, about 0.5 of the base of the bud is excised and the resultant tip is grafted to the etiolated rootstock.

Regardless of the method used to therapy the germplasm received in quarantine, it is important that the therapied germplasm be indexed for all known graft-transmissible pathogens before being released from quarantine. This usually requires both biological and laboratory testing. Table 2 summarizes the pathogens tested and the test used for each pathogen as required in California. If the germplasm has been obtained from a region where a virus or virus-like pathogen is present, that pathogen should be tested for. For example, germplasm originating from Florida should be tested for citrus blight

4.2. Clean Stock Programs

Clean stock programs provide for recovery of healthy sources of local cultivars and provide a mechanism whereby pathogen-tested propagating material is available to the local citrus industry. A clean stock program is often carried on by a research institution utilizing the expertise of specialists in plant pathology, horticulture, and possibly other departments, but the clean stock program may be operated by the regulatory agency having jurisdiction over the citrus production area.

There are several steps in a clean stock program (see generalized scheme in Fig. 1). The clean stock program selects mother trees from local cultivars, the selection of potential mother trees should be made with regard to horticultural characteristics without pathogen content being considered. However, it is usually recommended that a measurable trait or phenotype be used to make the best use of limited resources. Examples of measurable traits or phenotypes are earlier maturation, consistent production of more fruit, higher color score on fruit, etc. Following the selection of potential mother trees, the process for cleanup is similar to the quarantine program. The trees are indexed to determine what citrus viruses are present. Therapy follows, either shoot tip grafting or thermotherapy, depending on viruses present and expertise and facilities available. Following therapy, the resultant recovered potential mother trees must be completely indexed for all graft-transmissible pathogens present in the area.

An important component of a clean stock program is the horticultural evaluation. The availability of the data from the horticultural evaluation is valuable to the local industry in helping growers decide if they want to grow a particular cultivar or variety. In Florida, the productivity on a per tree basis over a 15–20 year period was increased by 10–15% because the growers could evaluate the publically available data obtained from the horticultural evaluation trials from the accessions being therapied in the Florida clean stock program (Lee, 2000). The clean stock program maintains a source of mother trees, usually under protected conditions to prevent possible insect vectors access, and makes budwood available upon request.

4.3. Certification Programs

Certification programs are important to guarantee the sanitary status and horticultural quality of propagating material for the commercial production of nursery plants (Lee, 2000; Lee et al., 1999; Navarro, 1993). Certification programs must have regulatory authority and are usually operated by a state or provincial agency having the legal authority to impose restrictions. All the propagation of citrus is done by commercial nurseries, but the primary foundation trees are usually maintained by the agency in charge of the certification program. In a certification program, testing of primary foundation trees on a recurring basis is done for commonly occurring citrus viruses or viroids and nursery increase blocks may be tested for pathogens known to be insect vectored in the region. A typical scheme for a certification program is shown in Fig. 2.

5. OTHER METHODS OF CONTROL

5.1. Pest management areas

Pest management areas for eradication or suppression of citrus viruses and other pathogens have been used effectively in California. As an example, in California pest management areas for the suppression of CTV were established in the Central Valley of California in 1963 (Gottwald et al., 2002). An epidemic of CTV decline on sour orange had occurred in the southern California production area resulting a wide spread incidence of tristeza, but CTV was rarely found in the Central Valley production area (Fresno, Kern, Madera, and Tulare counties). To control losses due to CTV, the Central California Citrus Pest Control Agency was charged with locating and removing CTV-infected trees. This agency was later renamed the Central California Tristeza Eradication Agency (CCTEA) with responsibility through an agreement of the Joint Operations of the Tristeza Eradication Program (Joint Powers Agreement or JPA) of what was then five citrus pest control districts in Fresno, Tulare, and Kern counties. Two of the districts withdrew in 1995 and 1996, leaving three pest management districts. The program is funded by a special assessment on growers depending on the acreage of citrus in the district.

The purpose of the CCTEA is to identify and eradicate CTV in a timely, orderly, and cost effective manner and to encourage and support appropriate research programs that pursue method to eliminate the threat of CTV (Gottwald et al., 2002). The program has been charged with ensuring that all trees planted in the suppressive areas remain virus-free by identifying reservoirs of CTV so the trees can be eliminated. While the approach for surveying and eliminating CTV has changed over the years and technology has been developed and applied, the area still has a very low incidence of CTV after five decades.

5.2. Mild strain cross-protection

MSCP is the phenomenon which occurs when a mild isolate of a virus is inoculated into a plant, and when that plant is later inoculated (or challenged) with a severe isolate of the same virus, the symptoms of the severe isolate are suppressed or delayed in expression (Lee, Brlansky, Garnsey, & Yokomi, 1987). MSCP is not the same as virus resistance. MSCP as a management strategy is useful only when (1) the disease is endemic with no other possibility of control; (2) there is confidence that the mild isolate being used will not spread and cause damage in other crops or act in combination with other viruses to cause more severe damage; (3) there is some level of confidence that the mild isolate being used is stable (especially if it has been derived by mutation) and will continue to be mild when purposefully inoculated into many plants.

MSCP of citrus viruses has been a useful method to control, or manage, disease losses due to severe viruses, especially with tristeza (Bederski, Roistacher, Silvestre, & Müller, 2010; da Graça & van Vuuren, 2010; Lee & Keremane, 2013; Roistacher et al., 2010). Empirically selected mild isolates of CTV have been used in several citrus production areas to maintain the ability to produce fruit despite the presence of severe stem pitting strains of CTV; Pera sweet orange in Brazil (Müller & Costa, 1987), grapefruit in Australia and South Africa, limes in India, Hassuku dwarf on pummelo in Japan. More recently, in Florida isolates of CTV were empirically selected which protected against CTV decline on sour orange rootstock (Lee & Keremane, 2013).

The mechanism of MSCP in citrus is not understood, and it is probable that several mechanisms are involved in the phenomena perhaps depending on the virus and/or host. The mechanisms of protein mediated protection, such as CP-mediated resistance, RNA-mediated resistance, and RNA silencing have been suggested as mechanisms of cross protection. The practical cross protection employed at the field level so far has involved selection of cross protection strains by empirical methods, but as a better understanding of the mechanisms(s) determine the efficiency and effectiveness of MSCP, better and quicker methods of selection of isolates for MSCP should develop in the foreseeable future. The ability to transform plants, including citrus, should enable the development of a better understanding of MSCP at the molecular level with specific viruses.

5.3. CP-mediated resistance

CP-mediated resistance to *Tobacco mosaic virus* (TMV) was demonstrated by Sherwood and Fulton (1982) and in *Cucumber mosaic virus* by Dodds, Lee, and Tiffany (1985). In transgenic plants, the level of expression of the CP seems to determine if resistance is expressed for not, with the higher expression of the CP having better protection against challenge (Powell, Sanders, Tumer, Fraley, & Beachy, 1990). CP-mediated resistance has been reported to be less strain specific than forms of resistance involving RNA (Lomonossoff, 1995). Dominguez et al. (2002) reported pathogen-derived resistance to CTV in 10–33% of the Mexican lime plants transformed with the CP gene of CTV.

5.4. RNA-mediated resistance

Using TMV as the model system, RNA-mediated resistance to explain cross protection was suggested by Palukaitis and Zaitlin (1984). In this model, the protecting virus produces excessive positive-sense RNA which hybridizes to the newly formed minus-strand RNA and this stops the further replication and translation of the challenge virus. This protection by RNA hybridization theory has been demonstrated at the cell level but not at the whole plant level.

Virus resistance due to an expressed transgene with sequence homology to the virus is a form of RNA-mediated resistance (Febres, Lee, & Moore, 2008; Lindbo, Silvarosales, Proebsting, & Dougherty, 1993; Ratcliff, Harrison, & Baulcombe, 1997; Waterhouse, Graham, & Wang, 1998). This RNA-mediated resistance, also called gene silencing, and has been termed RNA interference or RNAi. Silencing is initiated in plants by dsRNA that is processed into small interfering RNA (siRNA) of about 21–26 nt. The siRNAs then mediate RNA degradation of complementary sequences; a process known as posttransciptional gene silencing (PTGS). SiRNAs can also direct methylation of complementary DNA sequences (Mette, Aufsatz, van der Winden, Matzke, & Matzke, 2000; Wassenegger, Heimes, Riedel, & Sanger, 1994). Thus, methylation of the transcribed sequence is associated with PTGS. Methylation of promoter sequences inhibits transcription. RNAi has been proposed to be a natural plant defense mechanism against viruses and transposons and is also involved in the regulation of the expression levels of certain gene (Voinnet, 2002; Waterhouse, Wang, & Lough, 2001).

RNA-mediated protection against CTV has been reported in Duncan grapefruit transformed with the 3′ end of the CTV genome (Febres et al., 2008). The most resistant line showed no transgene mRNA accumulation and promoter methylation of cytosines; both characteristic of PTGS.

Most likely MSCP uses one of these two methods as suggested by experiments with transformed plants, either CP-mediated resistance or RNA-mediated resistance, or a combination. The virus and even strain of the virus probably determines the mechanism used. Regardless, at least with CTV, mild strain cross protection, using empirically selected strains, has been demonstrated to be a viable control mechanism. Further research will provide more information as to the nature of this protection.

REFERENCES

Aguilar, E., Villalobos, W., Moreira, L., Rodriguez, C. M., Kitajima, E. W., & Rivera, C. (2005). First report of *Xylella fastidiosa* infecting citrus in Costa Rica. *Plant Disease, 89*, 687.

Ahlawat, Y. S., Chenulu, V. V., Viswanath, S. M., Pandey, P. K., & Bhagabati, K. N. (1985). Mosaic disease of citrus in India. *Current Science, 54*, 873–874.

Ahlawat, Y. S., Pant, R. P., Lockhart, B. E. L., Srivastava, M., Charkaborty, N. K., & Varma, A. (1996). Association of a Badnavirus with citrus mosaic disease India. *Plant Disease, 80*, 590–592.

Barthe, G. A., Ceccardi, T. L., Manjunath, K. L., & Derrick, K. S. (1998). *Citrus psorosis virus*: Nucleotide sequencing of the coat protein gene and detection by hybridization and RT-PCR. *Journal of General Virology, 79*, 1531–1537.

Bassanezi, R. B., Montesino, L. H., Gimenes-Fernandes, N., Yamamoto, P. T., Gottwald, T. R., Amorim, L., et al. (2013). Efficacy of area-wide inoculum reduction and vector control on temporal progress of huanglongbing in your sweet orange plantings. *Plant Disease, 97*, 789–796.

Bederski, K., Roistacher, C. N., Silvestre, O. P., & Müller, G. W. (2010). Long-term cross protection of severe stem pitting *Citrus tristeza virus* in Peru. In M. E. Hilf, L. W. Timmer, R. G. Milne, & J. V. da Graça (Eds.), *Proceedings of the 17th conference of the International Organisation of Citrus Virologists* (pp. 67–79).

Bové, J. M. (1995). *Virus and virus-like diseases of citrus in the Near East region*. Rome: FAO, 518 p.

Bové, J. M. (2006). Huanglongbing: A destructive, newly-emerging, century-old disease of citrus. *Journal of Plant Pathology, 88*, 7–37.

Bové, J. M., Renaudin, J., Foissac, X., Gaurivaud, P., Carle, P., Laigret, F., et al. (2002). *Spiroplasma citri*: From functional genomics to … genomics!. In N. Duran-Vila, R. G. Milne, & J. V. da Graça (Eds.), *Proceedings of the 15th conference of the International Organisation of Citrus Virologists* (pp. 278–287).

Brlansky, R. H., & Timmer, L. W. (2014). *2014 Florida citrus pest management guide: Exocortis, cachexia, and other viroids* (p. 179). Gainesville, FL: University of Florida.

Calavan, E. C., & Bové, J. M. (1989). Ecology of *Spiroplasma citri*. In R. F. Whitcomb, & J. G. Tully (Eds.), *The mycoplasmas* (pp. 425–485). New York: Academic Press.

Calavan, E. C., Frolich, E. F., Carpenter, J. B., Roistacher, C. N., & Christiansen, D. W. (1964). Rapid indexing for exocortis of citrus. *Phytopathology, 54*, 1359–1362.

Caruso, A., Davino, M., & Terranova, G. (1993). Observation of impietratura disease symptoms in four citrus species. In P. Moreno, J. V. da Graça, & L. W. Timmer (Eds.), *Proceedings of the 12th conference of the International Organisation of Citrus Virologists* (pp. 379–382).

Castle, W. S. (1995). Rootstock as a fruit quality factor in citrus and deciduous tree crops. *New Zealand Journal of Crop and Horticultural Science, 23*, 383–394. http://dx.doi.org/10.1080/01140671.1995.9513914.

Castle, W. S., & Gottwald, T. R. (2005). Spatio-temporal analysis of tree decline losses among Navel orange trees on Swingle citrumelo rootstock in two central Florida citrus groves. In M. E. Hilf, N. Duran-Vila, & M. A. Rocha-Peña (Eds.), *Proceedings of the 16th conference of the International Organisation of Citrus Virologists* (pp. 375–380).

Changyong, Z., Xueyuan, Z., Yuanhui, J., & Xinhua, H. (1993). The occurrence of satsuma dwarf virus in China. In P. Moreno, J. V. da Graça, & L. W. Timmer (Eds.), *Proceedings of the 12th conference of the International Organisation of Citrus Virologists* (pp. 349–351).

Childs, J. F. L. (1950). The cachexia disease of Orlando tangelo. *The Plant Disease Reporter, 34*, 295–298.

Cui, P. F., Gu, C. F., & Roistacher, C. N. (1991). Occurrence of *Satsuma dwarf virus* in Zhejiang Province, China. *Plant Disease, 75*, 149–153.

da Graça, J. V., & van Vuuren, S. P. (2010). *Citrus tristeza virus*: Complex and tristeza disease. In A. V. Karasev, & M. E. Hilf (Eds.), *Managing* Citrus tristeza virus *losses using cross protection* (pp. 247–260). St. Paul: APS Press.

Dakshinamurti, V., & Reddy, G. S. (1975). Mosaic—A transmissible disorder of sweet oranges. *Indian Phytopathology, 28*, 398–399.

Dawson, W. O. (2010). *Citrus tristeza virus*: Complex and tristeza disease. In A. V. Karasev, & M. E. Hilf (Eds.), *Molecular genetics of* Citrus tristeza virus (pp. 53–72). St. Paul: APS Press.

Dawson, T. E., & Mooney, P. A. (2000). Evidence for trifoliate resistance breaking isolates of *Citrus tristeza virus* in New Zealand. In N. Duran-Vila, R. G. Milne, & J. V. da Graça (Eds.), *Proceedings of the 14th conference of the International Organisation of Citrus Virologists* (pp. 69–76).

Derrick, K. S., Brlansky, R. H., de Graca, J. V., Lee, R. F., Timmer, L. W., & Nguyen, T. K. (1988). Partial characterization of a virus associated with citrus ringspot. *Phytopathology, 78*, 1298–1301.

Derrick, K. S., Lee, R. F., Brlanksy, R. H., Timmer, L. W., Hewitt, B. G., & Barthe, G. A. (1990). Proteins associated with citrus blight. *Plant Disease, 74*, 168–170.

Derrick, K. S., & Timmer, L. W. (2000). Citrus blight and other disease of recalcitrant etiology. *Annual Review of Phytopathology, 38*, 181–205.

Dodds, J. A., Lee, S. Q., & Tiffany, M. (1985). Cross protection between strains of *Cucumber mosaic virus*: Effect of host and type of inoculum on accumulation of virions and double-stranded RNA of the challenge strain. *Virology, 144*, 301–309.

Dominguez, A., de Mendoza, A. H., Guerri, J., Cambra, M., Navarro, L., Moreno, P., et al. (2002). Pathogen-derived resistance *to Citrus tristeza virus* (CTV) in transgenic Mexican lime (*Citrus aurantifolia* (Christ.) Swing.) plants expressing its p25 coat protein gene. *Molecular Breeding, 10*, 1–10.

Duran-Vila, N., Roistacher, C. N., Rivera-Bustamante, R., & Semancik, J. S. (1988). A definition of citrus viroids groups and their relationship to the exocortis disease. *Journal of General Virology, 69*, 3069–3080.

Fawcett, H. S. (1934). Is psorosis of citrus a virus disease? *Phytopathology, 24*, 659–668.

Fawcett, H. S. (1936). *Citrus diseases and their control.* New York/London: McGraw-Hill Book Co., Inc, 656 p.

Febres, V. J., Lee, R. F., & Moore, G. A. (2008). Transgenic resistance to *Citrus tristeza virus* in grapefruit. *Plant Cell Reports, 27*, 93–104.

Flores, R., Duran-Vila, N., Pallas, V., & Semancik, J. S. (1985). Detection of viroid and viroid-like RNAs from grapevine. *Journal of General Virology, 66*, 2095–2102.

Garnsey, S. M., Barrett, H. C., & Hutchison, D. J. (1987). Identification of *Citrus tristeza virus* resistance in citrus relatives and its potential applications. *Phytophylactica, 19*, 187–191.

Garnsey, S. M., Civerolo, E. L., Gumpf, D. J., Paul, C., Hilf, M. E., Lee, R. F., et al. (2005). Biological characterization of an international collection of *Citrus tristeza virus* (CTV) isolates. In M. E. Hilf, N. Duran-Vila, & M. A. Rocha-Pena (Eds.), *Proceedings of the 16th conference of the International Organisation of Citrus Virologists* (pp. 75–93).

Garnsey, S. M., Gumpf, D. J., Roistacher, C. N., Civerolo, E. L., Lee, R. F., Yokomi, R. K., et al. (1987). Toward a standardized evaluation of the biological properties of *Citrus tristeza virus*. *Phytophylactica, 19*, 151–157.

Gottwald, T. R., Polek, M., & Riley, K. M. (2002). History, present incidence, and spatial distribution of *Citrus tristeza virus* in the California Central Valley. In N. Duran-Vila, R. G. Milne, & J. V. da Graça (Eds.), *Proceedings of the 15th conference of the International Organisation of Citrus Virologists* (pp. 83–94).

Grant, T. J. (1957). Effect of heat treatments on tristeza and psorosis viruses in citrus. *The Plant Disease Reporter, 41*, 232–234.

Guerra-Moreno, A. S., Manjunath, K. L., Brlanksy, R. H., & Lee, R. F. (2005). Citrus leprosis symptoms can be associated with the presence of two different viruses: Cytoplasmic and nuclear, the former having a multipartite RNA genome. In M. E. Hilf, N. Duran-Vila, & M. A. Rocha-Peña (Eds.), *Proceedings of the 15th conference of the International Organisation of Citrus Virologists* (pp. 230–239).

Hajeri, S., Ramadugu, C., Keremane, M., Vidalakis, G., & Lee, R. (2010). Nucleotide sequence and genome organization of *Dweet mottle virus* and its relationship to members of the family Betaflexiviridae. *Archives of Virology, 155*, 1523–1527.

Halbert, S. E., Genc, H., Cevik, B., Brown, L. G., Rosales, I. M., Manjunath, K. L., et al. (2004). Distribution and characterization of *Citrus tristeza virus* in South Florida following establishment of *Toxoptera citricida*. *Plant Disease*, *88*, 935–941.

Hartung, J. S., Beretta, J., Brlansky, R. H., Spisso, J., & Lee, R. F. (1994). Citrus variegated chlorosis bacterium: Axenic culture, pathogenicity, and serological relationships with other strains of *Xylella fastidiosa*. *Phytopathology*, *84*, 591–597.

Hiltabrand, W. F. (1957). Certification program for maintenance of virus-free propagation sources of citrus in California. In J. M. Wallace (Ed.), *Citrus virus diseases* (pp. 229–231). Richmond, CA: University of California, Division of Agricultural Sciences.

Hopkins, D. L., & Purcell, A. H. (2002). *Xylella fastidiosa*: Cause of Pierce's disease of grapevine and other emergent diseases. *Plant Disease*, *86*, 1056–1066.

Huang, Q., & Hartung, J. S. (2001). Cloning and sequence analysis of an infectious clone of *Citrus yellow mosaic virus* that can infect sweet orange via Agrobacterium-mediated inoculation. *Journal of General Virology*, *82*, 2549–2558.

Ito, T., Icki, H., Ozaki, K., & Ito, T. (2001). Characterization of a new citrus viroid species tentatively termed citrus viroid OS. *Archives of Virology*, *146*, 975–982.

Ito, T., Ito, T., Shiotani, H., Iwanami, T., Ozaki, K., & Muramoto, K. (2007). Genetic diversity and a heterogeneous population of *Citrus mosaic virus* within a single citrus tree. *Journal of General Plant Pathology*, *73*, 147–151.

Iwanami, T., Kano, T., & Koizumi, M. (1991). Pathogenic diversity of *Citrus tatter leaf virus* isolates. *Annals of Plant Phytopathological Society of Japan*, *45*, 712–720.

Jagoueix, S., Bové, J. M., & Garnier, M. (1994). The phloem-limited bacterium of greening disease of citrus is a member of the alpha subdivision of the proteo-bacteria. *International Journal of Systematic Bacteriology*, *44*, 379–386.

Karasev, A. V., Han, S. S., & Iwanamı, T. (2001). Satsuma dwarf and related viruses belong to a new lineage of plant picorna-like viruses. *Virus Genes*, *23*, 45–52.

Koizumi, M., Kano, T., Ieki, H., & Mae, H. (1988). *China laurestine*: A symptomless carrier of *Satsuma dwarf virus* which accelerates natural transmission in the fields. In L. W. Timmer, S. M. Garnsey, & L. Navarro (Eds.), *Proceedings of the 10th conference of the International Organisation of Citrus Virologists* (pp. 348–352).

Krueger, R. R., Bash, J. A., & Lee, R. F. (2005). *Dweet mottle virus* and *Citrus leaf blotch virus*. *Topics in Subtropics Newsletter*, *3*, 1.

Lee, R. F. (2000). Why have mandatory citrus certification programs? In J. V. da Graça, R. F. Lee, & R. K. Yokomi (Eds.), *Proceedings of the 14th conference of the International Organisation of Citrus Virologists* (pp. 311–325).

Lee, R. F., Baker, P. S., & Rocha-Pena, M. A. (1994). *The* Citrus tristeza virus *(CTV): An introduction to current priorities with special reference to the worsening situation in Central America and the Caribbean*. Rome: International Institute of Biological Control, Centre for Agriculture and BioSciences International, and Food and Agricultural Organization, 197 p.

Lee, R. F., Brlansky, R. H., Garnsey, S. M., & Yokomi, R. K. (1987). Traits of *Citrus tristeza virus* important for mild strain cross protection of citrus: The Florida approach. *Phytophylactica*, *19*, 215–218.

Lee, R. F., Derrick, K. S., Beretta, M. J. G., Chagas, C. M., & Rosetti, V. (1991). Citrus variegated chlorosis: A new destructive disease of citrus in Brazil. *Citrus Industry Magazine*, *12–13*, 15.

Lee, R. F., Derrick, K. S., Futch, S. H., & Tucker, D. P. H. (1993). Graft transmission of citrus measles. In P. Moreno, J. V. da Graça, & L. W. Timmer (Eds.), *Proceedings of the 12th conference of the International Organisation of Citrus Virologists* (pp. 371–374).

Lee, R. F., & Keremane, M. L. (2013). Mild strain cross protection of tristeza: A review of research to protect against decline on sour orange in Florida. *Frontiers in Microbiology*, *4*, 259. http://dx.doi.org/10.3389/fmicb.2013.00259.

Lee, R. F., Lehman, P. S., & Navarro, L. (1999). Nursery practices and certification programs for budwood and rootstocks. In L. W. Timmer, & L. W. Duncan (Eds.), *Citrus healthy management* (pp. 35–46). St. Paul, MN: APS Press.

Lee, R. F., Marais, L. J., Timmer, L. W., & Graham, J. H. (1984). Syringe injection of water into the trunk: A rapid diagnostic test for citrus blight. *Plant Disease*, *68*, 511–513.

Lee, R. F., Yokomi, R. K., & Vidalakis, G. (2008). First report of citrus measles in California. In *Proceedings of the 11th international citrus congress, 2008 October 26–30, Wuhan, China.*

Lindbo, J. A., Silvarosales, L., Proebsting, W. M., & Dougherty, W. G. (1993). Induction of a highly specific antiviral state in transgenic plants—Implications for regulation of gene-expression and virus-resistance. *Plant Cell*, *5*, 1749–1759.

Locali-Fabris, E. C., Freitas-Astu, J., Souza, A. A., Takita, M. A., Astua-Monge, G., Antonioli-Luizon, R., et al. (2006). Complete nucleotide sequence, genomic organization and phylogenetic analysis of *Citrus leprosis virus* cytoplasmic type. *Journal of General Virology*, *87*, 2721–2729.

Loconsole, G., Önelge, O., Potere, A., Giampetruzzi, O., Bozan, S., Satar, A., et al. (2012). Identification and characterization of *Citrus yellow vein clearing virus*, a putative new member of the genus *Mandarivirus*. *Phytopathology*, *102*, 1168–1175

Lomonossoff, G. P. (1995). Pathogen-derived resistance to plant viruses. *Annual Review of Phytopathology*, *33*, 323–343.

Lopes, S. A., Frare, G. F., Bertolini, E., Cambra, M., Fernandes, N. G., Ayres, A. J., et al. (2009). Liberibacters associated with citrus huanglongbing in Brazil: '*Candidatus* Liberibacter asiaticus' is heat tolerant, '*Candidatus* L. americanus' is heat sensitive. *Plant Disease*, *93*, 257–262.

Lopes, S. A., Frare, G. F., Camargo, L. E. A., Wulff, N. A., Teixeira, D. C., Bassanezi, R. B., et al. (2010). Liberibacters associated with orange jasmine in Brazil: Incidence in urban areas and relatedness to citrus Liberibacters. *Plant Pathology*, *59*, 1044–1053.

Maharaj, S. B., & da Graça, J. V. (1988). Observation of isometric virus-like particles associated with citrus vein enation-infected citrus and the viruliferous aphid vector *Toxoptera citricidus*. *Phytophylactica*, *20*, 357–360.

Manjunath, K. L., Halbert, S. E., Ramadugu, C., Webb, S., & Lee, R. F. (2008). Detection of '*Candidatus* Liberibacter asiaticus' in *Diaphorina citri* and its importance in the management of citrus huanglongbing in Florida. *Phytopathology*, *98*, 387–396.

Mette, M. F., Aufsatz, W., van der Winden, J., Matzke, M. A., & Matzke, A. J. (2000). Transcriptional silencing and promoter methylation triggered by double-stranded RNA. *EMBO Journal*, *19*, 5194–5201.

Milne, R. G., Garcia, M. L., & Grau, O. (2000). Genus *Ophiovirus*. In M. H. V. van Regenmortel, C. M. Fauquest, D. H. L. Bishop, E. B. Carstens, M. K. Estes, S. M. Lemon, J. Maniloff, M. A. Moyo, D. J. McGeoch, C. R. Pringle, & R. B. Wickner (Eds.), *Viruses taxonomy: Seventh report of the international committee on taxonomy of viruses* (pp. 626–631). San Diego: Academic Press.

Müller, G. W., & Costa, A. S. (1987). Search for outstanding plants in tristeza infected citrus orchards: The best approach to control the diseases by premmunization. *Phytophylactica*, *19*, 197–198.

Navarro, L. (1993). Citrus sanitation, quarantine and certification programs. In P. Moreno, J. V. da Graça, & L. W. Timmer (Eds.), *Proceedings of the 12th conference of the International Organisation of Citrus Virologists* (pp. 383–391).

Navarro, L., Civerolo, E. L., Juarez, J., & Garnsey, S. M. (1989). Improving therapy methods for citrus germplasm exchange. In R. H. Brlansky, R. F. Lee, & L. W. Timmer (Eds.), *Proceedings of the 11th conference of the International Organisation of Citrus Virologists* (pp. 400–408).

Navarro, L., Roistacher, C. N., & Mirashige, T. (1975). Improvement of shoot-tip graftting *in vitro* for virus-free citrus. *Journal of the American Society for Horticultural Science*, *100*, 471–479.

Palukaitis, P., & Zaitlin, M. (1984). A model to explain the "cross-protection" phenomenon shown by plant viruses and viroids. In T. Kosuge, & E. W. Nester (Eds.), *Plant–microbe interactions: Molecular and genetics perspectives* (pp. 213–222). New York: Macmillan Publishing Co.
Powell, C. A., Pelosi, R. R., Sonoda, R. M., & Lee, R. F. (1998). A psorosis-like agent prevalent in Florida's grapefruit groves and budwood sources. *Plant Disease*, *82*, 208–209.
Powell, P. A., Sanders, P. R., Tumer, N., Fraley, R. T., & Beachy, R. N. (1990). Protection against *Tobacco mosaic virus* infection in transgenic plants requires accumulation of coat protein rather than coat protein RNA sequences. *Virology*, *175*, 2124–2130.
Ratcliff, F., Harrison, B., & Baulcombe, D. (1997). A similarity between viral defense and gene silencing in plants. *Science*, *276*, 1558.
Rocha-Peña, M. A., Lee, R. F., Lastra, R., Niblett, C. L., Ochoa-Corona, F. M., Garnsey, S. M., et al. (1995). *Citrus tristeza virus* and its aphid vector *Toxoptera citricida*: Threats to citrus production in the Caribbean and Central and North America. *Plant Disease*, *79*, 437–445.
Rodrigues, J. C. V., Kitajima, E. W., Chilers, C. C., & Chagas, C. M. (2003). *Citrus leprosis virus* vectored by *Brevipalpus phoenicis* (Acari: Tenuipalpidae) on citrus in Brazil. *Experimental and Applied Acarology*, *30*, 161–179.
Roistacher, C. N. (1991). *Graft-transmissible disease of citrus. Handbook for detection and diagnosis.* Rome: FAO-IOCV, 286 p.
Roistacher, C. N. (1993). Psorosis—A review. In P. Moreno, J. V. da Graça, & L. W. Timmer (Eds.), *Proceedings of the 12th conference of the International Organisation of Citrus Virologists* (pp. 139–154).
Roistacher, C. N., da Graça, J. V., & Müller, G. W. (2010). Cross protection against *Citrus tristeza virus*—A review. In N. Duran-Vila, R. G. Milne, & J. V. da Graça (Eds.), *Proceedings of the 17th conference of the International Organisation of Citrus Virologists* (pp. 1–27).
Roistacher, C. N., Nauer, E. M., & Wagner, R. C. (1980). Transmissibility of cachexia, dweet mottle, psorosis, tatterleaf and infectious variegation viruses on knife blades and its prevention. In E. C. Calavan, S. M. Garnsey, & L. W. Timmer (Eds.), *Proceedings of the 8th conference of the International Organisation of Citrus Virologists* (pp. 225–229).
Roy, A., Choudhary, N., Guillermo, L. M., Shao, J., Govindarajulu, A., Achor, D., et al. (2013). A novel virus of the genus *Cilevirus* causing symptoms similar to citrus leprosis. *Phytopathology*, *103*, 488–500.
Roy, A., Stone, A., Otero-Colina, G., Wei, G., Chaurdhary, N., Achor, D., et al. (2013). Genome assembly of *Citrus leprosis virus* nuclear type reveals a close association with *Orchid fleck virus*. *Genome Announcements*, *4*, e00519-13.
Saglio, P., Hospital, M., Lafléche, D., Bové, J. M., Tully, J. G., & Freundt, E. A. (1973). *Spiroplasma citri* gen. and sp.n.: A mycoplasma-like organism associated with "stubborn" disease of citrus. *International Journal of Systematic Bacteriology*, *23*, 191–204.
Schwarz, R. E. (1965). Aphid-borne virus diseases of citrus and their vectors in South Africa. A. Investigations into the epidemiology of aphid transmissible virus diseases of citrus by means of trap plants. *South African Journal of Agricultural Science*, *8*, 839–852.
Semancik, J. S., Roistacher, C. N., Rivera-Bustamante, R., & Duran-Vila, N. (1988). Citrus cachexia viroid, a new viroid of citrus: Relationship to viroids of the exocortis disease complex. *Journal of General Virology*, *69*, 3059–3068.
Serra, P., Barbosa, C. J., Daròs, J. A., Flores, R., & Duran-Vila, N. (2007). Citrus viroid V: Molecular characterization and synergistic interactions with other members of the genus *Apscaviroid*. *Virology*, *370*, 102–112.
Sherwood, J. L., & Fulton, R. W. (1982). The specific involvement of coat protein in *Tobacco mosaic virus* cross protection. *Virology*, *119*, 150–158.
Swingle, W. T., & Reese, P. C. (1967). The botany of citrus and its wild relatives of the orange subfamily. In W. Reuther, H. J. Webber, & L. D. Batchlor (Eds.), *Vol. 1. The citrus industry* (revised ed., pp. 190–430). Berkley: University of California Press.

Tanaka, H. (1972). Mechanical transmission of viruses of satsuma dwarf and natsudaidai dwarf from citrus to citrus. *Annals of the Phytopathological Society of Japan, 38*, 156–160.

Teixeira, D. C., Ayres, J., Kitajima, E. W., Danet, L., Jagoueix, S., Saillard, C., et al. (2005). First report of a Huanglongbing-like disease of citrus in São Paulo state, Brazil and association of a new Liberibacter species, *Candidatus* Liberibacter americanus, with the disease. *Plant Disease, 89*, 107–111.

Tucker, D. P. H., Lee, R. F., Timmer, L. W., Albrigo, L. G., & Brlansky, R. H. (1984). Experimental transmission of citrus blight. *Plant Disease, 68*, 979–980.

Vives, M. C., Galipienso, L., Navarro, L., Moreno, P., & Guerri, J. (2002). *Citrus leaf blotch virus*: A new citrus virus associated with bud union crease on trifoliate rootstocks. In N. Duran-Vila, R. G. Milne, & J. V. da Graça (Eds.), *Proceedings of the 15th conference of the International Organisation of Citrus Virologists* (pp. 205–212).

Vives, M. C., Velázquez, K., Pina, J. A., Moreno, P., Guerri, J., & Navarro, L. (2013). Identification of a new enovovirus associated with citrus vein enation disease by deep sequencing of small RNAs. *Phytopathology, 103*, 1077–1086.

Vogel, R., & Bové, J. M. (1968). Cristacortis, a virus disease inducing stem pitting on sour orange and other citrus species. In J. F. L. Childs (Ed.), *Proceedings of the 4th conference of the International Organisation of Citrus Virologists* (pp. 221–228).

Voinnet, O. (2002). RNA silencing: Small RNAs as ubiquitous regulators of gene expression. *Current Opinion in Plant Biology, 5*, 444–451.

Volk, G. M., Bonnart, R., Krueger, R., & Lee, R. (2012). Cryopreservation of citrus shoot tips using micrografting for recovery. *CryoLetters, 33*, 418–426.

Wallace, J. M., & Drake, R. J. (1953). A virus-induced vein enation in citrus. *Citrus Leaves, 33*, 22–24.

Wallace, J. M., & Drake, R. J. (1962). Tatterleaf, a previously undescribed virus effect of citrus. *The Plant Disease Reporter, 46*, 211–212.

Wang, N., & Brlansky, R. (2013). Citrus blight: Update on research. *Citrus Industry Magazine*, (pp.28 and 30).

Wang, Q. C., Panis, B., Engelmann, F., Lambardi, M., & Valkonen, J. P. T. (2009). Cryotherapy of shoot tips: A technique for pathogen eradication to produce healthy planting materials and prepare healthy plant genetic resources for cryopreservation. *Annals of Applied Biology, 154*, 351–363.

Wassenegger, M., Heimes, S., Riedel, L., & Sanger, H. L. (1994). RNA-directed *de novo* methylation of genomic sequences in plants. *Cell, 76*, 567–576.

Waterhouse, P. M., Graham, H. W., & Wang, M. B. (1998). Virus resistance and gene silencing in plants can be induced by simultaneous expression of sense and antisense RNA. *Proceedings of the National Academy of Sciences of the United States of America, 95*, 13959–13964.

Waterhouse, P. M., Wang, M. B., & Lough, T. (2001). Gene silencing as an adaptive defence against viruses. *Nature, 411*, 834–842.

Weathers, L. G. (1960). Yellow-vein disease of citrus and studies on interactions between yellow-vein and other viruses of citrus. *Virology, 11*, 753–764.

Webber, H. J., Reuther, W., & Lawton, H. W. (1967). History and development of the citrus industry. In W. Reuther, H. J. Webber, & L. D. Batchelor (Eds.), *Vol. 1. The citrus industry* (pp. 1–39). Riverside, CA: University of California, Division of Agricultural Sciences (Chapter 1).

Ye, F., Laigret, F., Whitley, J. C., Citti, C., Finch, L. R., Carle, P., et al. (1992). A physical and genetic map of the *Spiroplasma citri* genome. *Nucleic Acids Research, 20*, 1559–1565.

Yokomi, R. K., Mello, A. F. S., Fletcher, J., & Saponari, M. (2010). Estimation of citrus stubborn incidence in citrus groves by real-time PCR. In N. Duran-Vila, R. G. Milne, & J. V. da Graça (Eds.), *Proceedings of the 17th conference of the International Organisation of Citrus Virologists* (pp. 131–141).

CHAPTER SIX

Control of Viruses Infecting Grapevine

Varvara I. Maliogka*[,1], Giovanni P. Martelli[†], Marc Fuchs[‡], Nikolaos I. Katis*

*Faculty of agriculture, Forestry and Natural Environment, School of Agriculture, Plant Pathology Lab, Aristotle University of Thessaloniki, Thessaloniki, Greece
[†]Università degli Studi di Bari "Aldo Moro", Bari, Italy
[‡]Department of Plant Pathology and Plant–Microbe Biology, Cornell University, New York State Agricultural Experiment Station, Geneva, New York, USA
[1]Corresponding author: e-mail address: vmaliogk@agro.auth.gr

Contents

Abstract

Grapevine is a high value vegetatively propagated fruit crop that suffers from numerous viruses, including some that seriously affect the profitability of vineyards. Nowadays, 64 viruses belonging to different genera and families have been reported in grapevines and new virus species will likely be described in the future. Three viral diseases namely leafroll, rugose wood, and infectious degeneration are of major economic importance worldwide. The viruses associated with these diseases are transmitted by mealybugs, scale and soft scale insects, or dagger nematodes. Here, we review control measures

Advances in Virus Research, Volume 91
ISSN 0065-3527
http://dx.doi.org/10.1016/bs.aivir.2014.11.002

of the major grapevine viral diseases. More specifically, emphasis is laid on (i) approaches for the production of clean stocks and propagative material through effective sanitation, robust diagnosis, as well as local and regional certification efforts, (ii) the management of vectors of viruses using cultural, biological, and chemical methods, and (iii) the production of resistant grapevines mainly through the application of genetic engineering. The benefits and limitations of the different control measures are discussed with regard to accomplishments and future research directions.

1. ORIGIN, BOTANY, AND ECONOMIC IMPORTANCE OF GRAPEVINE (*VITIS VINIFERA* L.)

Grapevine (*Vitis vinifera* L.) is one of the oldest fruit crop plants on the planet and due to its regenerative ability it was considered as a symbol of life, thus frequently referred to as the *tree of life* (Vivier & Pretorius, 2000). Grapevine is classified into the genus *Vitis*, which consists of the sub-genera *Euvitis* and *Muscadinia* and together with 11–13 other genera belongs to the family Vitaceae (Pongracz, 1978). It occurs mainly in Asia, North America, and Europe (tropical and subtropical areas). *V. vinifera* L. subsp. *vinifera* (or *sativa*) is of major economic significance worldwide, the majority of its cultivars deriving from wild forms [*V. vinifera* L. subsp. *sylvestris* (Gmelin) Hegi; Crespan, 2004; Rosseto, McNally, & Henry, 2002; Sefc et al., 2003; This et al., 2004]. Other species of the genus, such as the North American *Vitis rupestris*, *Vitis riparia*, *Vitis berlandieri*, and their hybrids, are commonly used as rootstocks because they are highly resistant to phylloxera (*Daktulosphaira vitifoliae*) and mildews. About 5000 true *V. vinifera* cultivars are utilized nowadays by the wine, table (fresh fruit), and raisin (dried fruit) grape industries of the world (Jackson, 1994).

Cultivation and domestication of the grapevine appear to have occurred between the seventh and fourth millennia BC in a region between the Black sea and Iran (Terral et al., 2010; Zohary & Hopf, 2000). From this area, the cultivated forms (types) of grapevine were eventually transferred by humans to the Near East, Middle East, and Central Europe, which may have constituted secondary domestication centers (Arroyo-Garcia et al., 2006; Grassi et al., 2003). An independent domestication may have happened in Spain as well (Nunez & Walker, 1989; Stevenson, 1985). Nowadays, there are about 8 million hectares of vineyards around the world (Vivier & Pretorius, 2000).

2. THE MAIN VIRUSES INFECTING GRAPEVINE

It is common knowledge that viruses and other infectious agents (viroids, phytoplasmas, and phloem- and xylem-limited bacteria) are primary plant pathogens, which are detrimental to agricultural crops as a whole, but are especially so to vegetatively propagated crops. Over time, these types of crops have undergone a severe sanitary deterioration on a worldwide basis (Hadidi, Barba, & Jelkmann, 2011). The grapevine makes no exception (Martelli, 2014), being the woody crop that hosts the highest number of intracellular infectious agents of all (Table 1). Among the causes that have created such an unhappy situation, a major role has been and is still being played by the international trading of infected nursery productions. These materials disseminate over long distance disease agents (and their vectors) so that, when planted in the field, they serve as inoculum sources for secondary vector-mediated spreading at a site.

So far, more than 60 viral species belonging to different genera have been reported to infect the grapevine (Martelli, 2014). Many of these viruses, usually in combinations, are responsible for serious diseases that might have detrimental effects on the vines. Leafroll, rugose wood (RW), and infectious degeneration are the three main disorders of grapevine, which are induced by virus species belonging to the families *Closteroviridae*, *Betafexiviridae*, and *Secoviridae*, respectively.

2.1. Closterovirids associated with grapevine leafroll disease

Leafroll is probably the most widespread virus disease of grapevine occurring in most viticultural areas of the world (Martelli & Boudon-Padieu, 2006). Disease symptoms develop during late summer and autumn and include reddening or yellowing of the leaves in red-berried and white-berried cultivars, respectively (Fig. 1). Usually, the veins remain green and the leaf margins roll downward, while decreased fruit quality and delayed maturation are often observed. The infection reduces vigor and yield of the vines by 15–20% on average (Martelli & Boudon-Padieu, 2006). Nevertheless, it is difficult to assess the actual economic impact of grapevine leafroll disease (GLRD). Recently, it was estimated that the cost of the disease ranges between $25,000 and $40,000 per hectare for vineyards with a 25-year lifespan in the absence of any control measure (Atallah, Gomez, Fuchs, & Martinson, 2012). Leafroll symptoms vary according to the grapevine cultivar and the viruses involved (Krake, 1993). The American *Vitis* species that are mainly used as rootstocks do not show any disease symptoms, except for a decrease in vigor (Martelli & Boudon-Padieu, 2006).

Several phloem-limited virus species, all members of the family *Closteroviridae*, have been reported as being related with GLRD (Table 2). They are designated grapevine leafroll-associated viruses (GLRaVs) followed by a

Table 1 Viruses of *Vitis* spp. and their taxonomic affiliation[a]

Family	Genus	Species
A. Viruses belonging to genera included into families		
Viruses with a single-stranded DNA genome		
Geminiviridae	Undetermined	Grapevine Cabernet franc-associated virus (GCFaV) (=Grapevine red blotch-associated virus)
Viruses with a double-stranded DNA genome		
Caulimoviridae	*Badnavirus*	*Grapevine vein clearing virus* (GVCV)
Viruses with a double-stranded RNA genome		
Reoviridae	*Oryzavirus*	Unnamed virus
Endornaviridae	*Endornavirus*	Two unnamed viruses
Partitiviridae	*Alphacryptovirus*	Raphanus sativus cryptic virus 3 (RsCV-3) like *Beet cryptic virus* 3(BCV-3) like
Viruses with a negative-sense single-stranded RNA genome		
Bunyaviridae	*Tospovirus*	*Tomato spotted wilt virus* (TSWV)
Viruses with a positive-sense single-stranded RNA genome (filamentous particles)		
Closteroviridae	*Closterovirus*	*Grapevine leafroll-associated virus 2* (GLRaV-2)
	Ampelovirus	*Grapevine leafroll-associated virus 1* (GLRaV-1) *Grapevine leafroll-associated virus 3* (GLRaV-3) *Grapevine leafroll-associated virus 4* (GLRaV-4) GLRaV-4 strain 5 GLRaV-4 strain 6 GLRaV-4 strain 9 GLRaV-4 strain Car
	Velarivirus	*Grapevine leafroll-associated virus* 7 (GLRaV-7)
Alphaflexiviridae	*Potexvirus*	*Potato virus X* (PVX)
Betaflexiviridae	*Foveavirus*	*Grapevine rupestris stem pitting-associated virus* (GRSPaV)
	Trichovirus	*Grapevine berry inner necrosis virus* (GINV) *Grapevine Pinot gris virus* (GPGV)
	Vitivirus	*Grapevine virus A* (GVA) *Grapevine virus B* (GVB) *Grapevine virus D* (GVD) *Grapevine virus E* (GVE) Grapevine virus F (GVF)

Table 1 Viruses of *Vitis* spp. and their taxonomic affiliation—cont'd

Family	Genus	Species
Potyviridae	*Potyvirus*	Unidentified potyvirus-like virus isolated in Japan from a Russian cv. *Bean common mosaic virus* (BCMV), peanut strain
Viruses with a positive-sense single-stranded RNA genome (rod-shaped particles)		
Virgaviridae	*Tobamovirus*	*Tobacco mosaic virus* (TMV) *Tomato mosaic virus* (ToMV)
Viruses with a positive-sense single-stranded RNA genome (isometric particles)		
Secoviridae	*Fabavirus*	*Broadbean wilt virus* (BBWV)
	Nepovirus	*Artichoke italian latent virus* (AILV) *Arabis mosaic virus* (ArMV) *Blueberry leaf mottle virus* (BBLMV) *Cherry leafroll virus* (CLRV) *Grapevine Bulgarian latent virus* (GBLV) *Grapevine Anatolian ringspot virus* (GARSV) *Grapevine deformation virus* (GDefV) *Grapevine chrome mosaic virus* (GCMV) *Grapevine fanleaf virus* (GFLV) *Grapevine Tunisian ringspot virus* (GTRV) *Peach rosette mosaic virus* (PRMV) *Raspberry ringspot virus* (RpRV) *Tobacco ringspot virus* (TRSV) *Tomato ringspot virus* (ToRSV) *Tomato blackring virus* (TBRV)
	Unassigned in the family	*Strawberry latent ringspot virus* (SLRSV)
Bromoviridae	*Alfamovirus*	*Alfalfa mosaic virus* (AMV)
	Cucumovirus	*Cucumber mosaic virus* (CMV)
	Ilarvirus	*Grapevine line pattern virus* (GLPV) Grapevine angular mosaic virus (GAMoV)
Tombusviridae	*Carmovirus*	*Carnation mottle virus* (CarMV)
	Necrovirus	*Tobacco necrosis virus D* (TNV-D)
	Tombusvirus	*Grapevine Algerian latent virus* (GALV) *Petunia asteroid mosaic virus* (PAMV)

Continued

Table 1 Viruses of *Vitis* spp. and their taxonomic affiliation—cont'd

Family	Genus	Species
Tymoviridae	*Marafivirus*	*Grapevine asteroid mosaic-associated virus* (GAMaV) *Grapevine redglobe virus* (GRGV) *Grapevine Syrah virus 1* (GSV-1) *Blackberry virus S* (BlVS) Unnamed putative marafivirus-like virus
	Maculavirus	*Grapevine fleck virus* (GFkV) *Grapevine rupestris vein feathering virus* (GRVFV)
B. Viruses belonging to genera unassigned to families		
	Idaeovirus	*Raspberry bushy dwarf virus* (RBDV)
	Sobemovirus	*Sowbane mosaic virus* (SoMV)
C. Taxonomically unassigned viruses		
		Unnamed filamentous virus *Grapevine Ajinashika virus* (GAgV) *Grapevine stunt virus* (GSV) *Grapevine labile rod-shaped virus* (GLRSV) *Southern tomato virus* (STV)

[a]Scientific names of definitive virus species are written in *italics*. The names of tentative species are written in Roman characters. The updated taxonomy of all classified grapevine viruses can be found in King, Adams, Carstens and Lefkowitz (2012). Virus Taxonomy. Ninth Report of the International Committee Virus Taxonomy. Elsevier-Academic Press, Amsterdam, the Netherlands. This table comprises also the new viruses reported from southeastern United States, a detailed description of which is yet to be published. In addition to the above-listed viruses, grapevines can host viroids (5), phytoplasmas (8), and insect-transmitted xylematic bacteria (1).
By permission from Martelli (2014).

number for their identification at the species level. Currently, there are five recognized GLRaVs, most of which (GLRaV-1, -3, and -4) are ampeloviruses, GLRaV-2 is a closterovirus and GLRaV-7 is a member of the newly proposed genus *Velarivirus* (Al Rwahnih et al., 2012). GLRaVs of the genus *Ampelovirus* are further separated into two phylogenetically distinct clades, subgroup I that includes GLRaV-1 and -3 and subgroup II consisting of GLRaV-4 and related strains, i.e., the former GLRaV-5, -6, -9, -Pr, -De, and -Car (Abou Ghanem-Sabanadzovic, Sabanadzovic, Gugerli, & Rowhani, 2010, Abou Ghanem-Sabanadzovic, Sabanadzovic, Uyemoto, Golino, & Rowhani, 2010; Maliogka, Dovas, & Katis, 2008, Maliogka, Dovas, Lotos, Efthimiou, & Katis, 2009; Martelli et al., 2012; Thompson, Fuchs, & Perry, 2012).

Figure 1 Leafroll symptoms on grapevine. (A) Mid vein reddening and curling of the leaves of an infected vine. (B) Close-up of a leafroll-infected leaf from a white-berried cultivar. (C) Close-up of a leafroll-infected leaf from a red-berried cultivar. (See the color plate.)

Table 2 Viruses associated with grapevine leafroll disease

Genus		Virus species	Geographic distribution	Vectors
Ampelovirus	Subgroup I	GLRaV-1	Worldwide	Mealybugs, soft scale insects
		GLRaV-3	Worldwide	Mealybugs, soft scale, and scale insects
	Subgroup II	GLRaV-4[a]	Europe, Near East, North and South America	Mealybugs
Closterovirus		GLRaV-2	Worldwide	Unknown
Velarivirus[b]		GLRaV-7	Europe, Near East, North and South America, China	Unknown

[a]This species includes variants of the formerly known GLRaV-4, -5, -6, -9, -Pr, and -Car (proposal awaiting International Committee on Taxonomy of Viruses (ICTV) approval, Martelli et al., 2012).
[b]Proposed genus in the family *Closteroviridae* (Al Rwahnih, Dolja, Daubert, Koonin, & Rowhani, 2012).

GLRaV-1, -3, and most strains of GLRaV-2 are strong inducers of leafroll symptoms in grapevine. However, GLRaV-3 is the most prevalent etiological agent of GLRD (Maree et al., 2013). The GLRaV-4 cluster generally elicits milder symptomatology, while infections with GLRaV-7 induce no or very mild symptoms (Al Rwahnih et al., 2012; Martelli et al., 2012). Some GLRaV-2 strains (e.g., GLRaV-2 RG) are involved in severe cases of graft incompatibility on certain rootstocks (Greif et al., 1995; Uyemoto, Rowhani, Luvisi, & Krag, 2001). Even though all GLRaVs are found in different regions of the world (Table 2), GLRaV-3 is the most widely distributed (Charles et al., 2006; Maree et al., 2013).

GLRaVs, as all grapevine-infecting viruses, are transmitted through vegetative propagation and grafting. In addition to this mode of dispersal, laboratory transmission assays and field surveys showed that several mealybugs and soft scale species are efficient vectors of these agents (Table 2). More specifically, mealybugs of the genera *Heliococcus*, *Phenacoccus*, *Planococcus*, and *Pseudococcus*, soft scale insects of the genera *Pulvinaria*, *Neopulvinaria*, *Parthenolecanium*, *Coccus*, *Saissetia*, and *Parasaissetia*, and scale insects of the genus *Ceroplastes* have been reported to transmit one or more of the leafroll-associated ampeloviruses (Le Maguet, Beuve, Herrbach, & Lemaire, 2012; Martelli & Boudon-Padieu, 2006; Martelli, Saldarelli, & Minafra, 2011; Tsai, Rowhani, Golino, Daane, & Almeida, 2010). Transmission seems to occur in a semi-persistent manner and instar nymphs are more efficient vectors of GLRaV-3 than adult mealybugs (Tsai et al., 2008). Nevertheless, there is no indication of virus-vector specificity in the transmission process (Le Maguet et al., 2012; Tsai et al., 2010). So far, no vector has been identified for GLRaV-7 and GLRaV-4 strain Car and none is suspected for GLRaV-2.

The contribution of vectors to the epidemiology of GLRaV-1, -3, and -4 can be important. A number of field studies on GLRaV-3 indicate annual spread rates of 8–12% (Cabaleiro & Segura, 1997, 2006; Charles, Froud, van den Brink, & Allan, 2009; Golino, Weber, Sim, & Rowhani, 2008). Infectious crawlers can either walk from vine to vine or be dispersed by the wind. In New Zealand, up to 15 m wind dispersal of mealybug crawlers was observed (Lo et al., 2006). In recent surveys, GLRaV-2 and -3 were detected in *Vitis californica* and natural *V. californica* × *V. vinifera* hybrids surrounding Napa Valley vineyards (Klaassen et al., 2011) and GLRaV-2 was found in symptomless vines of *Muscadinia rotundifolia* and *Vitis aestivalis* in Mississippi (USA; Abou Ghanem-Sabanadzovic & Sabanadzovic, 2014). However, there is no strong evidence that they could act as significant virus reservoirs. More studies are needed to address this issue.

2.2. Flexivirids related to the RW disease complex

RW is a disease complex occurring in most grapevine-growing regions of the world (Martelli & Boudon-Padieu, 2006). It consists of four different syndromes that are latent in self-rooted *V. vinifera* species and American rootstocks but develop symptoms on grafted grapevines. These syndromes can be distinguished by graft transmission to specific *Vitis* indicator plants and are named after the symptoms induced as Rupestris stem pitting, corky bark, Kober stem grooving, and LN33 stem grooving (Martelli, 1993; for a detailed description of the symptoms developed on the indicators, see Martelli & Boudon-Padieu, 2006). The most characteristic symptom of the RW is the appearance of alterations of the woody cylinder such as pits and grooves that may occur on the scion and rootstock, or on both (Fig. 2). A swelling above the bud union is also often observed on grafted grapevines. Infected plants show an overall decrease in vigor and yield (Fig. 2) and some of them might die a couple of years after planting (Martelli & Boudon-Padieu, 2006). Furthermore, occasions of graft incompatibility might occur. The intensity of symptoms depends on the scion/rootstock combination and the climatic conditions.

Phloem-restricted viruses belonging to two genera of the family *Betaflexiviridae* are associated with the RW complex (Table 3; Martelli, 1993). These are *Grapevine virus A*, *-B*, and *-D* (GVA, GVB, and GVD, respectively) that belong to the genus *Vitivirus*, and *Grapevine rupestris stem pitting-associated virus* (GRSPaV), which is a member of the genus

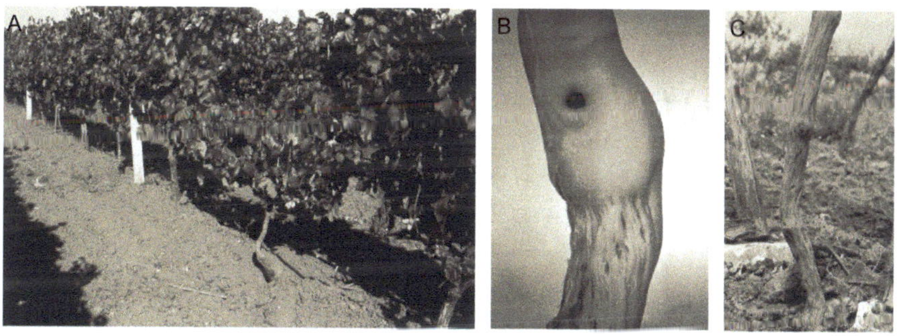

Figure 2 Rugose wood symptoms. (A) Infected plant on the field. (B and C) Pitting and grooving on the stem. (See the color plate.)

Table 3 Viruses associated with grapevine rugose wood complex

Genus	Virus species	Geographic distribution	Vectors
Vitivirus	GVA	Worldwide	Mealybugs, soft scale insects
	GVB	Worldwide	Mealybugs
	GVD	Worldwide	Unknown
Foveavirus	GRSPaV	Worldwide	Unknown

Foveavirus. GVA has been implicated with the etiology of Kober stem grooving (Chevalier, Greif, Clauzel, Walter, & Fritsch, 1995), while GVB is putatively responsible for the corky bark syndrome (Bonavia et al., 1996), and GVD may have an implication in the same disease (Abou Ghanem et al., 1997). The etiological role of *Grapevine virus E* (GVE) and Grapevine virus F (GVF) in RW is not known. On the other hand, GRSPaV is associated with Rupestris stem pitting (Zhang, Uyemoto, Golino, & Rowhani, 1998). From these viral species, it seems that GVA and GRSPaV are the most prevalent viruses of RW since they have been detected in most grapevine varieties all over the world (Goszczynski & Jooste, 2003; Meng & Gonsalves, 2003). At the same time, a wide range of sequence variants have been reported for both viruses, though it is still unknown how this high genetic variability correlates with pathogenicity. Interestingly, it has been suggested that GVA is also putatively implicated with Shiraz disease in South Africa and Australia (Goszczynski & Habili, 2012), whereas variants of GRSPaV are associated with vein necrosis of 110 Richter (Bouyahia et al., 2005). Their possible involvement with Syrah decline (Lima, Alkowni, et al., 2006) has recently been dismissed (Beuve et al., 2013).

RW-related viruses are transmitted through vegetative propagation and grafting. However, vector transmission plays also a role in the epidemiology of some of these virus species. Similarly to the GLRaVs, GVA, GVB, and GVE are semi-persistently vectored by several mealybugs and soft scale insects. More specifically, mealybugs of the genera *Heliococcus*, *Planococcus*, *Pseudococcus*, and *Phenacoccus* and soft scales of the genera *Parthenolecanium* and *Neopulinaria* can transmit GVA (Le Maguet et al., 2012; Martelli & Boudon-Padieu, 2006), while GVB is vectored by *Planococcus*, *Pseudococcus*, and *Phenacoccus* species (Le Maguet et al., 2012; Martelli & Boudon-Padieu, 2006). GVE isolates are transmitted by the mealybug *Pseudococcus comstocki* (Nakaune, Toda, Mochizuki, & Nakano, 2008). Studies conducted by various researchers showed that GVA transmission often occurs simultaneously with GLRaV-1 and GLRaV-3 (Engelbrecht & Kasdorf, 1990; Fortusini, Scattini, Prati, Cinquanta, & Belli, 1997; Hommay, Komar, Lemaire, & Herrbach, 2008; Zorloni, Prati, Bianco, & Belli, 2006), suggesting some level of assistance for transmission, but this needs further investigation. Furthermore, experiments with GVA showed that, similarly to the GLRaVs, first instar nymphs of mealybugs are also the most efficient vectors of this virus (Le Maguet et al., 2012). GVA and GVB were recently detected, along with GLRaV-2 and -3, in *V. californica* and in natural

V. californica × *V. vinifera* hybrids surrounding Napa Valley vineyards (Klaassen et al., 2011). However, the role of these hosts in the epidemiology of vitiviruses is unknown. Finally, GRSPaV has no known vector but it has been detected in pollen and seeds from infected vines as well as in young seedlings (Lima, Rosa, Golino, & Rowhani, 2006; Rowhani et al., 2000).

2.3. Nepoviruses responsible for grapevine fanleaf degeneration

Fanleaf degeneration is among the most severe and widespread virus diseases of grapevine (Martelli & Boudon-Padieu, 2006; Martelli & Savino, 1990). It is regarded as the oldest viral disease of this crop that existed in the Mediterranean and Near East since the very beginning of viticulture and has spread afterward all over the world. Two groups of degenerative conditions have been described, infectious degeneration and grapevine decline caused by a number of European and American nepoviruses of the family *Secoviridae*, respectively (Martelli, 1993). The degeneration induced by the European nepoviruses is collectively known as fanleaf.

The main etiological agent of fanleaf disease is *Grapevine fanleaf virus* (GFLV), which occurs in almost all regions where *V. vinifera* and hybrid rootstocks are cultivated, contrary to the other reported nepoviruses that show a more regional distribution and are therefore characterized as European and American viruses (Andret-Link, Laporte, et al., 2004; Digiaro, Elbeaino, & Martelli, 2007; Martelli, 1978). There are two distinct symptomatologies induced by different strains of GFLV. Malformations induced by strains causing distortions and yellow mosaic caused by chromogenic virus strains (Fig. 3; Martelli & Boudon-Padieu, 2006). The most typical foliar symptom is a distortion with toothed margins, closer primary veins, and widely open petiolar sinuses; thus, the leaves resemble a fan, hence the name of the virus. In the case of the chromogenic strains, the foliage develops bright yellow discolorations. In general, foliar symptoms appear early in the spring and might last throughout the vegetative period. Canes from infected vines also show short internodes, double nodes, and zig–zag growth between nodes (Hewitt, 1950). Crop losses by the virus can be high (up to 80%), while fruit quality and the productive life of the infected vineyards are significantly reduced (Andret-Link, Laporte, et al., 2004; Raski, Goheen, Lider, & Meredith, 1983).

GFLV is easily disseminated through grafting and use of infected propagation material, but it can also be transmitted semi-persistently from vine to

Figure 3 Symptoms of fanleaf disease. (A and B) Yellow mosaic on the leaves. (C) Canes from infected vines showing short internodes, double nodes, and zig-zag growth between nodes compared to cane from healthy plant (left). (D) Uneven development of the bunch on infected grapevine. (See the color plate.)

vine by the ectoparasitic nematode *Xiphinema index* (family *Longidoridae*) that feeds on growing root tips (Andret-Link, Schmitt-Keichinger, Demangeat, Komar, & Fuchs, 2004; Hewitt, Raski, & Goheen, 1958; Raski et al., 1983). The transmission process is characterized by a specific and complementary association between *X. index* and GFLV (Brown & Weischer, 1998). The virus can be acquired and transmitted by both juvenile stages and adults of the vector. After acquisition GFLV virions are retained at the esophageal tract of the nematode (Taylor & Robertson, 1970). The coat protein is the sole viral determinant of the specific transmission of GFLV by its vector (Andret-Link, Schmitt-Keichinger, et al., 2004). Virus particles can persist in viruliferous *X. index* for over 4 years (Demangeat et al., 2005). Of course, the virus can also be disseminated by human activities such as transfer of soil containing roots from infected plants and viruliferous *X. index*, whereas occasional transmission through seeds has also been reported (Martelli, Walter, & Pinck, 2003), but its epidemiological significance is unknown.

3. DIAGNOSIS OF GRAPEVINE VIRUSES

The application of appropriate and reliable methods for the detection of grapevine viruses is an important step toward their efficient control. Such methods are used both for monitoring the sanitary status of the vines in the field (Constable, Connellan, Nicholas, & Rodoni, 2012; Leonhardt, Wawrosch, Auer, & Kopp, 1998; Sharma et al., 2011; Thompson et al., 2014) and during certification schemes (EPPO, 2008; Martelli, 1999; Rowhani, Uyemoto, Golino, & Martelli, 2005). In these programs, detection methods are applied on the initial selected plant material to verify that it fulfills the sanitary requirements and also during virus elimination procedures to assess the health status of the treated material. Three main groups of diagnostic methods are implemented, i.e., biological indexing, serological, and molecular assays. Of these, biological indexing is mainly used in certification programs or when putative new diseases are studied. On the other hand, serological and molecular methods are more convenient and rapid and are therefore more widely applied.

3.1. Biological indexing

Grafting on woody indicators of the genus *Vitis* is a compulsory step in the certification schemes of grapevine mainly because some diseases cannot be identified otherwise (EPPO, 2008; Martelli, 1999). Grafting can be done by various techniques such as whip or cleft grafting, chip-bud grafting, machine grafting, or green grafting (Martelli, 1999; Walter et al., 1990). Mechanical inoculation of herbaceous hosts is also used for detecting sap-transmitted viruses such as those of the genus *Nepovirus* (Rowhani et al., 2005).

3.2. Serological assays

Serological methods, which are based on the specific recognition of viruses by homologous antibodies, are routinely used for rapid screening of plant material. Of these, the enzyme-linked immunosorbent assay (ELISA) is the most widely applied. Direct (Double antibody sandwich, DAS) and indirect (Triple antibody sandwich, TAS) ELISA assays (Boscia et al., 1997) have been developed and successfully applied for most of the known grapevine viruses. Nowadays, commercial ELISA kits from different companies are available. Different parts of the plant such as leaves, buds, stems, roots, and cortical scrapings can be used as source of antigens. Of course, it should be taken into consideration that viruses are not evenly distributed in an infected plant and that there is a seasonal fluctuation of their titer (Constable et al., 2012; Monis & Bestwick, 1996; Tsai, Daugherty, & Almeida, 2012). As these factors affect the reliability of the assay, it is crucial to sample the part of the plant and during the time of the year in which virus titers are the highest. Alternatively, bark scrapings can be used for a more reliable testing throughout the year (EPPO, 2008). According to EPPO directions (2008) for the production of certified material, the use of ELISA is recommended for GFLV and other European nepoviruses, for members of the family *Closteroviridae* (GLRaV-1 to -4), vitiviruses (GVA and GVB), and *Grapevine fleck virus* (GFkV) from the genus *Maculavirus*. It should be noted that ELISA is not used alone but complementarily to other diagnostic procedures.

3.3. Molecular assays

Molecular methods are based on the detection of the genomic nucleic acids of the viruses. They are widely used in diagnostics due to a higher sensitivity in comparison to serological techniques. Reverse transcription-polymerase chain reaction (RT-PCR) is probably the most frequently used molecular assay for the detection of grapevine viruses. The method is rapid and very sensitive, thus circumvents the problems of low virus concentration. The preparation of the template used for RT-PCR is of major importance. Grapevine, as most woody species, contains high quantities of polysaccharides and polyphenolic compounds that inhibit the enzymes used for the assay (Nassuth, Pollari, Helmeczy, Stewart, & Kofalvi, 2000). In order to minimize the impact of these substances, various approaches have been followed such as total RNA extraction, the most widely used, immunocapture (Koolivand, Sokhandan-Bashir, Akbar Behjatnia, & Jafazi Joozani, 2014; Sefc, Leonhardt, & Steinkellner, 2000), dilution of the plant extract (Habili, Fazeli, & Rezaian, 1997), and spotting of plant sap extract onto a nylon membrane that is used after thermal treatment (Dovas & Katis, 2003a; La Notte, Minafra, & Saldarelli, 1997). For the detection of grapevine viruses, different versions of RT-PCR have been developed such as nested RT-PCR (Dovas & Katis, 2003a, 2003b), multiplex (Dovas & Katis, 2003b; Gambino & Gribaudo, 2006), and more recently real-time PCR (Lopez-Fabuel et al., 2013; Osman, Leutenegger, Golino & Rowhani, 2007). Furthermore, generic RT-PCR assays have been successfully applied for the simultaneous identification of related viruses, reducing the time, and cost of detection (Dovas & Katis, 2003a, 2003b). More advanced technologies such as micro- or macroarrays are being used for the simultaneous detection of a broad spectrum of viruses (Engel et al., 2010; Thompson, Fuchs, Fischer, & Perry, 2012; Thompson et al., 2014). Finally, next-generation sequencing is a new promising technology that has been successfully used for the rapid identification and sequencing of all putative viruses present in a sample as well as for the identification of new agents of a disease (Al Rwahnih et al., 2013; Coetzee et al., 2010; Giampetruzzi et al., 2012; Poojari, Alabi, Fofanov, & Naidu, 2013; Zhang, Singh, Kaur, & Qiu, 2011).

4. CONTROL

4.1. Introduction

To mitigate the impact of viruses, prophylactic and curative measures are utilized. Prophylactic measures should always be considered wherever grapevines are grown and whenever new vineyards are established, while curative measures should be tailored to diseased vineyards.

Prophylactic measures such as sanitary selection and certification help prevent the presence of viruses in the propagation material, and thus, reduce their long-distance dispersal via exchange of carelessly selected nursery productions (budwood and planting materials). They also lessen the rate of primary infection in vineyards. Prophylactic strategies facilitate the production of high quality planting material and the establishment of healthy vines that are paramount for high-quality production. The impact of prophylactic measures is somewhat limited in new plantings for which secondary infection of viruses can be problematic. This is the case, for example, in clean vineyards that are surrounded by diseased vineyards in areas where vectors, i.e., dagger nematodes, mealybugs, and scale and soft scale insects, are present. These conditions are conducive to an influx of viruses in the clean vineyard from surrounding diseased vineyards.

Cultural, chemical, biological, and sanitary measures are the curative measures deployed to manage virus diseases, as recently reported for leafroll (Almeida et al., 2013). The selection and implementation of the most appropriate cultural approach, i.e., roguing versus extensive or total elimination in vineyards can be facilitated by economic analyses and profitable thresholds (Atallah et al., 2012). Agrochemicals can substantially reduce vector populations and slow virus spread in diseased vineyards (Almeida et al., 2013). However, concerns have been raised on their use in terms of adverse effects on the environment and human health, and there is an increasing demand for sustainable and environmentally safe viticulture practices. Some of the curative measures can be difficult to implement and, altogether, are often limited at offering a satisfactory management of viruses.

4.2. Production and use of certified propagative material

4.2.1 The principles

The production of propagative material (cuttings, rooted cuttings, and grafted plants) with improved sanitary traits can reduce the inoculum potential. In perspective, this will have a beneficial impact on the health conditions of the viticultural industry of those countries that pursue and implement appropriate strategies for restraining the production and distribution of infected stocks. These strategies take into account the sanitary improvement of the crops (sanitary selection and sanitation) and the certification of nursery productions.

Certification can be defined as: (i) "A procedure whereby true-to-type candidate mother plants to be used as source of material for propagation, undergo controls and, whenever necessary, treatments to secure absence from any number of pathogens, as specified by regulations officially issued, or endorsed, by competent governmental agencies" (Martelli & Walter, 1998) or (ii) "A system for the production of vegetatively propagated plants for planting, intended for further propagation or for sale, obtained from selected candidate material after several propagation stages under conditions ensuring that stated health standards are met. The filiation of the material is considered throughout the scheme" (EPPO, 2008).

These definitions imply that certification is an interdisciplinary endeavor requiring phytopathological (primarily virological), pomological, and, when required, technological (e.g., oenological) competences. Certification schemes enforced in the European Union (EU) are typically based upon: (i) pomological (clonal) and sanitary selection in the field for the identification of desirable accessions (putative clones) of any given cultivar, based on the evaluation of the quantity and quality of the yield; (ii) assessment of the varietal conformity (trueness-to-type); (iii) assessment of the sanitary status of putative clones and their sanitation (when needed); and (iv) technological evaluation of the produce (when needed; Fig. 4). The identification of clones is a lengthy procedure regulated by EU Directive 72/169, as outlined by the Office International de la Vigne et du Vin (Anonymous, 1991). It encompasses the comparative performance in the field of selected accessions of any given cultivar, among themselves and with the already registered clones of that cultivar.

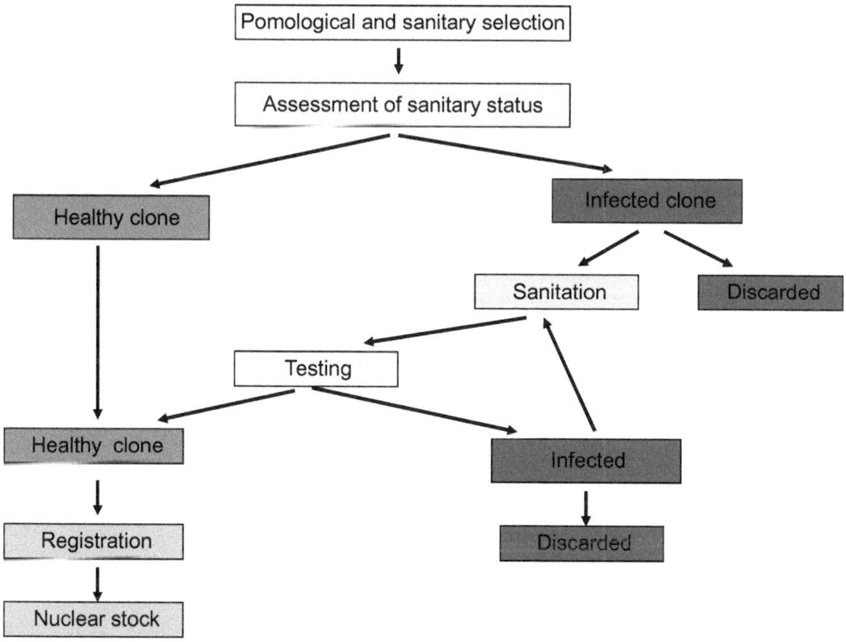

Figure 4 Flowchart of the procedure implemented in Italy for the generation of certified nuclear stocks, in accordance with the certification program enforced in the European Union. (See the color plate.)

The outcome of these activities is a "registrable stock," i.e., a clonal accession true-to-type possessing a well-established sanitary status, from which, upon registration, a set of materials of different categories will derive, which EPPO (2008) defines as follows: (i) *Nuclear stock*: Material individually tested by the most rigorous procedure in the scheme. Material propagated from a nuclear stock may remain nuclear stock under appropriate conditions. (ii) *Propagation stock*: Material derived from the multiplication of a nuclear stock, under conditions ensuring freedom from infection. Pathogen freedom is checked by an appropriate procedure. Material derived from propagation stock under the same conditions remains propagation stock, but, according to the plant species concerned, a maximum number of generations of propagation may be fixed at this stage. (iii) *Certified stock*: Material which is produced from propagation stock under appropriate conditions. In the EU certification system, propagation stocks and certified stocks are labeled with a white and a blue tags, respectively.

Certification can be either voluntary or compulsory. The first type is regimented by regulations issued by public agencies to which participating physical or juridical persons subscribe voluntarily. It cannot be forcefully imposed. By contrast, compulsory certification is enforced whenever the dissemination through propagative materials of a severe disease agent liable to give rise to destructive outbreaks (e.g., *Xylella fastidiosa*) is to be prevented.

The EU certification protocol differs from those implemented in several third countries, which can basically be regarded as mere clean stock programs, for they guarantee the health status and the trueness-to-type (not always) of certified stocks, but not their clonal origin, a property which is highly valued in some EU Member States with a long tradition in this field (Martelli, 1992; Martelli & Walter, 1998).

4.2.2 The International Council for the Study of Virus and Virus-like Diseases of the Grapevine

The International Council for the Study of Virus and Virus-like Diseases of the Grapevine (ICVG) was established in 1962 by a group of American and European plant pathologists who realized the importance of creating an international organization for promoting research on grapevine virology and favoring the exchange of information among researchers (Bovey & Gugerli, 2003). From the very beginning, ICVG has been instrumented in fostering basic and applied research in grapevine virology, attracting the attention of scientists, growers, nurserymen, and regulators on the detrimental effects of infectious diseases on the well-being of the industry, and supporting initiatives for the establishment and implementation of clean stock programs and certification schemes. To this aim, an EPPO/NAPPO workshop on grapevine certification was held during the 14th Meeting (ICVG, 2003), and three different "Recommendations" were produced in 1997, 2003, and 2012. These documents were intended to inform regulators on the current status of the knowledge on infectious diseases of grapevines, in the hope that they could serve as guidelines when sanitary provisions for the production and marketing of propagative material (nursery productions) were to be issued by countries' hosting relevant viticultural industries.

4.2.3 Certification in non-EU countries

4.2.3.1 United States of America

In the early 1960s, the recognition that the California grape industry was affected to a large extent by virus diseases, prompted the University of California at Davis, to initiate a "Clean grape stock program" for sanitizing locally grown and imported grape cultivars, through heat treatment and/or meristem tip culture (Goheen, 1980). This was the beginning of a successful endeavor, which has developed into the current National Grape Clean Plant Network (NCPN Grapes, http://nationalcleanplantnetwork.org/Grape_CPN/).

The grapevine network is a part of the NCPN established by United States Department of Agriculture (USDA) within the Farm Bill—H.R. 6124 Food, Conservation, and Energy Act of 2008, whose mission is to "Provide high quality asexually propagated plant material free of targeted plant pathogens and pests that cause economic loss, to protect the environment and ensure the global competitiveness of specialty crop producers."

In 2010, the Grape CPN was consolidated in a single national network with five centers: (i) The Grapevine, Fruit Tree, and Nut Tree Clean Plant Program at Foundation Plant Services, University of California, Davis, CA (UCD); (ii) The Southeast Vine Improvement and Distribution Program, Florida A&M University, Tallahassee, FL; (iii) The Midwest Grapevine Tissue Culture and Virus Testing Program, Missouri State University, Mountain Grove, MO; (iv) The Eastern Regional Grape Importation, Indexing and Clean Plant Program, Cornell University—New York State Agricultural Experiment Station, Geneva, NY; and (v) Clean Plant Center Northwest-Grapes, Prosser, WA.

At the UCD Foundation Plant Services, certified, true-to-type but not clonally selected grapevine accessions are analyzed according to a testing scheme denoted "Protocol 2010," for the presence of a large number of infectious agents comprising 38 members of the following virus genera: *Nepovirus*, *Closterovirus*, *Vitivirus*, *Foveavirus*, *Maculavirus*, *Marafivirus*, *Trichovirus*, plus phytoplasmas, *X. fastidiosa* (Rowhani & Golino, 2010), and, lately, *Agrobacterium vitis* (A. Rowhani, personal communication). Infected accessions are sanitized (the presence of GRSPaV is tolerated) and grown in a foundation vineyard on the UCD campus. Propagating material from this plot qualifies for the California Department of Food and Agriculture certification program and can be distributed to growers.

4.2.3.2 Canada

The Canadian Plant Protection Export Certification Program (PPECP) for grapevine nursery stocks, established in 1997, was primarily developed for meeting foreign import requirements. However, the plant material produced under this scheme can also be used within Canada (Johnson, 2003) The PPECP is a voluntary and typical clean stock program that deals with phytosanitary certification issues only and does not take into account trueness-to-type and quality standards. Nursery stocks accepted for PPECP purposes belong to one or the other of different categories (elite, foundation, certified) all deriving from nuclear stocks defined as "Original mother plants tested in Canada for the viruses of concern by the Centre for Plant Health (Sidney, BC) of the Canadian Food Inspection Agency (CFIA) or a laboratory approved by it." The pathogens certified absent under PPECP are viruses [12 nepoviruses, 10 "graft-transmissible agents" that include a couple of leafroll-associated viruses (GLRaV-1 and GLRaV-3) and some unspecified RW-associated agents], four phytoplasmas, two bacteria (*X. fastidiosa* and *Xanthomonas ampelinus*), and eight fungi (CFIA, 1997).

4.2.3.3 Argentina and Chile

Certification procedures with the characteristics of clean stock programs have been established in Argentina in 2001 (http://www.infoleg.gob.ar/infolegInternet/anexos/65000-69999/69288/norma.htm) and in Chile in 2007 (Anonymous, 2007; N. Fiore, personal communication). In both countries, nuclear stocks, which do not undergo clonal selection, are certified for the absence of a limited number of viruses, i.e., GFLV, GLRaV-1, -2, and -3, GFkV, and plus RW, vein mosaic and vein necrosis in Argentina and GFLV, GLRaV-1, -2, and -3, GVA, and GVB in Chile.

4.2.3.4 South Africa

In South Africa, a Vine Improvement Association was established in 1986 with the following mission: "To promote the interests of members through plant improvement and plant certification of vine propagation material in the interest of the South African Wine Industry" and, in 1992, a certification scheme for wine grapes was licensed under the Plant Improvement Act No. 53 of 1976 of the South African Department of Agriculture, Forestry, and Fisheries (http://www.plantsa.co.za/wvv.php). The South African certification protocol encompasses clonal selection and requires, for both rootstocks and scions, freedom from the following virus diseases: fanleaf, fleck, leafroll, corky bark, stem pitting/grooving, Shiraz disease, plus visual freedom from the bacterial diseases crown gall (*A. vitis*), and bacterial blight (*Xylophilus ampelinus*), the fungi *Pythium* spp. and *Phythophthora* spp. and a group of nematodes and insects, among which the pseudococcid mealybugs (*Pseudococcus longispinus* and *Planococcus ficus*), known as vectors of some viruses, associated with GLRD and RW.

4.2.3.5 New Zealand and Australia

In both countries, grape certification is a private undertaking (Bonfiglioli, Hoskins, Kelly, Edwards, & Thorpe, 2003; D. Nitschke, personal communication). In New Zealand, the certification scheme is now supported by the New Zealand Winegrowers Board, which has developed a Grafted Grapevine Standard and an associated certification program with the objective of minimizing the chances that infected material be released to the industry. Certified mother plants are true-to-type, clonally selected, and free from GLRaV-3, the only virus taken into account by the certification protocol (Anonymous, 2011).

The Australian Standard for Grapevine propagation material is a clean stock program recently announced by the nongovernment organization Standards Australia (http://www.standards.org.au/Pages/default.aspx) with the publication "AS 5588-2013 Grapevine propagation material." As specified by Standard Australia: "The objective of the standard is to provide the purchaser of grapevine propagation material with assurance as to the origin, varietal identity, physical specifications, and health status of grapevine propagation material." To meet the requirements of the standard, vines must undergo hot water treatment targeting the agent of crown gall (*A. vitis*) and some fungi of trunk diseases and must test negative for two GLRD (GLRaV-1 and GLRaV-3) and two RW (GVA and GVB) viruses.

4.2.4 Certification in the EU

In Europe, the selection and propagation of clones from a single vine date back to the early 1900s, when a program was initiated in Germany for improving the performance of locally grown cultivars (Kassemeyer, 1992). A similar approach was followed in the mid 1900s by France and Italy, which explains why the current EU grape amelioration protocols, in addition to varietal conformity, guarantee the clonal origin of the accessions submitted to selection (Martelli & Walter, 1998)

In the late 1960s, the harmonization of the programs that were running in several Member States was attempted by the EU Commission through the Council Directive 68/93/EC (1968) on "Marketing of material for the vegetative propagation of the grapevine." This Directive classified this material in three categories "basic," "certified," and "standard," which are still retained and, with reference to sanitary provisions, stated that (i) when nurseries of mother vine plots for the production of "basic" and "certified" propagating material are established, the highest possible guarantee must exist that the soil is not infected by harmful organisms, viruses in particular; (ii) in these vineyards, the presence of harmful organisms which reduce the value of propagative material is tolerated only within the "narrowest possible limit"; and (iii) these vineyards must be kept free from plants showing symptoms of virus diseases.

This Directive was amended shortly afterward (1971) by Directive 71/140/EC, which reported that "In the vineyards producing 'basic' material, harmful virus diseases, notably fanleaf and leafroll, must be eliminated. Vineyards producing material of other categories must be kept free from plants showing symptoms of virus diseases."

The vagueness in phytosanitary matters of these Directives, which failed to recognize the advances made in the knowledge of the complexity of the grapevine virus world and connected diseases, reflected in the legislation of the Member States (Portugal, Spain, France, Germany, Italy, and Greece) in which they had been incorporated. The risk was that differences in the sanitary quality of certified nursery productions of diverse Member States could be prejudicial to their export and free circulation within the Community. To face this problem, an international group of experts composed by ICVG members was established in 1991 (Martelli, 1992). This panel produced a proposal (Martelli et al., 1993) which, together with a recommendation voted by ICVG in 2003, was circulated among representatives of the Ministries of Agriculture of EU countries and forwarded to the officials who were in charge of the negotiations for a further amendment of Directive 68/93/EC.

The annex to the latest document (Directive 2005/43/EC) states that: (i) The soil or, if necessary, the substrate of culture gives sufficient guarantees regarding the absence of harmful organisms or their vectors, in particular nematodes which carry viral diseases. The stock nurseries and the cutting nurseries shall be established under appropriate conditions to avoid any risk of contamination by harmful organisms. (ii) The presence of harmful organisms, which reduce the usefulness of the propagation material, shall be at the lowest possible level.

Ironically, and contrary to the recommendation by grapevine virologists and their organization (ICVG), this "lowest possible level" consisted in the absence from (i) complex of infectious degeneration (GFLV and ArMV), (ii) GLRD (GLRaV-1 and GLRaV-3), and (iii) GFkV (only for rootstocks).

Besides the bizarre tolerance of the presence of GFkV in the scions, as if this virus could not translocate to the rootstock of a grafted vine, no mention was made of GLRaV-2 and its RG strain, which is unanimously recognized as most insidious inducer of graft incompatibility, nor of any of the viruses of the RW complex.

The ultimate result is that, by virtue of Directive 2005/43/EC, the EU grapevine nursery industry is allowed to produce and release "certified" material with a deplorably low sanitary level. However, since this Directive sets minimal sanitary standards, Member States are allowed to introduce stricter sanitary provisions. Taking advantage of this, the Italian nurserymen and conservative breeders (MIVA, Moltiplicatori Italiani Viticoli Associati) have signed in 2008 an agreement, endorsed by the Ministry of Agriculture, Food, and Forestry Policies, whereby GLRaV-2 and the RW-associated agents, GVA and GVB, were added to the list of viruses whose absence from nursery productions must be certified.

4.3. Methods used for virus elimination

A crucial step in the production of certified plant material is the selection of clean, virus-tested stocks that will be vegetatively propagated. However, it is not always easy to identify such plants. In some cases, most of the tested vines of certain grapevine varieties are infected by a number of viruses, thus making the application of sanitation techniques necessary in order to obtain healthy plant material. During the past 20 years or so different procedures have been applied to achieve virus elimination with variable results (Panattoni, Luvisi, & Triolo, 2013). A description of the most widely used techniques is given below.

4.3.1 Thermotherapy

Thermotherapy is the most frequently applied therapeutics methodology in sanitation protocols (Panattoni et al., 2013). During heat treatment, plants are gradually grown at moderately high to high temperatures for a certain period of time. The temperature should be within the survival limits of the plant species and usually ranges from 35 to 38 °C. *In vitro* or *in vivo* thermotherapies, either alone or in combination with other techniques such as meristem and shoot tip culture, are successfully applied against a panel of grapevine viruses belonging to different genera (Table 4).

It has been suggested that thermal treatment is more effective against viruses located in the parenchymatic tissue such as nepoviruses (Gribaudo et al., 2006; Grout, 1990). In fact, GFLV has shown strong susceptibility to heat stress in different *Vitis* species (Panattoni & Triolo, 2010). However, studies on viruses with different tissue tropism (e.g., phloem-restricted ampelo- and vitiviruses) showed their variable susceptibility to high temperatures, suggesting that other factors apart from localization are also influencing the outcome of the treatment (Panattoni & Triolo, 2010). Even though

Table 4 Sanitation methods applied against major viral pathogens of grapevine

Treatment	Genus	Virus	References
Thermotherapy	*Ampelovirus*	GLRaV-1, GLRaV-3, GLRaV-4 strain Pr	Leonhardt et al. (1998), Valero, Ibanez, and Mortes (2003), Skiada, Grigoriadou, Maliogka, Katis, and Eleftheriou (2009), Maliogka, Skiada, Eleftheriou, and Katis (2009), and Panattoni and Triolo (2010)
	Foveavirus	GRSPaV	Gribaudo, Gambino, Cuozzo, and Mannini (2006), Skiada et al. (2009), and Maliogka, Skiada, et al. (2009)
	Vitivirus	GVA	Guidoni, Mannini, Ferrandino, Argamante, and Di Stefano (1997), Panattoni, D'Anna, Cristani, and Triolo (2007), and Panattoni and Triolo (2010)
	Nepovirus	GFLV, ArMV	Leonhardt et al. (1998), Valero et al. (2003), Salami, Ebadi, Zamani, and Habibi (2009), and Panattoni and Triolo (2010)
	Maculavirus	GFkV	Panattoni and Triolo (2010)
Meristem tip and shoot tip culture	*Ampelovirus*	GLRaV-1, GLRaV-4 strain Pr	Skiada et al. (2009), Maliogka, Skiada, et al. (2009), and Youssef, Al-Dhaher, and Shalaby (2009)
	Foveavirus	GRSPaV	Gribaudo et al. (2006), Skiada et al. (2009), and Maliogka, Skiada, et al. (2009)
	Vitivirus	GVA	Guidoni et al. (1997) and Bottalico, Savino, and Campanale (2000)
	Nepovirus	GFLV	Youssef et al. (2009)
Somatic embryogenesis	*Ampelovirus*	GLRaV-1, GLRaV-3	Goussard, Wiid, and Kasdorf (1991) and Gambino, Bondaz, and Gribaudo (2006)
	Foveavirus	GRSPaV	Gribaudo et al. (2006) and Gambino et al. (2006)

Table 4 Sanitation methods applied against major viral pathogens of grapevine—cont'd

Treatment	Genus	Virus	References
	Vitivirus	GVA	Gambino et al. (2006)
	Nepovirus	ArMV, GFLV	Goussard et al. (1991), Borroto-Fernández, Sommerbauer, Popwich, Schartl, and Laimer (2009), Gambino, Di Matteo, and Gribaudo (2009), and Gambino, Vallania, and Gribaudo (2010)
	Maculavirus	GFkV	Popescu, Buciumeanu, and Visoiu (2003)
Chemotherapy	*Ampelovirus*	GLRaV-1, GLRaV-3	Panattoni, D'Anna, and Triolo (2006, 2007), Guta, Buciumeanu, Gheorghe, and Teodorescu (2010), and Panattoni, Luvisi, and Triolo (2011)
	Foveavirus	GRSPaV	Skiada, Maliogka, Katis, and Eleftheriou (2013)
	Vitivirus	GVA	Panattoni, D'Anna, Cristani, et al. (2007)
	Nepovirus	GFLV	Weiland, Cantos, Troncoso, and Perez-Camacho (2004)
Cryotherapy	*Vitivirus*	GVA	Wang et al. (2003) and Bayati, Shams-Bakhsh, and Moieni (2011)
Electrotherapy	*Ampelovirus*	GLRaV-1, GLRaV-3	Guta et al. (2010)
	Vitivirus	GVA	Bayati et al. (2011)

its exact mechanism is not known, it has been hypothesized that heat treatment increases viral degradation in the plant cell and might slow down its replication and movement toward the newly grown plant tissue (Cooper & Walkey, 1978). More recent studies associate the effect of heat treatment with RNA silencing, an antiviral immune-like defense system (Voinnet, 2001). It has been observed that RNA silencing is temperature-dependent and that it is significantly enhanced at higher temperatures (Chellappan, Vanitharani, Ogbe, & Fauquet, 2005; Wang, Cuellar, Rajamaki, Hirata, & Valkonen, 2008), hence leading to increased degradation rates of viral RNA.

4.3.2 Meristem tip, shoot tip culture, and somatic embryogenesis

The *in vitro* culture of meristems, shoot tips, and somatic embryos has also been used for the regeneration of virus-free plantlets. Extremely small tissue explants corresponding to the meristem (0.2–0.7 mm) or shoot (5.0–10.0 mm) tip (Panattoni et al., 2013) are excised from virus-infected plants and grown *in vitro* in appropriate media for a period, which lasts several months in the case of meristem culture due to the smaller size of the explant, while the regeneration time for shoot tips is shorter (1–2 months; Fig. 5). Of course, the smaller sized meristem tip stands a better chance to be virus-free but, at the same time, the survival/regeneration rates are lower than those of shoot tips (Maliogka, Skiada, et al., 2009; Skiada et al., 2009). For somatic embryogenesis, tissues from anthers, ovaries, or leaves are used (Gribaudo et al., 2006; Gribaudo, Gambino, & Vallania, 2004; Panattoni et al., 2013). The explants are cultivated on a callus-inducing medium and, a few months later, calli are transferred on a differentiation medium for the generation of somatic embryos (Gribaudo et al., 2006).

These techniques have successfully been applied against several of the most important grapevine viruses (Table 4). The way whereby meristem or shoot tip culture methods are leading to the production of virus-free explants is mainly associated with the fact that virus concentration and localization is not uniform in plant tissue. Meristem tip culture is particularly effective against phloem-located viruses. This is attributed to their inability to invade the meristem since this part of the plant has no differentiated

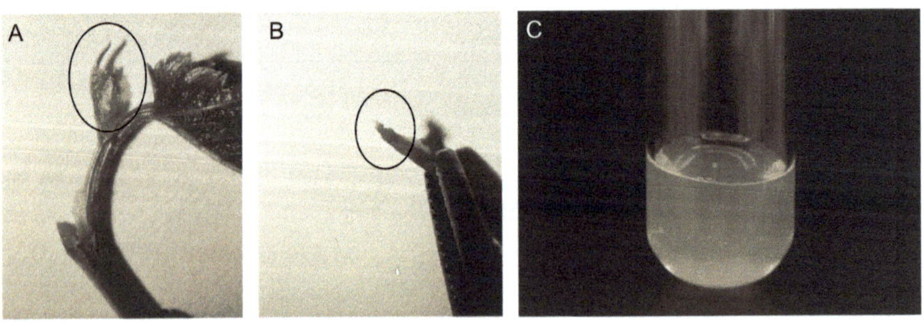

Figure 5 Procedure of meristem tip isolation. (A) Apical bud before the meristem isolation. (B) Meristem tip after the removal of leaf primordia. (C) Apical meristem after a 4-week culture *in vitro*. (See the color plate.)

vascular tissue (Grout, 1990). Likewise, somatic embryogenesis is highly effective at eradicating phloem-limited viruses (Gambino et al., 2006; Goussard et al., 1991; Popescu et al., 2003) but also nepoviruses (Table 4). Interestingly, this method gave very good results even with GRSPaV, which is recalcitrant to elimination with the traditional sanitation techniques (Gribaudo et al., 2006; Minafra & Boscia, 2003; Skiada et al., 2009). The exact mechanism of action of somatic embryogenesis is still not clear. A recent work of Gambino, Vallania, et al. (2010) showed differences in the ability of phloem-limited viruses to spread in callus tissues compared to the nepovirus GFLV, which could have implications on the sanitation efficiency. Nevertheless, it should be taken into consideration that this technique is more time consuming and cultivar dependent as compared to the other procedures with a risk of inducing somaclonal variation in the regenerated plantlets (Gambino et al., 2006; Gribaudo et al., 2006).

4.3.3 Chemotherapy

Even though it has fewer applications in comparison to the great bulk of work conducted on thermotherapy, a number of important grapevine viruses have been eliminated with chemotherapy during the past 10 years (Table 4). Very often, the application of this method came as an alternative to the more traditional procedures in order to eliminate more recalcitrant viruses such as GVA or GRSPaV (Panattoni, D'Anna, Cristani, et al., 2007; Skiada et al., 2013). The method is based on the *in vitro* administration of chemical compounds displaying antiviral activity to the explants. These are mainly drugs that have been largely used against human viruses in medical research and were then applied for the elimination of plant viruses.

Among the most widely used antiviral compounds in grapevine are the synthetic nucleotide analogues ribavirin (1-®-D-ribofuranosyl-1,2,4-triazole-3-carboxamide) and tiazofurin (2-®-D-ribofuranosylthiazole-4-carboxamide) that inhibit IMPDH (inosine 5-monophosphate dehydrogenase), a key enzyme in the *de novo* biosynthesis of guanine nucleotides (Shu & Nair, 2008), as well as mycophenolic acid that is a nonnucleotide IMPDH inhibitor. Other substances that showed significant therapeutic potential against grapevine viruses are the dihydroxypropyladenine (DHPA) and oseltamivir, which belong to the groups of *S*-adenosilhomocysteine hydrolase (key enzyme in methylation reactions) and neuraminidase inhibitors, respectively (Guta et al., 2010; Panattoni, D'Anna, Cristani, et al., 2007; Panattoni et al., 2006). The efficiency of chemotherapy depends on the grapevine cultivar tested and the chemical substance used. Nevertheless, this method could be more effective than *in vitro* thermotherapy combined with meristem or shoot tip culture (Skiada et al., 2013). Interestingly, the combined administration of drugs proved to be, in some cases, more effective in virus elimination than single drug treatments. More specifically, a combination of ribavirin and DHPA led to GVA-free grapevine explants (Panattoni, D'Anna, Cristani, et al., 2007).

The action mechanism of antiviral compounds in plants has been poorly studied. Even though it is known how several of these chemicals function against animal viruses, mainly by interfering with viral replication (for a review, see Panattoni et al., 2013), it is still a mystery how drugs such as neuraminidase inhibitors can interfere with plant viruses. As all methods, chemotherapy has drawbacks. The most severe disadvantage of the technique is the induction of phytotoxicity, the intensity of which was found to depend on the tested cultivar, the antiviral compound applied, and the dose used (Panattoni, D'Anna, Cristani, et al., 2007, Panattoni, D'Anna, & Triolo, 2007; Skiada et al., 2013). There are also other concerns such as the putative accumulation of residues and the presumed mutagenic effects in plants that need to be addressed in future studies.

4.3.4 Cryotherapy

Cryotherapy is a promising new method to eradicate pathogens from infected plant tissue (Wang, Panis, Engelmann, Lambardi, & Valkonen, 2009). It is based on the application of cryopreservation techniques that include the storage of living cells, tissues, and organs at ultra-low temperature, usually in liquid nitrogen (−196 °C). Cryopreservation is considered an ideal means of long-term storage of germplasm and it has been applied to a high number of plant species (Engelmann, 1997). Briefly, a cryopreservation protocol involves encapsulation of shoot tips, followed by dehydration and immersion in liquid nitrogen (Wang et al., 2003). After some time at this low temperature, the encapsulated shoot tips are thawed and regenerated by *in vitro* culture in appropriate media. Alternatively, naked or encapsulated shoot tips may be cryopreserved by vitrification procedures.

Since its initial application to eliminate *Plum pox virus* (PPV) from an interspecific *Prunus* rootstock (Brison, de Boucaud, Pierronnet, & Dosba, 1997), cryotherapy has been used against a range of pathogens (viruses, phytoplasmas, and bacteria) in different plant species (for a review, see Wang et al., 2009). In grapevine, cryotherapy of shoot tips was successfully used for the eradication of GVA (Table 4; Bayati et al., 2011; Wang et al., 2003). Interestingly, very high rates (97%) of GVA-free plants were obtained with this method in comparison with meristem culture (12%; Wang et al., 2003). As previously mentioned, the apical part of the meristematic tissue (dome) is often virus-free or contains low virus concentration. The aim of cryotherapy is to kill infected cells that are more distantly located to this apical part by injuring them through cryo-treatment and hence lead to the regeneration of healthy shoots from the surviving cells of the meristematic dome. By the studies on other plant–virus combinations, in which the virus enters almost all tissues of the meristem, it was shown that the application of thermotherapy followed by cryotherapy of shoot tips may enhance virus eradication (Wang et al., 2008).

Cryotherapy is proposed as an alternative sanitation method that allows treatment of many samples. It has high efficiency and avoids problems related with meristem tip culture (low regeneration rates, difficulties in tissue excision). The main drawback of the technique is the genotype specificity. Finally as in all methods involving tissue culture, it is suggested to check the true-to-typeness of the regenerated plantlets following cryotherapy (Wang et al., 2009).

4.3.5 Electrotherapy

The application of continuous electric current on plant tissue is a novel approach for sanitation purposes. With this treatment, high survival and elimination rates of GLRaV-1, GLRaV-3, and GVA have been achieved (Table 4; Bayati et al., 2011; Guta et al., 2010). Electrotherapy was applied either on potted grapevine plants followed by *in vitro* culture of shoot tips or directly on shoot tips. It was suggested that the plants regenerated from this treatment should be tested for genetic stability, fidelity, and uniformity (Guta et al., 2010).

4.4. Control of virus vectors

As already mentioned, nepoviruses and many ampeloviruses and vitiviruses that infect grapevine are transmitted by nematodes, mealybugs, and scale insects, respectively (Engelbrecht & Kasdorf, 1990; Tsai et al., 2010; van Zyl, Vivier, & Walker, 2012). The control of vectors has been considered one of the basic approaches used against the viruses they transmit. However, nematodes are among the most difficult crop pests to control, due to their large distribution in the soil (D'Addabbo et al., 2011). It should be emphasized that some of the control methods described against nematodes and mealybugs have yet to be evaluated in vineyards and their effect on virus transmission is often unknown. Therefore, although they might be helpful at reducing vector populations, they might have little effect, if any, on virus spread.

4.4.1 Cultural control

4.4.1.1 Crop rotation and fallow period

The use of certified planting material is the first and crucial step when planting a new vineyard. However, when replanting an old vineyard, which is possibly infested with nematodes and infected with GFLV or any other nepovirus, special attention should be taken to protect the newly established vines. Earlier studies suggested 3 years crop rotation before replanting old vineyards (Raski, 1955). Later studies detected GFLV in viruliferous nematodes in soil without host roots for up to 4 years (Demangeat et al., 2005). However, *X. index*-infested fields should be left fallow or cultivated with plants other than grapevine and fig for 6–10 years to avoid infection by GFLV (Golino, Uyemoto, & Goheen, 1992; McKerry, 2000).This is impractical and therefore alternative measures should be considered in order to avoid contamination/infection of the newly established vines.

When vines are pulled out before replanting an old vineyard, all the above-ground plant material is removed and destroyed while some of the root system remains. It is estimated that after vine removal, 70–80% of the root mass remain viable in the soil for a long time depending on the vine age, the rootstock, and soil type (Bell et al., 2009). Since some of the mealybugs vectoring GLRaVs can survive on roots, a mixture of systemic herbicides (e.g., glyphosate + triclopyr) should be applied before elimination to destroy the roots. However, the effectiveness of this approach in terms of management of GLRaVs and vitiviruses remains to be investigated.

4.4.1.2 Other cultural methods

Various cultural measures can also be applied to reduce mealybug populations. Mealybugs hide under the bark of the trunk, cordon, spurs, and canes and become sheltered from insecticides, natural enemies, and environmental conditions (see Daane et al., 2012). Therefore, it is recommended to strip off mealybug-infested bark and to prevent movement of the mealybugs from the trunk to the clusters. Application of an insecticide, as well as flaming to kill the mealybugs is also recommended. The mealybug-infested bark should be destroyed instead of leaving it between the rows in order to avoid movement of the mealybugs back to the vines.

Cover crops can also be used to protect vines from mealybugs by increasing the number of natural enemies present under the cover crop itself.

Parasitoids, generalist predators such as lacewings and some ladybeetle species, might use the floral food resources in the cover crop as alternative food (see Daane et al., 2012). However, many mealybugs species can also feed on ground vegetation and therefore, in some cases, the cover crops may provide an alternative habitat for the mealybugs themselves.

Controlling vine vigor may also result in improved control of mealybugs (Daane et al., 2012). Vigorous vines may increase mealybug populations in two ways. First, excess nitrogen increases the size of mealybug females and the number of eggs in each ovisac. Second, the excess foliage in the vigorous vines provides better shelter to the mealybugs, by decreasing the temperature inside the leaf canopy.

All these methods developed to reduce mealybugs populations need to be evaluated for their potential at reducing spread of GRLaVs or vitiviruses.

4.4.2 Biological control of the nematode and mealybug vectors

4.4.2.1 Nematodes

Various *Trichoderma* species have been successfully used for the control of *X. index* (Darago, Szabo, Hracs, Takacs, & Nagy, 2013). Also, some rhizobacteria isolated from grapevines protect roots from damage caused by *X. index* (Aballay, Martensson, & Persson, 2011; Aballay, Prodan, Martensson, & Persson, 2012), suggesting that they can be used in biological control programs.

Alternative methods were also adopted to reduce populations of *X. index*. For example, marigold and hairy vetch were used as cover crops in vineyards to reduce *X. index* populations (Villate, Morin, Demangeat, Van Helden, & Esmenjaud, 2012). Also, saponins from *Medicago sativa* L. can be effective against *X. index* (D'Addabbo et al., 2011). Different antagonist plants such as *Brassica juncea* substantially reduce *X. index* populations in the soil (Aballay, Sepulveda, & Insunza, 2004).

The effect of these biological approaches on the spread of GFLV remains to be investigated.

4.4.2.2 Mealybugs/soft scale insects

Natural enemies that attack mealybugs and help reducing their populations in the field include the predator *Cryptolaemus montrozieri* Mulsant (Daane et al., 2012) and some lady beetle species in the subfamily Scyminae, lacewing larvae, and Cecidomyiid flies (i.e., predators midges; Abbas, 1999). Knowledge on the impact of these predators on mealybugs is rather limited but studies in New Zealand showed that *Diadiplosis kaebelei* (Koebele) reduced the population of *P. longispinus* by about 30% (Charles, 1985). The most successful biological control program against mealybugs is based primarily on encyrtid parasitoids that are mealybug specialists (Daane et al., 2012; Noyes & Hayat, 1994). The parasitoid *Anagyrus pseudococci* (Girauld) is a well-studied and widely distributed natural enemy (Berlinger, 1977; Duso, 1989; Walton & Pringle, 2004) that has been utilized to control *Planococcus citri* and *P. ficus* (Blumberg, Klein, & Mendel, 1995; Daane, Malakar-Kuenen, & Walton, 2004; Islam & Copland, 2000).

More research is needed to determine whether biological control of mealybugs and soft scale vectors has any effect on the spread of GLRaVs and vitiviruses.

4.4.3 *Chemical control*

4.4.3.1 Nematodes

Over the last 30–40 years, agrochemicals have been extensively used for the control of dagger nematodes and to restrict field spread of nepoviruses. However, soil fumigation is usually ineffective as the nematodes move deep in the soil (Lear, Goheen, & Raski, 1981). In any case, it is recommended to remove the root debris and dry the soil as deep as possible before the application of nematicides (van Zyl et al., 2012). Also, the use of agrochemicals is being limited for environmental safety and human and animal health. Therefore, research is focusing on alternative strategies for controlling *X. index*.

4.4.3.2 Mealybugs/soft scales

Organophosphate compounds are effectively used to control mealybugs and soft scales (Gonzalez, Poblete, & Barria, 2001; Sazo, Araya, & Cerda, 2008; Walton & Pringle, 2001), but new insecticides such as neonicotinoids, insect plant regulators, and biosynthesis inhibitors with a novel mode of action are gaining part of the market (Daane et al., 2006; Lo & Walker, 2010; Sunitha, Jagginavar, & Birodar, 2009). One of the major advantages of some of the newly marketed insecticides is that they have systemic properties, irrespectively of being used through the irrigation system or as a foliar spray, and therefore they kill mealybugs even when sheltered either under the bark or on vine roots.

The timing of applications is critical with most insecticides as the exposed insects are more easily reached than those hidden under the bark, and the immature stages are more susceptible than the adults (Daane et al., 2012). For these reasons, research has been focusing on the development of new insecticides that penetrate into the protected habitats of the mealybugs and on defining the proper application times for an effective control.

4.5. Resistance to viruses and vectors

The use of grapevines that are resistant or tolerant to viruses would be ideal to manage virus diseases. Tolerance is defined by the ability of grapevines to reduce damage caused by viruses, while resistance is defined by their ability to limit virus multiplication. In spite of a long history of cultivation and breeding efforts (Burger, Bouquet, & Striem, 2009), virus resistance has not been achieved in most elite cultivars because sources of resistance have not been found in *Vitis* species (Laimer et al., 2009; Oliver & Fuchs, 2011).

4.5.1 Resistance to viruses and their vectors in Vitis *spp*

Considering the many viruses that can be a major constraint on grapevine production (Table 1) and the long history of grapevine cultivation, little useable resistance to viruses, and their vectors has been identified in *Vitis* species, except against *X. index*, the ectoparasitic nematode vector of GFLV (Oliver & Fuchs, 2011).

The most prominent success in the development of resistance in *Vitis* sp. is against *X. index*. Indeed, *M. rotundifolia* and hybrid crosses between *M. rotundifolia* and *Vitis* species show high resistance to *X. index* with few or no root-associated nematodes (Esmenjaud et al., 2010). Other rootstocks, which are resistant to this nematode species, include *Vitis arizonica*, *Vitis rufotomentosa*, and Börner, among others (Oliver & Fuchs, 2011). Resistance can appear as a hypersensitive reaction to nematode feeding (Staudt & Weischer, 1992). Recently, the first major genetic locus responsible for resistance to an ectoparasitic nematode in grapevine [*X. index* resistance locus 1 (*XiR1*)] was identified (Hwang et al., 2010). This finding should facilitate resistance breeding efforts in the future. Several rootstocks with resistance to *X. index* such as Dog Ridge, Ramsey, Börner, Schwarzmann, O39-16, the Ramsey x Schwarzmann (RS) and RS series, Nemadex, and the Grapevine rootstocks for nematode resistance (GRN) series have been released over time to the wine and grape industries (Oliver & Fuchs, 2011). However, none of the rootstocks that are highly resistant to *X. index* prevent replication of GFLV, nor do they prevent virus translocation to scions (Laimer et al., 2009). Even *M. rotundifolia*, *V. arizonica*, and hybrids derived thereof can be infected with GFLV although at a low rate but they substantially delay infection, possibly through a reduction in nematode-feeding events (Walker, Wolpert, & Weber, 1994). While a substantial delay

(~10 years) in infection may provide economic benefits to growers dealing with *X. index* and GFLV, ultimately, if the vines are infected, *X. index*-resistant rootstocks are not fully satisfactory to manage GFLV. In addition, some of the *X. index*-resistant rootstocks have undesired viticultural characteristics. For example, O39-16 is highly vigorous and has poor rooting ability, while Börner and other promising species and muscadine hybrids are susceptible to lime-induced chlorosis (Oliver & Fuchs, 2011).

For viruses other than GFLV, *Vitis labrusca* is resistant to *Tomato ringspot virus* (ToRSV) and *Tobacco ringspot virus* (TRSV), and some interspecific hybrids show some resistance to TRSV but are susceptible to ToRSV (Oliver & Fuchs, 2011). Most rootstocks also show field resistance to ToRSV; this resistance is likely against the dagger nematode *Xiphinema americanum* rather than the virus itself (Oliver & Fuchs, 2011).

For leafroll, no source of resistant material has been identified in any *Vitis* sp. and infection is latent in *V. labrusca*, interspecific hybrids and rootstocks (Oliver & Fuchs, 2011). Resistance or tolerance to mealybugs and soft scales is not known (Daane et al., 2012; Oliver & Fuchs, 2011). Similarly, there is no recognized source of useful resistance to viruses associated with the RW disease complex and their vectors (Oliver & Fuchs, 2011).

Based on the severity of viral epidemics, difficulties at implementing efficient control strategies in vineyards, there is a need to develop elite cultivars and rootstocks that are resistant or tolerant to viruses and/or their vectors.

4.5.2 Engineered resistance to viruses and their vectors

The advent of plant biotechnologies in the mid 1980s opened new avenues for the development of virus-resistant plants. Based on a lack of recognized sources of resistance in *Vitis* species, genetic engineering is likely the only approach to achieve resistance to viruses in grapevines. Since the pioneering work of Mullins and colleagues who were the first to stably transform grape plants (Mullins, Tang, & Facciotti, 1990), several reports described the development of transgenic grapevines for virus resistance (Gambino, Gribaudo, Leopold, Schartl, & Laimer, 2005; Gambino, Perrone, et al., 2010; Jardak-Jamoussi et al., 2009; Krastanova et al., 1995; Le Gall, Torregrosa, Danglot, Candresse, & Bouquet, 1994; Maghuly et al., 2006; Martinelli, Candioli, Costa, & Minafra, 2002; Mauro et al., 1995; Radian-Sade et al., 2000; Spielmann, Krastanova, Douet-Orhant, & Gugerli, 2000; Winterhagen, Dubois, Sinn, Wetzel, & Reustle, 2009; Xue et al., 1999). Efforts to engineer resistance are primarily under way for *Arabis mosaic virus* (ArMV), GFLV, GVA, GVB, and GLRaV-3 (Laimer et al., 2009).

Several approaches have been selected to achieve virus resistance in otherwise susceptible scions and rootstocks. Most are based on the overexpression of virus-derived genetic constructs, as an application of the concept of pathogen-derived resistance (Gottula & Fuchs, 2009). Virus-derived genes, mainly the coat protein, movement protein, or RNA-dependent RNA polymerase genes, have been engineered and transferred into the genome of rootstocks, *V. vinifera* or interspecific hybrids as complete, truncated, sense, antisense, translatable, or untranslatable versions (Gambino et al., 2005; Krastanova et al., 1995; Le Gall et al., 1994; Maghuly et al., 2006; Martinelli et al., 2002; Mauro et al., 1995; Radian-Sade et al., 2000; Spielmann et al., 2000; Valat, Fuchs, & Burrus, 2006; Xue et al., 1999). Virus-derived gene fragments are also used as hairpin constructs to activate RNA silencing, a potent immune defense against viruses (Gambino, Perrone, et al., 2010; Jardak-Jamoussi et al., 2009; Winterhagen et al., 2009). This antiviral defense is stimulated by double-stranded RNA (dsRNA), which can arise from self-complementary RNA produced by inverted-repeat genetic elements or by bidirectional convergent transcription. The dsRNA molecules are processed by ribonuclease III-type Dicer-like enzymes into small interfering RNAs' (siRNAs') duplexes that are 21–24 nucleotides in size. Following processing, stabilized siRNAs' duplexes are incorporated into Argonaute proteins to form an RNA-induced silencing complex that targets mRNAs complementary to siRNAs, i.e., viral RNAs, and induce their posttranscriptional gene silencing processing by endonucleolytic cleavage (Pumplin & Voinnet, 2013). Exploiting the mechanisms of RNA silencing is providing unprecedented opportunities to engineer virus resistance in grapevines through activation of an innate defense mechanism that results in a targeted and specific nucleolytic degradation of viral RNAs.

Another pathway of RNA silencing has been recently exploited to engineer resistance to grapevine viruses. This approach is based on the use of precursors of eukaryotic posttranscriptional small RNA regulators called microRNAs (miRNAs) and their tailored modification for virus resistance. Modified miRNAs are referred to as artificial miRNAs (amiRNAs). Two artificial miRNAs were produced by replacing nucleotides of the *V. vinifera* miR166f pre-miRNA with a 21 nucleotide stretch from two open reading frames of the GVA genome. Their *in planta* transient expression elicited various levels of resistance to GVA (Roumi et al., 2012). For GFLV, the pre-miR319a of *Arabidopsis thaliana* was modified with a short nucleotide sequence of the coat protein coding region. Transient ectopic expression of these amiRNAs was determined in grapevine somatic embryos following agroinfection, suggesting promise for resistance (Jelly, Schellenbaum, Walter, & Maillot, 2012).

Expression of specific antibodies raised against viruses such as GFLV (Nölke et al., 2009) or GLRaV-3 (Orecchia et al., 2008) is also explored to achieve resistance in grapevines. Similar efforts are under way using nanobodies against GFLV (C. Ritzenthaler, personal communication). In principle, this approach should facilitate broad-spectrum resistance in comparison with the nucleotide sequence-specific antiviral pathways of RNA silencing.

The proof of principle of the different genetic constructs at achieving virus resistance is often validated first in herbaceous hosts prior to their application in grapevines. Herbaceous hosts offer the benefits of mechanical inoculation for resistance evaluation, short time to achieve systemic infection, and high throughput testing. High levels of resistance are reported for GFLV (Bardonnet, Hans, Serghini, & Pinck, 1994; Jardak-Jamoussi et al., 2009; Nölke et al., 2009), ArMV (Spielmann et al., 2000), *Tomato black ring virus* (Pacot-Hiriart, Le Gall, Candresse, Delbos, & Dunez, 1999), GVA (Brumin et al., 2009; Minafra et al., 1998; Radian-Sade et al., 2000), GVB (Minafra et al., 1998), and *Grapevine berry inner necrosis virus* (Yoshikawa et al., 2000) in genetically modified (GM) *Nicotiana* (Laimer et al., 2009).

For grapevines, evaluating the transgenic material for resistance can be challenging. Screens for resistance mostly rely on graft-inoculation in the laboratory or greenhouse (Laimer et al., 2009). Transfection of protoplasts was also explored to rapidly identify resistant candidate individuals within a population of transgenic rootstocks (Valat et al., 2006). In spite of promising results (Laimer et al., 2009), it is not known whether resistance data obtained in the laboratory translate into similar data in a vineyard. Though little work has been done in vineyards, a 3-year field trial of transgenic rootstocks indicated a few GFLV-resistant clones (Vigne, Komar, & Fuchs, 2004). However, 3 years of field testing are not adequate to validate resistance in a perennial crop like grapevines; longer term vineyard evaluations are needed. To date, the translation of laboratory knowledge to real world solutions has been slow and no virus-resistant transgenic grapevine has been released yet (Laimer et al., 2009). It will be interesting to see whether transgenic grapevines with practical resistance to viruses will be made available to the grape and wine industries in the future.

Staking host resistance to a vector and RNA silencing-based resistance to a virus would be an elegant approach to achieve durable resistance against a disease, as discussed for fanleaf degeneration (Bouquet, Torregrosa, & Chatelet, 2004). Also, transgenic grapevines carrying siRNA constructs designed against viruses and their vectors could have potential to control not only viral diseases but also their spread, given RNA silencing-based resistance is a powerful tool to engineer crops for resistance against nematodes (Baum et al., 2007) and insects (Gu & Knipple, 2013).

4.5.3 Virus-resistant transgenic grapevines and environmental safety issues

Environmental and human safety issues have been expressed on the release of virus-resistant transgenic plants, including virus-resistant transgenic grapevines. Issues relate to recombination, heteroencapsidation, transgene dissemination through pollen flow, and allergenicity. Some of these issues are particularly relevant in the case of perennial crops like grapevines that are grown for many years.

Recombinant viruses were not found in transgenic grapevines expressing the coat protein gene of GVA or GVB that were challenge inoculated with the homologous or heterologous virus (Fuchs et al., 2007). Similarly, a vineyard study with transgenic grapevine rootstocks engineered for GFLV resistance suggested no detectable environmental impact beyond natural background events regarding the emergence of recombinant GFLV species. Indeed, recombinant GFLV variants were found in conventional grapevines but not in those grafted onto transgenic rootstocks expressing the coat protein gene of GFLV (Vigne et al., 2004). However, heterologous encapsidation was demonstrated in transgenic *Nicotiana* carrying the coat protein gene of GVA or GVB that were challenged with the heterologous virus, as well as in coinfected nontransgenic *Nicotiana* (Buzkan et al., 2001). The epidemiological significance of the occurrence of heterologous encapsidation in transgenic *Nicotiana* is unknown.

Transgene dissemination through pollen dispersal is not considered as an important environmental safety issue for grapevines because most cultivated cultivars are hermaphrodites and generally considered as self-pollinators. Nonetheless, transgene dispersal was studied with transgenic *V. vinifera* cv. Dornfelder expressing a *β-glucuronidase* (GUS) gene construct that contains an intron sequence. The average cross pollination rate of conventional vines surrounding the transgenic vines was 2.7% at a distance of 20 m, as determined by GUS staining of seedlings during a 3-year experiment (Harst, Cobanov, Hausmann, Eibach, & Töpfer, 2009). These results suggested that a physical barrier of a few dozen meters should limit inadvertent transgene dispersal through pollen flow. Also, the use of a transgenic rootstock that can be grafted to standard conventional scions' cultivars to produce conventional fruits is another strategy to address concerns about transgene flow in grapevines.

Another area of potential concern is the possibility of introducing allergenic proteins and toxic compounds in virus-resistant transgenic crops, including grapevines (Oliver, Tennant, & Fuchs, 2011). In general, viral proteins expressed in transgenic grapevines are not regarded as potential allergens or toxins given their physicochemical and structural properties, a long history of safe consumption and the low exposure levels, if the corresponding transgenic grapevines were to be released in the future, due to low or undetectable transgene protein accumulation because of RNA silencing-driven degradation of the corresponding mRNAs. There is also a wealth of information that clearly indicates no real risks associated with released virus-resistant transgenic fruit and vegetable crops in terms of allergenicity and toxicity (Oliver et al., 2011).

Based on the studies with transgenic grapevines and other perennial fruit crops, including some that have been evaluated for more than a decade in the field, the environmental and human safety issues raised on their release have limited, if any, significance beyond background events (Fuchs et al., 2007; Fuchs & Gonsalves, 2007; Oliver et al., 2011). Thus, safety concerns should not be considered as an impetus to hinder research efforts on the development and release of virus-resistant transgenic grapevines.

4.5.4 Virus-resistant transgenic grapevines and social perception

Social factors are playing a role in stymieing the adoption of virus-resistant transgenic grapevines and concerns have sometimes created a strong climate of opposition, alongside a rejection of new technologies, and certain viticulture innovations and practices. Concerns are often stronger in areas where opportunities for regular interactions between researchers, extension educators, the wine and grape industries, and local communities are limited. Similarly, heightened concerns are often more expressed in areas, where there is a perceived disconnect between research priorities and industry challenges.

To facilitate a forum of discussion for researchers, the wine and grape industries, and local communities, several initiatives have been taken. For instance, focus groups have been convened for targeted dialogs and participation in interactive technology assessment (The Local Monitoring Committee, Lemaire, Moneyron, & Masson, 2010). A 7-year effort to debate why and how research on engineered resistance to GFLV in transgenic grapevine rootstocks should be conducted in France captured divergent and conflicting views among the various participants. Curiously, while these commensurate efforts to reach out to stakeholders and engage them in participative discussions were under way, discovery-based research was put on hold albeit a field trial of previously tested transgenic rootstocks was established with oversight from a local management committee comprised of growers, vintners, environmentalists, consumers, regulators, politicians, union members, and citizens representing the local community. Although the field trial was progressing nicely for some time, outreach efforts did not prevent the destruction of the transgenic vines first by an isolated activist and subsequently by an organized group of activists (The Local Monitoring Committee et al., 2010).

In areas where transparent partnerships among all actors, i.e., researchers, extension educators, the wine and grape industries, regulators, and local communities, are well established and nurtured, often through the land-grant mission of local research institutions, there is a culture of mutual understanding, tolerance, and respect. Such a culture is conducive for a continued dialog and innovative research efforts addressing challenges faced by the local grape and wine industries, including those incited by viruses and providing practical solutions.

5. CONCLUDING REMARKS

Since the first "historical" isolation on herbaceous hosts of a grapevine-infecting virus (GFLV; Cadman, Dias, and Harrison, 1960), the number of such viruses has progressively increased (Table 1) and is still on the raise. In fact, novelties can reasonably be expected from the extensive use of next-generation sequencing platforms, the novel powerful detection tools. These have already disclosed the existence in grapevines of six hitherto unknown viruses and are likely to continue picking up new infectious agents, some of which may be of economic relevance, as is the case of *Grapevine Pinot gris virus* (Giampetruzzi et al., 2012) and *Grapevine red blotch virus* (Al Rwahnih et al., 2013; Krenz, Thompson, Fuchs, & Perry, 2012). Thus, it can reasonably be assumed that, in perspective, the infectious diseases of unknown origin, which one steps on every once in a while, will no longer remain undetermined

Equally relevant advances have been made in diagnosis. The battery of tools currently available enables the dependable identification of all known viruses, while better performing detection protocols are likely to be developed in the future.

Sanitation procedures have also followed a positive improvement trend to the point that the absence of a virus can be ascertained. However, recalcitrant viruses can be knocked out upon exposure, repeatedly if required, to one or another of the currently available techniques (heat therapy, cryotherapy, electrotherapy, meristem tip culture, and somatic embryogenesis). Thus, the clean stock and certification programs under implementation in many countries, if appropriately constructed, are able to turn out clean propagative material, hence nursery productions, with a high sanitary standard.

What will the fate be of these expensively produced materials when planted in the field? How can the happy sanitary status of newly established vineyards be preserved over time? The problem rests with the contaminations operated by vectors that transmit viruses which are associated with, or are recognized agents of the three major disease complexes: infectious degeneration/decline, leafroll, and RW.

Nepovirus-transmitting nematodes are virtually impossible to eradicate. Fumigations are no longer allowed and long fallow periods can decrease the level of their population but are unable to extinguish them. Thus, growers whose soils host viruliferous nematodes are forced to move to another crop or to coexist with the problem. Introgression of resistance/tolerance through conventional breeding is an approach that has been pursued with limited, though encouraging success. Much more interesting and, perhaps, efficient results may be achieved through pathogen-derived transgenic resistance. The trials underway will provide an answer. However, should this be positive, transgenic grapevines will have a hard time to find their way into production vineyards, especially in the countries of old Europe where the anti-GM feeling is still profoundly entrenched. The French experience is enlightening in this regard.

Field protection of sanitized vines from mealybug-transmitted diseases (leafroll and RW) is also a serious, still unsolved issue. In this case, no natural sources of resistance/tolerance to any of the closteroviruses and vitiviruses involved in disease etiology have been identified so far, while the attempts to introduce transgenic resistance have apparently been discontinued. It ensures that the main, if not the only approach to preserve the health status of disease-free stocks rests with an appropriate management of mealybug populations. To this aim, much work is in progress in countries (e.g., South Africa and United States) where some viticultural districts are suffering much because of the rapid spread of these diseases, leafroll in particular.

In summary, the challenge and target of future research is not so much the development of more refined and highly performing techniques for the recognition or the elimination of viruses but, rather, the design of dependable strategies for preventing a quick sanitary deterioration of vineyards planted with costly certified materials.

ACKNOWLEDGMENTS

The authors would like to express their gratitude to Dr. N. Fiore, University of Chile, Santiago; Dr. A. Rowhani, University of California, Davis, USA; Dr. H.J. Maree, University of Stellenbosch, South Africa; and Mr. D. Nitschke, Riverland Vine Improvement Committee Inc., Monash, SA, Australia, for supplying most valuable information.

REFERENCES

Aballay, E., Martensson, A., & Persson, P. (2011). Screening of rhizosphere bacteria from grapevine for their suppressive effect on *Xiphinema index* Thorne & Allen on *in vitro* grape plants. *Plant and Soil*, *347*, 313–325.

Aballay, E., Prodan, S., Martensson, A., & Persson, P. (2012). Assessment of rhizobacteria from grapevine for their suppressive effect on the parasitic nematode *Xiphinema index*. *Crop Protection*, *42*, 36–41.

Aballay, E., Sepulveda, R., & Insunza, V. (2004). Evaluation of five nematode-antagonistic plants used as green manure to control *Xiphinema index* Thorne et Allen on *Vitis vinifera* L. *Nematropica*, *34*, 45–51.

Abbas, M. S. T. (1999). Studies on *Dicrodiplosis manihoti* Harris (Diptera: Cecidomyiidae), a common predator of mealybugs. *Journal of Pesticide Science*, *72*, 133–134.

Abou Ghanem, N., Saldarelli, P., Minafra, A., Buzkan, N., Castellano, M. A., & Martelli, G. P. (1997). Properties of *Grapevine virus D*, a novel putative trichovirus. *Journal of Plant Pathology*, *79*, 15–25.

Abou Ghanem-Sabanadzovic, N., & Sabanadzovic, S. (2014). First report of *Grapevine leafroll-associated virus 2* infecting muscadine (*Vitis rotundifolia*) and summer grape (*Vitis aestivalis*) in the United States. *Plant Disease*. http://dx.doi.org/10.1094/PDIS-12-13-1252-PDN. In press.

Abou Ghanem-Sabanadzovic, N., Sabanadzovic, S., Gugerli, P., & Rowhani, A. (2010). Genome organization, serology and phylogeny of grapevine-leafroll associated viruses 4 and 6: Taxonomic implications. *Virus Research*, *163*, 120–128.

Abou Ghanem-Sabanadzovic, N., Sabanadzovic, S., Uyemoto, J. K., Golino, D., & Rowhani, A. (2010). A putative new ampelovirus associated with grapevine leafroll disease. *Archives of Virology*, *155*, 1871–1876.

Al Rwahnih, M., Dave, A., Anderson, M., Rowhani, A., Uyemoto, J. K., & Sudarshana, M. R. (2013). Association of a DNA virus with grapevines affected by red blotch disease in California. *Phytopathology*, *103*, 1069–1076.

Al Rwahnih, M., Dolja, V. V., Daubert, S., Koonin, E. V., & Rowhani, A. (2012). Genomic and biological analyses of a virus from a symptomless grapevine support a new genus within the family *Closteroviridae*. *Virus Research*, *163*, 302–309.

Almeida, R. P. P., Daane, K. M., Bell, V. A., Blaisdell, G. K., Cooper, M. L., Herrbach, E., et al. (2013). Ecology and management of grapevine leafroll disease. *Frontiers in Microbiology*, *4*, 94. http://dx.doi.org/10.3389/fmicb.2013.00094.

Andret-Link, P., Laporte, C., Valat, L., Ritzenthaler, C., Demangeat, G., Vigne, E., et al. (2004). *Grapevine fanleaf virus*: Still a major threat to the grapevine industry. *Journal of Plant Pathology*, *86*, 183–195.

Andret-Link, P., Schmitt-Keichinger, C., Demangeat, G., Komar, V., & Fuchs, M. (2004). The specific transmission of *Grapevine fanleaf virus* by its nematode vector *Xiphinema index* is solely determined by the viral coat protein. *Virology, 320*, 12–22.

Anonymous. (1991). Programme type pour la réalisation de la selection clonale de la vigne. *Bulletin de l'OIV, 727*(728), 752–763.

Anonymous. (2007). *Establece norma especifica de certifcacion de material de propagacion de vides (*Vitis *spp.) y deroga resolucion n° 2.411 de 2007*. Santiago, Chile: Ministerio de Agricultura, Servicio Agricola y Ganadero, Division Semillas, 10 pp.

Anonymous. (2011). *Grafted grapevine standard*. New Zealand Winegrower's Board. http://www.nzwine.com.

Arroyo-Garcia, R., Ruiz-Garcia, L., Bolling, L., Ocete, R., Lopez, M. A., Arnold, C., et al. (2006). Multiple origins of cultivated grapevine (*Vitis vinifera* L. ssp. *sativa*) based on chloroplast DNA polymorphisms. *Molecular Ecology, 15*, 3707–3714.

Atallah, S., Gomez, M., Fuchs, M., & Martinson, T. (2012). Economic impact of grapevine leafroll disease on *Vitis vinifera* cv. Cabernet franc in Finger Lakes vineyards of New York. *American Journal of Enology and Viticulture, 63*, 73–79.

Bardonnet, N., Hans, F., Serghini, M. A., & Pinck, L. (1994). Protection against virus infection in tobacco plants expressing the coat protein of grapevine fanleaf nepovirus. *Plant Cell Reports, 13*, 357–360.

Baum, J. A., Bogaert, T., Clinton, W., Heck, G. R., Feldmann, P., Ilagan, O., et al. (2007). Control of coleopteran insect pests through RNA interference. *Nature Biotechnology, 25*, 1322–1326.

Bayati, S., Shams-Bakhsh, M., & Moieni, A. (2011). Elimination of *Grapevine virus A* (GVA) by cryotherapy and electrotherapy. *Journal of Agricultural Science and Technology, 13*, 443–450.

Bell, V. A., Bonfiglioli, R. G. E., Walker, J. T. S., Lo, P. L., Mackay, J. F., & McGregor, S. E. (2009). *Grapevine leafroll-associated virus 3* persistence in *Vitis vinifera* remnant roots. *Journal of Plant Pathology, 91*, 527–533.

Berlinger, M. J. (1977). The Mediterranean vine mealybug and its natural in southern Israel. *Phytoparasitica, 5*, 3–14.

Beuve, M., Moury, B., Spilmont, A. S., Sempé-Ignatovic, L., Hemmer, C., & Lemaire, O. (2013). Viral sanitary status of declining grapevine Syrah clones and genetic diversity of *Grapevine rupestris stem pitting-associated virus*. *European Journal of Plant Pathology, 135*, 439–452.

Blumberg, D., Klein, M., & Mendel, Z. (1995). Response by encapsulation of four mealybug species (Homoptera: Pseudococcidae) to parasitization by *Anagyrus pseudococci*. *Phytoparasitica, 23*, 157–163.

Bonavia, M., Digiaro, M., Boscia, D., Boari, A., Bottalico, G., Savino, V., et al. (1996). Studies on "corky rugose wood" of grapevine and on the diagnosis of *Grapevine virus B*. *Vitis, 35*, 53–58.

Bonfiglioli, R., Hoskins, N., Kelly, M., Edwards, F., & Thorpe, G. (2003). Introduction of a private certification scheme into an unregulated national industry: The New Zealand experience. In *Extended abstracts 14th meeting of ICVG, Locorotondo, Italy* (p. 152).

Borroto-Fernández, E. G., Sommerbauer, T., Popwich, E., Schartl, A., & Laimer, M. (2009). Somatic embryogenesis from anthers of the autochthonous *Vitis vinifera* cv. Domina leads to *Arabis mosaic virus*-free plants. *European Journal of Plant Pathology, 124*, 171–174.

Boscia, D., Digiaro, M., Fresno, J., Greif, C., Grenan, S., Kassemeyer, H. H., et al. (1997). ELISA for the detection and identification of grapevine viruses. In B. Walter (Ed.), *Sanitary selection of the grapevine. Protocols for detection of viruses and virus-like diseases* (pp. 129–155). Paris, France: INRA (Les Colloques no 86).

Bottalico, G., Savino, V., & Campanale, A. (2000). Improvements in grapevine sanitation protocols. In *Extended abstracts 13th meeting of ICVG, Adelaide, Australia* (p. 167).

Bouquet, A., Torregrosa, L., & Chatelet, P. (2004). Combination of biotechnological and conventional approaches to rootstock selection presenting a sustainable resistance to grape fanleaf disease transmission. *Bulletin de l'O.I.V.*, 77, 361–376.

Bouyahia, H., Boscia, D., Savino, V., La Notte, P., Pirolo, C., Castellano, M. A., et al. (2005). *Grapevine rupestris stem pitting-associated virus* is linked with grapevine vein necrosis. *Vitis*, *44*, 133–137.

Bovey, R., & Gugerli, P. (2003). A short history of ICVG. In *Extended abstracts 14th meeting of ICVG, Locorotondo, Italy* (pp. 1–2).

Brison, M., de Boucaud, M. T., Pierronnet, A., & Dosba, F. (1997). Effect of cryopreservation on the sanitary state of a cv. Prunus rootstock experimentally contaminated with plum pox potyvirus. *Plant Science*, *123*, 189–196.

Brown, D. J. F., & Weischer, B. (1998). Specificity, exclusivity and complementarity in the transmission of plant virus parasitic nematodes: An annotated terminology. *Fundamental and Applied Nematology*, *21*, 1–11.

Brumin, M., Stukalov, S., Haviv, S., Muruganantham, M., Moskovitz, Y., Batuman, O., et al. (2009). Post-transcriptional gene silencing and virus resistance in *Nicotiana* expressing a *Grapevine virus A* minireplicon. *Transgenic Research*, *18*, 331–345.

Burger, P., Bouquet, A., & Striem, M. J. (2009). Grape breeding. In S. M. Jain & P. M. Priyadarshan (Eds.), *Breeding plantation tree crops: Tropical species* (pp. 161–189). New York, USA: Springer Science and Business Media.

Buzkan, N., Minafra, A., Saldarelli, P., Castellano, A., Dell'Orco, M., Martelli, G. P., et al. (2001). Heterologous encapsidation in non-transgenic and transgenic *Nicotiana* plants infected by *Grapevine viruses A* and *B*. *Journal of Plant Pathology*, *83*, 37–43.

Cabaleiro, C., & Segura, A. (1997). Field transmission of *Grapevine leafroll associated virus 3* (GLRaV-3) by the mealybug *Planococcus citri*. *Plant Diseases*, *81*, 283–287.

Cabaleiro, C., & Segura, A. (2006). Temporal analysis of *Grapevine leafroll-associated virus 3* epidemics. *European Journal of Plant Pathology*, *114*, 441–446.

Cadman, C. H., Dias, H. F., & Harrison, B. D. (1960). Sap transmissible viruses associated with diseases of grape vines in Europe and North America. *Nature (London)*, *187*, 577–579.

CFIA (1997). *Plant protection export certification program for grapevine nursery stock,* Vitis *spp.* Nepean, Ontario, Canada: Canadian Food Inspection Agency, 11 pp.

Charles, J. G. (1985). *Diadiplosis koebelei* new record (Diptera. Cecidomyiidae) a predator of *Pseudococcus longispinus* (Homoptera: Pseudococcidae) from New Zealand. *New Zealand Journal of Zoology*, *8*, 331–334.

Charles, J. G., Cohen, D., Walker, J. T. S., Forgie, S. A., Bell, V. A., & Breen, K. C. (2006). *A review of* Grapevine leafroll-associated virus type 3 *(GLRaV-3) for the New Zealand wine industry: Report to New Zealand Wine growers. Hort Research Client Report. 18447.* (p. 79). Available at, http://wineinf.nzwine.com/downloads_process.asp?sIDD&cIDD&fIDD900&dIDD389.

Charles, J. G., Froud, K. J., van den Brink, R., & Allan, D. J. (2009). Mealybugs and the spread of *Grapevine leafroll-associated virus 3* (GLRsV-3) in a New Zealand vineyard. *Australasian Plant Pathology*, *38*, 576–583.

Chellappan, P., Vanitharani, R., Ogbe, F., & Fauquet, C. M. (2005). Effect of temperature on geminivirus-induced RNA silencing in plants. *Plant Physiology*, *138*, 1828–1841.

Chevalier, S., Greif, C., Clauzel, J. M., Walter, B., & Fritsch, C. (1995). Use of an immunocapture polymerase chain reaction procedure for the detection of *Grapevine virus A* in Kober stem grooving-infected grapevines. *Journal of Phytopathology*, *143*, 369–373.

Coetzee, B., Freeborough, M.-J., Maree, H. J., Celton, J.-M., Jasper, D., Rees, G., et al. (2010). Deep sequencing analysis of viruses infecting grapevines: Virome of a vineyard. *Virology*, *400*, 157–163.

Constable, F. E., Connellan, J., Nicholas, P., & Rodoni, B. C. (2012). Comparison of enzyme-linked immunosorbent assays and reverse transcription-polymerase chain reaction for the reliable detection of Australian grapevine viruses in two climates during three growing seasons. *Australian Journal of Grape and Wine Research*, *18*, 239–244.

Cooper, V. C., & Walkey, D. G. A. (1978). Thermal inactivation of *Cherry leaf roll virus* in tissue cultures of *Nicotiana rustica* raised from seeds and meristem tips. *The Annals of Applied Biology*, *88*, 273–278.

Crespan, M. (2004). Evidence on the evolution of polymorphism of microsatellite markers in varieties of *Vitis vinifera* L. *Theoretical and Applied Genetics*, *108*, 231–237.

Daane, K. M., Almeida, R. P. P., Bell, V. A., Walker, J. T. S., Botton, M., Fallahzadeh, M., et al. (2012). Biology and management of mealybugs in vineyards. In N. J. Bostanian, C. Vincent, & R. Isaacs (Eds.), *Arthropod management in vineyards: Pests, approaches and future directions* (pp. 271–307): The Netherlands: Springer Science.

Daane, K. M., Bentley, W. J., Walton, V. M., Malakar-Kuenen, R., Millar, J. G., Ingels, C. A., et al. (2006). New controls investigated vine mealybugs. *California Agriculture*, *60*, 31–38.

Daane, K. M., Malakar-Kuenen, R., & Walton, V. M. (2004). New controls investigated vine mealybugs. *California Agriculture*, *60*, 31–38.

D'Addabbo, T., Carbonara, T., Leonetti, P., Radicci, V., Tava, A., & Avato, P. (2011). Control of plant parasitic nematodes with active saponins and biomass from *Medicago sativa*. *Phytochemistry Reviews*, *10*, 503–519.

Darago, A., Szabo, M., Hracs, K., Takacs, A., & Nagy, P. I. (2013). *In vitro* investigations on the biological control of *Xiphinema index* with Trichoderma species. *Helminthologia*, *50*, 132–137.

Demangeat, G., Voisin, R., Minot, J. C., Bosselut, N., Fuchs, M., & Esmenjaud, D. (2005). Survival of *Xiphinema index* in vineyard soil and retention of *Grapevine fanleaf virus* over extended time in the absence of host plants. *Phytopathology*, *95*, 1151–1156.

Digiaro, M., Elbeaino, T., & Martelli, G. P. (2007). Development of degenerate and species-specific primers for the differential and simultaneous RT-PCR detection of grapevine-infecting nepoviruses of subgroups A, B and C. *Journal of Virological Methods*, *141*, 34–40.

Dovas, C. I., & Katis, N. I. (2003a). A spot nested RT-PCR method for the simultaneous detection of members of the *Vitivirus* and *Foveavirus* genera in grapevine. *Journal of Virological Methods*, *170*, 99–106.

Dovas, C. I., & Katis, N. I. (2003b). A spot multiplex nested RT-PCR for the simultaneous and generic detection of viruses involved in the etiology of grapevine leafroll and rugose wood of grapevine. *Journal of Virological Methods*, *109*, 217–226.

Duso, C. (1989). Bioecological study on *Planococcus ficus* (Sign.) in Veneto. *Bollettino del Laboratorio di Entomologia Agraria 'Filippo Silvestri'*, *46*, 3–20.

Engel, E. A., Escobar, P. F., Rojas, L. A., Rivera, P. A., Fiore, N., & Valenzuela, P. D. T. (2010). A diagnostic oligonucleotide microarray for simultaneous detection of grapevine viruses. *Journal of Virological Methods*, *163*, 445–451.

Engelbrecht, D. J., & Kasdorf, G. G. F. (1990). Transmission of grapevine leafroll disease and associated closteroviruses by the vine mealybug, *Planococcus ficus*. *Phytophylactica*, *22*, 341–346.

Engelmann, F. (1997). *In vitro* conservation methods. In J. A. Callow, B. V. Ford-Lloyd, & H. J. Newbury (Eds.), *Biotechnology and plant genetic resources* (pp. 119–161). Oxford, UK: CAB International.

EPPO. (2008). Pathogen-tested material of grapevine varieties and rootstocks (Certification scheme). *EPPO Bulletin/Bulletin OEPP*, *38*, 422–429.

Esmenjaud, D., Van Ghelder, C., Voisin, R., Bordenave, L., Decroocq, S., Bouquet, A., et al. (2010). Host suitability of *Vitis* and *Vitis-Muscadina* material to the nematode *Xiphinema index* over one to four years. *American Journal of Enology and Viticulture*, *6*, 96–101.

Fortusini, A., Scattini, G., Prati, S., Cinquanta, S., & Belli, G. (1997). Transmission of *Grapevine leafroll virus* 1 (GLRaV-1) and Grapevine virus A (GVA) by scale insects. In *Extended abstracts 12th meeting of ICVG, Lisbon, Portugal* (pp. 121–122).

Fuchs, M., Cambra, M., Capote, N., Jelkmann, W., Laval, V., Martelli, G. P., et al. (2007). Safety assessment of transgenic plums and grapevines expressing viral coat protein genes: New insights into real environmental impact of perennial plants engineered for virus resistance. *Journal of Plant Pathology*, *89*, 5–12.

Fuchs, M., & Gonsalves, D. (2007). Safety of virus-resistant transgenic plants two decades after their introduction: Lessons from realistic field risk assessment studies. *Annual Review of Phytopathology*, *45*, 173–202.

Gambino, G., Bondaz, J., & Gribaudo, I. (2006). Detection and elimination of viruses in callus, somatic embryos and regenerated plantlets of grapevine. *European Journal of Plant Pathology*, *114*, 397–404.

Gambino, G., Di Matteo, D., & Gribaudo, I. (2009). Elimination of *Grapevine fanleaf virus* from three *Vitis vinifera* cultivars by somatic embryogenesis. *European Journal of Plant Pathology*, *123*, 57–60.

Gambino, G., & Gribaudo, I. (2006). Simultaneous detection of nine grapevine viruses by multiplex reverse transcription-polymerase chain reaction with co-amplification of a plant RNA as internal control. *Phytopathology*, *96*, 1223–1229.

Gambino, G., Gribaudo, I., Leopold, S., Schartl, A., & Laimer, M. (2005). Molecular characterization of grapevine plants transformed with GFLV resistance genes: I. *Plant Cell Reports*, *25*, 546–553.

Gambino, G., Perrone, I., Carra, A., Chitarra, W., Boccacci, P., Torello Marinoni, D., et al. (2010). Transgene silencing in grapevines transformed with GFLV resistance genes: Analysis of variable expression of transgene, siRNAs production and cytosine methylation. *Transgenic Research*, *19*, 17–27.

Gambino, G., Vallania, R., & Gribaudo, I. (2010). *In situ* localization of *Grapevine fanleaf virus* and phloem-restricted viruses in embryogenic callus of *Vitis vinifera*. *European Journal of Plant Pathology*, *127*, 557–570.

Giampetruzzi, A., Roumi, V., Roberto, R., Malossini, U., Yoshikawa, N., La Notte, P., et al. (2012). A new grapevine virus discovered by deep sequencing of virus- and viroid-derived small RNAs in cv Pinot gris. *Virus Research*, *163*, 262–268.

Goheen, A. C. (1980). The California clean stock program. *California Agriculture*, *34*, 15–16.

Golino, D. A., Uyemoto, J. K., & Goheen, A. C. (1992). Grape virus diseases. In D. L. Flaherty, L. P. Christensen, W. T. Lanini, J. J. Marois, P. A. Phillips, & L. J. Wilson (Eds.), *Grape pest management* (2nd ed., pp. 101–109): Oakland, CA: University of California, Division of Agriculture & Natural Resources Publication, 3343.

Golino, D. A., Weber, E., Sim, S., & Rowhani, A. (2008). Leafroll disease is spreading rapidly in a Napa Valley vineyard. *California Agriculture*, *62*, 156–160.

Gonzalez, R. H., Poblete, J. G., & Barria, P. G. (2001). The tree fruit mealybugs in Chile. *Pseudococcus viburni* (Signoret). (Homoptera: Pseudococcidae). *Revista Fruticola*, *22*, 17–26.

Goszczynski, D. E., & Habili, N. (2012). *Grapevine virus A* variants of group II associated with Shiraz disease in South Africa are present in plants affected by Australian Shiraz disease, and have also been detected in the USA. *Plant Pathology*, *61*, 205–214

Goszczynski, D. E., & Jooste, A. E. C. (2003). Identification of divergent variants of *Grapevine virus A*. *European Journal of Plant Pathology*, *109*, 397–403.

Gottula, J., & Fuchs, M. (2009). Toward a quarter century of pathogen-derived resistance and practical approaches to engineered virus resistance in crops. In G. Loebenstein & J. P. Carr (Eds.), *Advances in Virus Research: Vol. 75. Natural and engineered resistance to plant viruses* (pp. 161–183): Elsevier.

Goussard, P. G., Wiid, J., & Kasdorf, G. G. F. (1991). The effectiveness of *in vitro* somatic embryogenesis in eliminating fanleaf virus and leafroll associated viruses from grapevines. *South African Journal of Enology and Viticulture, 12*, 77–81.

Grassi, F., Labra, M., Imazio, S., Spada, A., Sgorbati, S., & Sala, F. (2003). Evidence of a secondary grapevine domestication centre detected by SSR analysis. *Theoretical and Applied Genetics, 107*, 1315–1320.

Greif, C., Garau, R., Boscia, D., Prota, V. A., Fiori, M., Bass, P., et al. (1995). The relationship of Grapevine leafroll-associated closterovirus 2 with a graft-incompatibility condition of grapevines. *Phytopathologia Mediterranea, 34*, 167–173.

Gribaudo, I., Gambino, G., Cuozzo, D., & Mannini, F. (2006). Attempts to eliminate *Grapevine rupestris stem pitting associated virus* from grapevine clones. *Journal of Plant Pathology, 88*, 293–299.

Gribaudo, I., Gambino, G., & Vallania, R. (2004). Somatic embryogenesis from grapevine anthers: The optimal developmental stage for collecting explants. *American Journal of Enology and Viticulture, 55*, 427–430.

Grout, B. W. W. (1990). Meristem-tip culture for propagation and virus elimination. In R. D. Hall (Ed.), *Plant cell culture protocols* (pp. 115–125). Totowa, USA: Humana Press Inc.

Gu, L., & Knipple, D. C. (2013). Recent advances in RNA interference research in insects: Implications for future insect pest management strategies. *Crop Protection, 45*, 36–40.

Guidoni, S., Mannini, F., Ferrandino, A., Argamante, N., & Di Stefano, R. (1997). The effect of grapevine leafroll and rugose wood sanitation on agronomic performance and berry and leaf phenolic content of a Nebbiolo clone. *American Journal of Enology and Viticulture, 48*, 438–442.

Guta, I. C., Buciumeanu, E. C., Gheorghe, R. N., & Teodorescu, A. (2010). Solutions to eliminate *Grapevine leafroll associated virus* serotype 1 + 3 from *V. vinifera* L. cv. Ranai Magaraci. *Romanian Biotechnology Letters, 15*, 72–78.

Habili, N., Fazeli, C. F., & Rezaian, M. A. (1997). Identification of a cDNA clone specific to *Grapevine leafroll-associated virus 1*, and occurrence of the virus in Australia. *Plant Pathology, 46*, 516–522.

Hadidi, A., Barba, M., & Jelkmann, W. (2011). *Virus and virus-like diseases of pome and stone fruits*. St. Paul, MN, USA: APS Press.

Harst, M., Cobanov, B.-A., Hausmann, L., Eibach, R., & Töpfer, R. (2009). Evaluation of pollen dispersal and cross pollination using transgenic grapevine plants. *Environmental Biosafety Research, 8*, 87–99.

Hewitt, W. B. (1950). Fanleaf—Another vine disease found in California. *The Bulletin of the California Department of Agriculture, 39*, 62–63.

Hewitt, W. B., Raski, D. J., & Goheen, A. C. (1958). Nematode vector of soil-borne fanleaf virus of grapevines. *Phytopathology, 48*, 586–595.

Hommay, G., Komar, V., Lemaire, O., & Herrbach, E. (2008). *Grapevine virus A* transmission by larvae of *Parthenolecanium corni*. *European Journal of Plant Pathology, 121*, 185–188.

Hwang, C. F., Xu, K., Hu, R., Zhou, R., Riaz, S., & Walker, M. A. (2010). Cloning and characterization of XiR1, a locus responsible for dagger nematode resistance in grape. *Theoretical and Applied Genetics, 121*, 789–799.

Islam, K. S., & Copland, M. J. W. (2000). Influence of egg load and oviposition time interval on the host discrimination and offspring survival of *Anargyrus pseudococci* (Hymenoptera: Encyrtidae), a solitary endoparasitoid of citrus mealybug, *Planococcus citri* (Hemiptera: Pseudococcidae). *Bulletin of Entomological Research, 90*, 69–75.

Jackson, R. S. (1994). Grapevine species and varieties. In *Wine science: Principles and applications* (pp. 11–31): Academic Press.

Jardak-Jamoussi, R., Winterhagen, P., Bouamama, B., Dubois, C., Mliki, A., Wetzel, T., et al. (2009). Development and evaluation of a GFLV inverted repeat construct for genetic transformation of grapevine. *Plant Cell, Tissue and Organ Culture, 97*, 187–196.

Jelly, N. S., Schellenbaum, P., Walter, B., & Maillot, P. (2012). Transient expression of artificial microRNAs targeting *Grapevine fanleaf virus* and evidence for RNA silencing in grapevine somatic embryos. *Transgenic Research, 21*, 1319–1327.

Johnson, R. C. (2003). Grapevine certification and importation of grapevines into the member countries of the North American Plant Protection Organization (NAPPO). In *Extended abstracts 14th meeting of ICVG, Locorotondo, Italy* (pp. 147–148).

Kassemeyer, H. H. (1992). Certification of grapevines in Germany. In G. P. Martelli (Ed.), *Grapevine viruses and certification in EEC countries* (pp. 67–73).Bari, Italy: Istituto Agronomico Mediterraneo, Quaderno No. 3.

King, A. M. Q., Adams, M. J., Carstens, E. B., & Lefkowitz, E. J. (2012). *Ninth report of the International Committee Virus Taxonomy. Virus taxonomy.* Amsterdam, the Netherlands: Elsevier-Academic Press.

Klaassen, V. A., Sim, S. T., Dangl, G. S., Osman, F., Al Rwahnih, M., Rowhani, A., et al. (2011). *Vitis californica* and *Vitis californica* × *Vitis vinifera* hybrids are hosts for *Grapevine leafroll-associated virus-2* and *-3* and *Grapevine virus A* and *B*. *Plant Diseases, 95*, 657–665.

Koolivand, D., Sokhandan-Bashir, N., Akbar Behjatnia, S. A., & Jafazi Joozani, R. A. (2014). Detection of *Grapevine fanleaf virus* by immunocapture reverse transcription-polymerase chain reaction (IC-RT-PCR) with recombinant antibody. *Archives of Phytopathology and Plant Protection, 47*, 2070–2077. http://dx.doi.org/10.1080/03235408.2013.868697.

Krake, L. R. (1993). Characterization of grapevine leafroll disease by symptomatology. *The Australian and New Zealand Wine Industry Journal, 8*, 40–44.

Krastanova, S., Perrin, M., Barbier, P., Demangeat, G., Cornuet, P., Bardonnet, N., et al. (1995). Transformation of grapevine rootstocks with the coat protein gene of grapevine fanleaf nepovirus. *Plant Cell Reports, 14*, 550–554.

Krenz, B., Thompson, J. R., Fuchs, M., & Perry, K. L. (2012). Complete genome sequence of a new circular DNA virus from grapevine. *Journal of Virology, 86*, 7715.

La Notte, P., Minafra, A., & Saldarelli, P. (1997). A spot PCR technique for the detection of phloem-limited grapevine viruses. *Journal of Virological Methods, 66*, 103–108.

Laimer, M., Lemaire, O., Herrbach, E., Goldschmidt, V., Minafra, A., Bianco, P., et al. (2009). Resistance to viruses, phytoplasmas and their vectors in the grapevine in Europe: A review. *Journal of Plant Pathology, 91*, 7–23.

Le Gall, O., Torregrosa, L., Danglot, Y., Candresse, T., & Bouquet, A. (1994). *Agrobacterium*-mediated genetic transformation of grapevine somatic embryos and regeneration of transgenic plants expressing the coat protein of grapevine chrome mosaic nepovirus (GCMV). *Plant Science, 2*, 161–170.

Le Maguet, J., Beuve, M., Herrbach, E., & Lemaire, O. (2012). Transmission of six ampeloviruses and two vitiviruses to grapevine by *Phenacoccus aceris*. *Phytopathology, 102*, 717–723.

Lear, B., Goheen, A. C., & Raski, D. J. (1981). Effectiveness of soil fumigation for control of fanleaf nematode complex in grapevine. *American Journal of Enology and Viticulture, 32*, 208–211

Leonhardt, W., Wawrosch, C., Auer, A., & Kopp, B. (1998). Monitoring of virus diseases in Austrian grapevine varieties and virus elimination using *in vitro* thermotherapy. *Plant Cell, Tissue and Organ Culture, 52*, 71–74.

Lima, M. F., Alkowni, R., Uyemoto, J. K., Golino, D. A., Osman, F., & Rowhani, A. (2006). Molecular analysis of a California strain of Rupestris stem pitting-associated virus isolated from declining Syrah grapevines. *Archives of Virology, 151*, 1889–1894.

Lima, M. F., Rosa, C., Golino, D. A., & Rowhani, A. (2006). Detection of *Rupestris stem pitting associated virus* in seedlings of virus-infected maternal grapevine plants. In *Extended abstracts 15th meeting of ICVG, Stellenbosch* (pp. 244–245).

Lo, P. L., Bell, V. A., Walker, J. T. S., Cole, L. C., Rogers, D. J., & Charles, J. G. (2006). *Ecology and management of mealybugs in vineyards, 2005–2006. HortResearch Client Report No 19636* Auckland, New Zealand: The Horticulture and Food Research Institute of New Zealand.

Lo, P. L., & Walker, J. T. S. (2010). Good results from a soil-applied insecticide against mealybugs. *New Zealand Winegrower, 14,* 125–127.

Lopez-Fabuel, I., Wetzel, T., Bertolini, E., Bassler, A., Vidal, E., Torres, L. B., et al. (2013). Real-time multiplex RT-PCR for the simultaneous detection of the five main grapevine viruses. *Journal of Virological Methods, 188,* 21–24.

Maghuly, F., Leopold, S., Machado da Câmara, A., Borroto Fernandez, E., Ali Khan, M., Gambino, G., et al. (2006). Molecular characterization of grapevine plants transformed with GFLV resistance genes: II. *Plant Cell Reports, 25,* 546–553.

Maliogka, V. I., Dovas, C. I., & Katis, N. I. (2008). Evolutionary relationships of virus species belonging to a distinct lineage within the Ampelovirus genus. *Virus Research, 135,* 125–135.

Maliogka, V. I., Dovas, C. I., Lotos, L., Efthimiou, K., & Katis, N. I. (2009). Complete genome analysis and immunodetection of a member of a novel virus species belonging to the genus Ampelovirus. *Archives of Virology, 154,* 209–218.

Maliogka, V. I., Skiada, F. G., Eleftheriou, E. P., & Katis, N. I. (2009). Elimination of a new ampelovirus (GLRaV-Pr) and *Grapevine rupestris stem pitting associated virus* (GRSPaV) from two *Vitis vinifera* cultivars combining *in vitro* thermotherapy with shoot tip culture. *Scientia Horticulturae, 123,* 280–282.

Maree, H. J., Almeida, R. P. P., Bester, R., Chooi, K. M., Cohen, D., Dolja, V. V., et al. (2013). Grapevine leafroll associated virus 3. *Frontiers in Microbiology, 4,* 82. http://dx.doi.org/10.3389/fmicb.2013.00082.

Martelli, G. P. (1978). Nematode-borne viruses of grapevine, their epidemiology and control. *Nematologia Mediterranea, 6,* 1–27.

Martelli, G. P. (1992). *Grapevine viruses and certification in EEC countries.* Bari, Italy: Quaderno No. 3, Istituto Agronomico Mediterraneo.

Martelli, G. P. (1993). Rugose wood complex. In G. P. Martelli (Ed.), *Graft transmissible diseases of grapevines: Handbook for detection and diagnosis* (pp. 45–53). Rome: FAO Publication Division.

Martelli, G. P. (1999). Infectious diseases and certification of grapevines. In G. P. Martelli & M. Digiaro (Eds.) *Proceedings of the Mediterranean network on grapevine closteroviruses 1992–1997 and the viroses and virus-like diseases of the grapevine a bibliographic report, 1985–1997.* Bari: CIHEAM, p. 47–64 (Options Méditerranéennes: Série B. Etudes et Recherches; n. 29)

Martelli, G. P. (2014). Directory of virus and virus-like diseases of the grapevine and their agents. *Journal of Plant Pathology, 96*(Suppl. 1), 1–136.

Martelli, G. P., Abou Ghanem-Sabanadzovic, N., Agranovsky, A. A., Al Rwahnih, M., Dolja, V. V., Dovas, C. I., et al. (2012). Taxonomic revision of the family *Closteroviridae* with special reference to the grapevine leafroll-associated members of the genus *Ampelovirus* and the putative species unassigned to the family. *Journal of Plant Pathology, 94,* 7–19.

Martelli, G. P., & Boudon-Padieu, E. (2006). Directory of infectious diseases of grapevines. In G. P. Martelli & E. Boudon-Padieu (Eds.), *Options Méditerranéennes Série B: Vol. 55.* (pp. 9–194). Bari, Italy: CIHEAM-IAMB.

Martelli, G. P., De Sequeira, O. A., Kassemeyer, H. H., Padilla, V., Prota, U., Quacquarelli, A., et al. (1993). A scheme for grapevine certification in the European Economic Community. *British Crop Protection Council Monograph, 54,* 279–284.

Martelli, G. P., Saldarelli, P., & Minafra, A. (2011). Grapevine leafroll associated virus 3. *AAB descriptions of plant viruses,* No. 422.

Martelli, G. P., & Savino, V. (1990). Fanleaf degeneration. In R. C. Pearson & A. Goheen (Eds.), *Compendium of grape diseases* (pp. 48–49). St. Paul, MN, USA: APS Press.

Martelli, G. P., & Walter, B. (1998). Virus certification of grapevines. In A. Hadidi, R. K. Khetarpal, & H. Koganezawa (Eds.), *Plant virus disease control* (pp. 261–276). St. Paul, MN, USA: APS Press.

Martelli, G. P., Walter, B., & Pinck, L. (2003). Grapevine fanleaf virus. *AAB descriptions plant viruses,* No. 385.

Martinelli, L., Candioli, E., Costa, D., & Minafra, A. (2002). Stable insertion and expression of the movement protein gene of *Grapevine virus A* (GVA) in grape (*Vitis rupestris S.*). *Vitis, 41*, 189–193.

Mauro, M. C., Toutain, S., Walter, B., Pick, L., Otten, L., Coutos-Thevenot, P., et al. (1995). High efficiency regeneration of grapevine plants transformed with GFLV coat protein gene. *Plant Science, 112*, 97–106.

McKerry, M. V. (2000). Soil pests. In L. P. Christensen (Ed.), *Raisin production manual* (pp. 154–159): Oakland, CA: University of California, Division of Agriculture & Natural Resources Publication, 3393.

Meng, B., & Gonsalves, D. (2003). Rupestris stem pitting associated virus of grapevines: Genome structure, genetic diversity, detection and phylogenetic relationship to other plant viruses. *Current Opinion in Virology, 3*, 125–135.

Minafra, A., & Boscia, D. (2003). An overview of rugose-wood associated viruses: 2000–2003. In *Extended abstracts. 14th meeting of ICVG, Locorotondo 2003* (pp. 116–120).

Minafra, A., Gölles, R., Da Camara Machado, A., Saldarelli, P., Buzkan, N., Savino, V., et al. (1998). Expression of the coat protein gene of grapevine virus A and B in *Nicotiana* species and evaluation of the resistance conferred on transgenic plants. *Journal of Plant Pathology, 80*, 197–202.

Monis, J., & Bestwick, R. K. (1996). Detection and localization of grapevine leafroll associated closteroviruses in greenhouse and tissue culture grown plants. *American Journal of Enology and Viticulture, 47*, 199–205.

Mullins, M. G., Tang, F. C. A., & Facciotti, D. (1990). *Agrobacterium*-mediated genetic transformation of grapevines: Transgenic plants of *Vitis ruspestris* Scheele and buds of *Vitis vinifera* L. *Biotechnology, 8*, 1041–1045.

Nakaune, R., Toda, S., Mochizuki, M., & Nakano, M. (2008). Identification and characterization of a new vitivirus from grapevine. *Archives of Virology, 153*, 1827–1832.

Nassuth, A., Pollari, E., Helmeczy, K., Stewart, S., & Kofalvi, S. A. (2000). Improved RNA extraction and one-tube RT-PCR assay for simultaneous detection of control plant RNA plus several viruses in plant extracts. *Journal of Virological Methods, 90*, 37–49.

Nölke, G., Cobanov, P., Uhde-Holzem, K., Reustle, G., Fischer, R., & Schillberg, S. (2009). *Grapevine fanleaf virus* (GFLV)-specific antibodies confer GFLV and *Arabis mosaic virus* (ArMV) resistance in *Nicotiana benthamiana*. *Molecular Plant Pathology, 10*, 41–49.

Noyes, J. S., & Hayat, M. S. (1994). *Oriental mealybug parasitoids of the* Anagyrini *(Hymenoptera: Encyrtidae).* Wallingford: CAB International Press.

Nunez, D. R., & Walker, M. J. (1989). A review of palaeobotanical findings of early *Vitis* in the Mediterranean and of the origins of cultivated grapevines, with special reference to prehistoric exploitations in the western Mediterranean. *Review of Paleobotany and Polynology, 61*, 205–237.

Oliver, J. E., & Fuchs, M. (2011). Tolerance and resistance to viruses and their vectors in *Vitis* sp.: A virologist's perspective of the literature. *American Journal of Enology and Viticulture, 62*, 438–451.

Oliver, J. E., Tennant, P. F., & Fuchs, M. (2011). Virus-resistant transgenic horticultural crops: Safety issues and risk assessment. In B. Mou & R. Scorza (Eds.), *Transgenic horticultural crops: Challenges and opportunities* (pp. 263–287). Boca Raton, FL: CRC Press.

Orecchia, M., Nolke, G., Saldarelli, P., Dell'Orco, M., Uhde-Holzem, K., Sack, M., et al. (2008). Generation and characterization of a recombinant antibody fragment that binds to the coat protein of *Grapevine leafroll-associated virus 3*. *Archives of Virology, 153*, 1075–1084.

Osman, F., Leutenegger, C., Golino, D., & Rowhani, A. (2007). Real-time RT-PCR (TaqMan) assays for the detection of grapevine leafroll associated viruses 1–5 and 9. *Journal of Virological Methods, 141*, 22–29.

Pacot-Hiriart, C., Le Gall, O., Candresse, T., Delbos, R. P., & Dunez, J. (1999). Transgenic tobaccos transformed with a gene encoding a truncated form of the coat protein of Tomato black ring nepovirus are resistant to viral infection. *Plant Cell Reports, 19*, 203–209.

Panattoni, A., D'Anna, F., Cristani, C., & Triolo, E. (2007). *Grapevine vitivirus A* elimination in *Vitis vinifera* explants by antiviral drugs and thermotherapy. *Journal of Virological Methods, 146*, 129–135.

Panattoni, A., D'Anna, F., & Triolo, E. (2006). Improvement in grapevine chemotherapy. In *Extended abstracts 15th meeting of ICVG, Stellenbosh, South Africa* (pp. 139–140).

Panattoni, A., D'Anna, F., & Triolo, E. (2007). Antiviral activity of tiazofurin and mycophenolic acid against *Grapevine leafroll associated virus 3* in *Vitis vinifera* explants. *Antiviral Research, 73*, 206–211.

Panattoni, A., Luvisi, A., & Triolo, E. (2011). Selective chemotherapy on *Grapevine leafroll-associated virus-1* and *-3*. *Phytoparasitica, 39*, 503–508.

Panattoni, A., Luvisi, A., & Triolo, E. (2013). Review. Elimination of viruses in plants: Twenty years of progress. *Spanish Journal of Agricultural Research, 11*, 173–188.

Panattoni, A., & Triolo, E. (2010). Susceptibility of grapevine viruses to thermotherapy on *in vitro* collection of Kober 5BB. *Scientia Horticulturae, 125*, 63–67.

Pongracz, D. P. (1978). *Practical viticulture*. Claremont: David Philip Publisher Ltd.

Poojari, S., Alabi, O. J., Fofanov, V. Y., & Naidu, R. A. (2013). A leafhopper-transmissible DNA virus with novel evolutionary lineage in the family *Geminiviridae* implicated in grapevine redleaf disease by next-generation sequencing. *PLoS One, 8*(6), e64194.

Popescu, C. F., Buciumeanu, E., & Visoiu, E. (2003). Somatic embryogenesis a reliable method for *Grapevine fleck virus*-free regeneration. In *Extended abstracts 14th meeting of ICVG, Bari, Italy* (p. 243).

Pumplin, N., & Voinnet, O. (2013). RNA silencing suppression by plant pathogens: Defence, counter-defence and counter–counter defence. *Nature Reviews, 11*, 745–760.

Radian-Sade, S., Perl, A., Edelbaum, O., Kuznetsova, L., Gafny, R., Sela, I., et al. (2000). Transgenic *Nicotiana benthamiana* and grapevine plants transformed with *Grapevine virus A* (GVA) sequences. *Phytoparasitica, 28*, 79–86.

Raski, D. J. (1955). Additional observations on the nematodes attacking grapevines and their control. *American Journal of Enology and Viticulture, 6*, 29–31.

Raski, D. J., Goheen, A. C., Lider, L. A., & Meredith, C. P. (1983). Strategies against grapevine fanleaf virus and its nematode vector. *Plant Diseases, 67*, 335–339.

Rosseto, M., McNally, J., & Henry, R. J. (2002). Evaluating the potential of SSR flanking regions for examining relationships in Vitaceae. *Theoretical and Applied Genetics, 104*, 61–66.

Roumi, V., Afsharifar, A., Saldarelli, P., Niazi, A., Martelli, G. P., & Izadpanah, K. (2012). Transient expression of artificial microRNAs confers resistance to *Grapevine virus A* in *Nicotiana benthamiana*. *Journal of Plant Pathology, 94*, 643–649.

Rowhani, A., & Golino, D. (2010). Grapevine disease testing protocol 2010. *FPS Grape Program Newsletter*, (October), 10–11.

Rowhani, A., Uyemoto, J. K., Golino, D. A., & Martelli, G. P. (2005). Pathogen testing and certification of *Vitis* and *Prunus* species. *Annual Review of Phytopathology, 43*, 261–278.

Rowhani, A., Zhang, Y. P., Chin, J., Minafra, A., Golino, D. A., & Uyemoto, J. K. (2000). Grapevine rupestris stem pitting associated virus: Population diversity, titer in the host, and possible transmission vector. In *Extended abstracts 13th meeting of ICVG, Adelaide* (p. 37).

Salami, S. A., Ebadi, A., Zamani, Z., & Habibi, M. K. (2009). Incidence of *Grapevine fanleaf virus* in Iran: A survey study and production of virus-free material using meristem culture and thermotherapy. *European Journal of Horticultural Science, 74*, 42–46.

Sazo, L., Araya, J. E., & Cerda, J. D. L. (2008). Effect of a siliconate coadjuvant and insecticides in the control of mealybug of grapevines, *Pseudococcus viburni* (Hemiptera: Pseudococcidae). *Ciencia e Investigacion Agraria, 35*, 177–184.

Sefc, K. M., Leonhardt, W., & Steinkellner, H. (2000). Partial sequence identification of *Grapevine leafroll associated virus-1* and development of a highly sensitive IC-RT-PCR detection method. *Journal of Virological Methods, 86*, 101–106.

Sefc, K. M., Steinkellner, H., Lefort, F., Botta, R., Machado, A. D., Borrego, J., et al. (2003). Evaluation of the genetic contribution of local wild vines to European grapevine cultivars. *American Journal of Enology and Viticulture, 54*, 15–21.

Sharma, A. M., Wang, J., Duffy, S., Zhang, S., Wong, M. K., Rashed, A., et al. (2011). Occurrence of grapevine leafroll-associated virus complex in Napa Valley. *PLoS One, 6*(10), e26227.

Shu, Q., & Nair, V. (2008). Inosine monophosphate dehydrogenase (IMPDH) as a target in drug discovery. *Medicinal Research Reviews, 28*, 219–232.

Skiada, F. G., Grigoriadou, K., Maliogka, V. I., Katis, N. I., & Eleftheriou, E. P. (2009). Elimination of *Grapevine leafroll-associated virus* 1 and *Grapevine rupestris stem pitting-associated virus* from grapevine cv. Agiorgitiko, and a micropropagation protocol for mass production of virus-free plantlets. *Journal of Plant Pathology, 91*, 177–184.

Skiada, F. G., Maliogka, V. I., Katis, N. I., & Eleftheriou, E. P. (2013). Elimination of *Grapevine rupestris stem pitting-associated virus* (GRSPaV) from two *Vitis vinifera* cultivars by *in vitro* chemotherapy. *European Journal of Plant Pathology, 135*, 407–414.

Spielmann, A., Krastanova, S., Douet-Orhant, V., & Gugerli, P. (2000). Analysis of transgenic grapevine (*Vitis rupestris*) and *Nicotiana benthamiana* plants expressing an *Arabis mosaic virus* coat protein gene. *Plant Science, 156*, 235–244.

Staudt, G., & Weischer, B. (1992). Resistance to transmission of *Grapevine fanleaf virus* by *Xiphinema index* in *Vitis rotundifolia* and *Vitis munsoniana*. *Wein Wissenschaft, 47*, 56–61.

Stevenson, A. C. (1985). Studies in the vegetational history of S.W. Spain. II. Playnological investigations at Laguna de los Madres, Spain. *Journal of Biogeography, 12*, 293–314.

Sunitha, N. D., Jagginavar, S. B., & Birodar, A. P. (2009). Bioefficacy botanicals and newer insecticides against grape vine mealybugs, *Maconelllicoccus hirsutus* (Green). *Karnataka Journal of Agricultural Science, 22*, 710–711.

Taylor, C. E., & Robertson, W. M. (1970). Sites of virus retention in the alimentary tract of the nematode vectors, *Xiphinema diversicaudatum* (Micol.) and *X. index* Thorne and Allen. *The Annals of Applied Biology, 66*, 375–380.

Terral, J. F., Tabard, E., Bouby, L., Ivorra, S., Pastor, T., Figueiral, I., et al. (2010). Evolution and history of grapevine (Vitis vinifera) under domestication: New morphometric perspectives to understand seed domestication syndrome and reveal origins of ancient European cultivars. *Annals of Botany, 105*, 443–455.

The Local Monitoring Committee, Lemaire, O., Moneyron, A., & Masson, J. E. (2010). "Interactive technology assessment" and beyond: The field trial of genetically modified grapevines at INRA-Colmar. *PLoS Biology, 8*, e100551.

This, P., Jung, A., Boccacci, P., Borrego, J., Botta, R., Costantini, L., et al. (2004). Development of a common set of standard varieties and standardized method of scoring microsatellites markers for the analysis of grapevine genetic resources. *Theoretical and Applied Genetics, 109*, 1448–1458.

Thompson, J. R., Fuchs, M., Fischer, K. F., & Perry, K. L. (2012). Macroarray detection of grapevine leafroll-associated viruses. *Journal of Virological Methods, 183*, 161–169.

Thompson, J. R., Fuchs, M., McLane, H., Celebi-Toprak, F., Fischer, K. F., Potter, J. L., et al. (2014). Profiling viral infections in grapevine using a randomly primed reverse transcription-polymerase chain reaction/macroarray multiplex platform. *Phytopathology, 104*, 211–219.

Thompson, J. R., Fuchs, M., & Perry, K. (2012). Genomic analysis of *Grapevine-leafroll associated virus 5* and related viruses. *Virus Research, 163*, 19–27.

Tsai, C. W., Chau, J., Fernandez, L., Bosco, D., Daane, K. M., & Almeida, R. P. P. (2008). Transmission of *Grapevine leafroll-associated virus 3* by the vine mealybug (*Planococcus ficus*). *Phytopathology, 98*, 1093–1098.

Tsai, C. W., Daugherty, M. P., & Almeida, R. P. P. (2012). Seasonal dynamics and virus translocation of *Grapevine leafroll-associated virus 3* in grapevine cultivars. *Plant Pathology, 61*, 977–985.

Tsai, C. W., Rowhani, A., Golino, D. A., Daane, K. M., & Almeida, R. P. P. (2010). Mealybug transmission of grapevine leafroll viruses: An analysis of virus-vector specificity. *Phytopathology, 100*, 830–834.

Uyemoto, J. K., Rowhani, A., Luvisi, D., & Krag, R. (2001). New closterovirus in 'Redglobe' grape causes decline of grafted plants. *California Agriculture, 55*, 28–31.

Valat, L., Fuchs, M., & Burrus, M. (2006). Transgenic grapevine rootstock clones expressing the coat protein or movement protein genes of *Grapevine fanleaf virus*: Characterization and reaction to virus infection upon protoplast electroporation. *Plant Science, 170*, 739–747.

Valero, M., Ibanez, A., & Mortes, A. (2003). Effects of high vineyard temperatures on the *Grapevine leafroll associated virus* elimination from *Vitis vinifera* L. cv. Napoleon tissue cultures. *Scientia Horticulturae, 97*, 289–296.

van Zyl, S., Vivier, M. A., & Walker, M. A. (2012). *Xiphinema index* and its relationship to grapevines: A review. *South African Journal of Enology and Viticulture, 33*, 21–32.

Vigne, E., Komar, V., & Fuchs, M. (2004). Field safety assessment of recombination in transgenic grapevines expressing the coat protein gene of *Grapevine fanleaf virus*. *Transgenic Research, 13*, 165–179.

Villate, L., Morin, E., Demangeat, G., Van Helden, M., & Esmenjaud, D. (2012). Control of *Xiphinema index* populations by fallow plants under greenhouse and field conditions. *Phytopathology, 102*, 627–634.

Vivier, M. A., & Pretorius, I. S. (2000). Genetic improvement of grapevine: Tailoring grape varieties for the third millennium—A review. *South African Journal of Enology and Viticulture, 21*, 5–26.

Voinnet, O. (2001). RNA silencing as a plant immune system against viruses. *Trends in Genetics, 17*, 449–459.

Walker, M. A., Wolpert, J. A., & Weber, E. (1994). Viticultural characteristics of VR hybrid rootstocks in a vineyard site infected with *Grapevine fanleaf virus*. *Vitis, 33*, 19–23.

Walter, B., Bass, P., Legin, R., Vernoy, R., Collas, A., & Vesselle, G. (1990). The use of a green grafting technique for the detection of virus-like disease of the grapevine. *Journal of Phytopathology, 128*, 137–145.

Walton, V. M., & Pringle, K. L. (2001). Effects of pesticides and fungicides used on grapevines on the mealybug predatory beetle Nephus 'boschianus' (Coccinelidae, Scymnini). *South African Journal of Enology and Viticulture, 22*, 107–110.

Walton, V. M., & Pringle, K. L. (2004). Vine mealybug, *Planococcus ficus* (Signoret) (Hemiptera: Pseudococcidae), a key pest in South African vineyards. A review. *South African Journal of Enology and Viticulture, 25*, 54–62.

Wang, Q., Cuellar, W. J., Rajamaki, M. L., Hirata, Y., & Valkonen, J. P. T. (2008). Combined thermotherapy and cryotherapy for efficient virus eradication: Relation of virus distribution, subcellular changes, cell survival and RNA degradation in shoot tips. *Molecular Plant Pathology*, *9*, 237–250.

Wang, Q., Mawassi, M., Li, P., Gafny, R., Sela, I., & Tanne, E. (2003). Elimination of *Grapevine virus A* (GVA) by cryopreservation of *in vitro*-grown shoot tips of *Vitis vinifera* L. *Plant Science*, *165*, 321–327.

Wang, Q. C., Panis, B., Engelmann, F., Lambardi, M., & Valkonen, J. P. T. (2009). Cryotherapy of shoot tips: A technique for pathogen eradication to produce healthy planting materials and prepare healthy genetic resources for cryopreservation. *The Annals of Applied Biology*, *154*, 351–363.

Weiland, C. M., Cantos, M., Troncoso, A., & Perez-Camacho, F. (2004). Regeneration of virus-free plants by *in vitro* chemotherapy of GFLV (*Grapevine fanleaf virus*) infected explants of *Vitis vinifera* L. cv "Zalema". *Acta Horticulturae*, *652*, 463–466.

Winterhagen, P., Dubois, C., Sinn, M., Wetzel, T., & Reustle, G. M. (2009). Gene silencing and virus resistance based on defective interfering constructs in transgenic *Nicotiana benthamiana* is not linked to accumulation of siRNA. *Plant Physiology and Biochemistry*, *47*, 739–742.

Xue, B., Ling, K. S., Reid, C. L., Krastanova, S., Sekiya, M., Momol, E. A., et al. (1999). Transformation of five rootstocks with plant virus genes and a *vir*E2 gene from *Agrobacterium tumefaciens*. *In Vitro Cellular & Developmental Biology Plant*, *35*, 226–231.

Yoshikawa, N., Gotoh, S., Umezawa, M., Satoh, N., Satoh, H., Takabashi, T., et al. (2000). Transgenic *Nicotiana occidentalis* plants expressing the 50-kDa protein of *Apple chlorotic leaf spot virus* display increased susceptibility to homologous virus, but strong resistance to *Grapevine berry inner necrosis virus*. *Phytopathology*, *90*, 311–316.

Youssef, S. A., Al-Dhaher, M. M. A., & Shalaby, A. A. (2009). Elimination of *Grapevine fanleaf virus* (GFLV) and *Grapevine leaf roll-associated virus*-1 (GLRaV-1) from infected grapevine plants using meristem tip culture. *International Journal of Virology*, *5*, 89–99.

Zhang, Y., Singh, K., Kaur, R., & Qiu, W. (2011). Association of a novel DNA virus with the grapevine vein-clearing and vine decline syndrome. *Phytopathology*, *101*, 1081–1090.

Zhang, Y. P., Uyemoto, J. K., Golino, D. A., & Rowhani, A. (1998). Nucleotide sequence and RT-PCR detection of a virus associated with grapevine rupestris stem-pitting disease. *Phytopathology*, *88*, 1231–1237.

Zohary, D., & Hopf, M. (2000). *Domestication of plants in the old World* (3rd ed., pp. 151). New York: Oxford University Press.

Zorloni, A., Prati, S., Bianco, P., & Belli, G. (2006). Transmission of *Grapevine virus A* and *Grapevine leafroll-associated virus 3* by *Heliococcus bohemicus*. *Journal of Plant Pathology*, *88*, 325–328.

CHAPTER SEVEN

Biology, Etiology, and Control of Virus Diseases of Banana and Plantain

P. Lava Kumar*[,1], Ramasamy Selvarajan[†], Marie-Line Iskra-Caruana[‡], Matthieu Chabannes[‡], Rachid Hanna[§]

*International Institute of Tropical Agriculture (IITA), Oyo Road, PMB 5320, Ibadan, Nigeria
[†]National Research Center for Banana, Tiruchirapalli, PIN # 620102, TN, India
[‡]CIRAD, UMR BGPI, F-34398, Montpellier, France
[§]IITA, P.O Box. 2008 (Messa), Yaoundé, Cameroon
[1]Corresponding author: e-mail address: L.kumar@cgiar.org

Contents

Abstract

Banana and plantain (*Musa* spp.), produced in 10.3 million ha in the tropics, are among the world's top 10 food crops. They are vegetatively propagated using suckers or tissue culture plants and grown almost as perennial plantations. These are prone to the accumulation of pests and pathogens, especially viruses which contribute to yield reduction and are also barriers to the international exchange of germplasm. The most economically important viruses of banana and plantain are *Banana bunchy top virus* (BBTV), a complex of banana streak viruses (BSVs) and *Banana bract mosaic virus* (BBrMV). BBTV is known to cause the most serious economic losses in the "Old World," contributing to a yield reduction of up to 100% and responsible for a dramatic reduction in cropping area. The BSVs exist as episomal and endogenous forms are known to be worldwide in distribution. In India and the Philippines, BBrMV is known to be economically important

Advances in Virus Research, Volume 91
ISSN 0065-3527
http://dx.doi.org/10.1016/bs.aivir.2014.10.006

but recently the virus was discovered in Colombia and Costa Rica, thus signaling its spread into the "New World." Banana and plantain are also known to be susceptible to five other viruses of minor significance, such as *Abaca mosaic virus*, *Abaca bunchy top virus*, *Banana mild mosaic virus*, *Banana virus X*, and *Cucumber mosaic virus*. Studies over the past 100 years have contributed to important knowledge on disease biology, distribution, and spread. Research during the last 25 years have led to a better understanding of the virus–vector–host interactions, virus diversity, disease etiology, and epidemiology. In addition, new diagnostic tools were developed which were used for surveillance and the certification of planting material. Due to a lack of durable host resistance in the *Musa* spp., phytosanitary measures and the use of virus-free planting material are the major methods of virus control. The state of knowledge on BBTV, BBrMV, and BSVs, and other minor viruses, disease spread, and control are summarized in this review.

1. INTRODUCTION

Banana and plantain in the genus *Musa* (*Musicale*, Zingiberales) are among the major food crops and important horticultural crops produced in over 130 countries. They provide staple food and income for millions of smallholder farmers across the tropics in Africa, Asia, America, Oceania, and the Pacific (Fig. 1; Table 1). Total production was estimated at 139 million tons from 10.3 million ha in 2012; the two crops together rank no. 6 after maize, rice, wheat, potato, and cassava (FAOStat, 2014). All widely cultivated varieties are parthenocarpic, resulting from intra- and inter-crosses of the two seedy species, *M. acuminata* ($2n=2\times=AA$) and *M. balbisiana* ($2n=2\times=BB$) (Ortiz, 2013). Many of the domesticated cultivars are natural mutants with triploid genome ($2n=3\times=33$) such as dessert banana (AAA), cooking banana, and plantain (AAB or ABB), and seedless cultivars of both diploids (AA and AB) and synthetic tetraploids ($2n=4\times=44$) with genome constitutions of AAAA, AAAB, AABB, and ABBB (Heslop-Harrison & Schwarzacher, 2007). Human migration and trade, especially in the Asia-Pacific, have played a major role in the wide dissemination of cultivars into Africa and Latin America (De Langhe & De Maret, 1999; Perrier et al., 2011). Currently over 1000 *Musa* accessions with a wide range of morphological traits and genome constitutions have been identified in collections worldwide (Ortiz, 2013; Pollefeys, Sharrock, & Arnaud, 2004).

Botanical distinction between banana and plantain is not straightforward. In general, banana are eaten uncooked as a dessert, and unripe starchy fruits

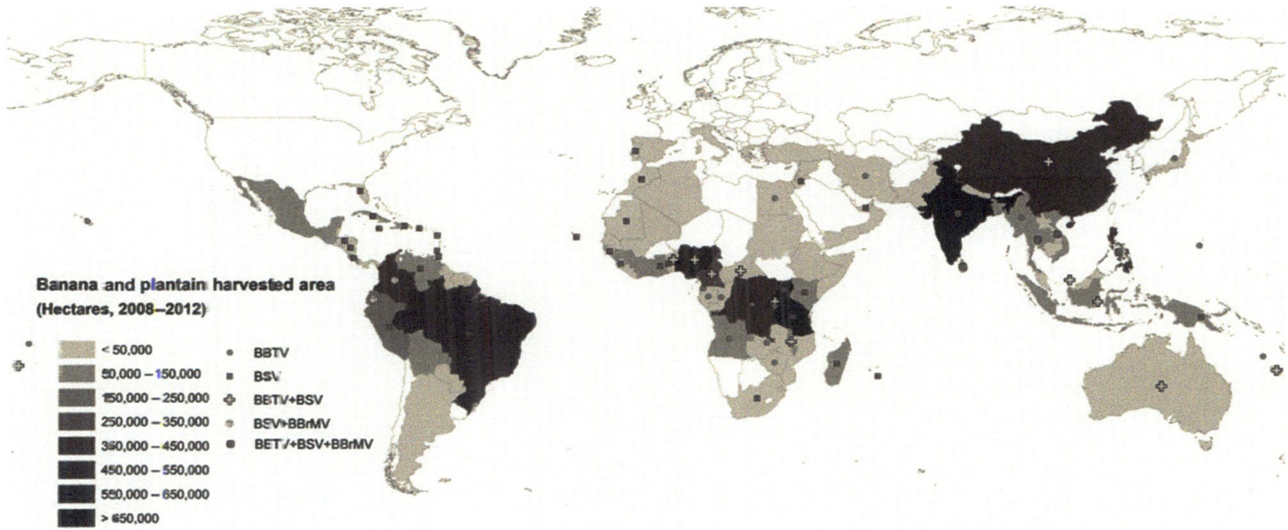

Figure 1 Banana and plantain production in various countries and distribution of the three major banana viruses, BBTV, BSVs, and BBrMV. (Average production data for banana and plantain for the years 2006–08. http://faostat.fao.org/) (See the color plate.)

Table 1 Combined area, production, and productivity of banana and plantain in 2012 and production trend from 2002–2012

	Area		**Production**		**Yield**	
	ha (×1000)	**% Change**	**t (×1000)**	**% Change**	**t/ha**	**% Change**
World	10,360.7	7.8	139,154.9	27.0	13.4	20.8
Africa	5862.2	4.1	42,408.1	20.3	7.2	16.9
Americas	2133.1	0.6	36,089.2	14.1	16.9	13.6
Asia	2258.0	24.2	58,731.0	40.3	26.0	21.3
Europe	10.5	0.8	399.9	−11.2	38.2	−12.1
Oceania	97.0	12.3	1526.7	19.4	15.7	8.1

% Change = Percentage increase or decrease compared with 2002 data.
Data source: FAO banana and plantain production statistics for 2012 (FAOStat, 2014).

that are cooked and eaten are referred to as plantain and cooking banana. Other cultivars are "beer banana" used for fermentation of the juice (Heslop-Harrison & Schwarzacher, 2007). Dessert banana are a major export commodity and a significant GDP earner for Ecuador, Costa Rica, Cameroon, and several other countries (FAOStat, 2012). Cooking banana and plantain are the most important sources of food and rural income in SSA (Nweke, Njoku, & Wilson, 1988). With 56.5% of the global area, Africa dominates the global *Musa* production area compared with 21.7% in Asia and 20.5% in Latin America. However, in terms of crop production and productivity, Africa is last, after Asia and America (Table 1). Dessert banana are grown mainly for commercial trade in Latin America and the Caribbean; cooking banana are widely grown in Asia-Pacific; plantain are dominant in Central and West Africa and Latin America; whereas the highland banana are mainly produced in East Africa (Ortiz & Swennen, 2014).

Banana and plantain grow, mature, and fruit without seasonality throughout the year. Suckers spring up from the underground rhizome to replace the main shoot that withers after fruiting, and this process of succession continues indefinitely (Morton, 1987). However, this exposes plantations to the effects of adverse environmental factors, pests, and pathogens. Farmers generally use young suckers removed from the old plantations to establish new fields. This practice has been among the major causes of outbreaks of several banana diseases and pests around the world (Jones, 2002), especially viruses, which are perpetuated along with the planting material.

About 20 virus species belonging to five families have been reported to infect banana and plantain worldwide (Fig. 1). *Banana bunchy top virus* (BBTV, genus *Babuvirus*, family *Nanoviridae*); several species of Banana streak viruses (BSVs, genus *Badnavirus*, family *Caulimoviridae*) are responsible for streak disease; and *Banana bract mosaic virus* (BBrMV, genus *Potyvirus*, family *Potyviridae*) are the most economically important viruses. *Abaca bunchy top virus* (ABTV, genus *Babuvirus*), abaca mosaic disease caused by a distinct strain of *Sugarcane mosaic virus* (SCMV) designated as SCMV-Ab (genus *Potyvirus*), *Banana mild mosaic virus*, and *Banana virus X* (BVX) both were unassigned members in the family *Betaflexiviridae*, and *Cucumber mosaic virus* (CMV, genus *Cucumovirus*, family *Bromoviridae*) are of minor significance. Viruses in general are a major concern to production because of their effects on yield and quality, and as constraints to the international exchange of *Musa* germplasm. Direct losses are incurred from reduced production, and indirect losses are associated with maintaining plant health, including the production of virus-free planting material. This review summarizes the present status of knowledge on the distribution, diversity diagnostics, and control of banana virus diseases.

2. MAJOR VIRUS DISEASES OF BANANA AND PLANTAIN

2.1. Banana bunchy top disease

2.1.1 Disease discovery and biology

Banana bunchy top disease (BBTD) is the most devastating virus disease of banana and plantain and is prevalent in the Old World (Dale, 1987). The disease was first recorded during an epidemic in Cavendish banana (AAA) in 1889 in Fiji (Magee, 1927). The origin of the virus in Fiji epidemic is not clearly known; however, it is supposed to have been introduced through infected suckers from Tanna (Vanuatu) (Simmonds, 1931). Available records indicate the wide dissemination of BBTD in the Old World along with the movement of planting material by humans, traders and returning soldiers in the early part of the twentieth century (Fahmy, 1927; Magee, 1927, 1953; Wardlaw, 1961). At present, BBTD occurs in 36 countries in Africa, Asia, and Oceania (Blomme et al., 2013; Diekmann & Putter, 1996; Jones, 2013; Kumar et al., 2011). Except for Hawaii (USA) (Conant, 1992) there are no records of BBTD in the New World.

Based on the symptoms, cytology, observation and features of vector transmission, the causal agent was initially established as a virus (Magee, 1940). Based on its localization within phloem tissue and the persistent mode of transmission by *Pentalonia nigronervosa* Coquerel (Hemiptera, *Aphididae*), the putative virus in the BBTD etiology was suspected to be a member of the luteovirus group (Mathews, 1982; Wu & Su, 1990a). In 1991, studies (Harding, Burns, & Dale, 1991; Iskra, Garnier, & Bove, 1989; Thomas & Dietzgen, 1991) reported the purification of virus particles of 18 nm diameter consisting of a coat protein of 20.5 kDa and a single-stranded DNA of about 1 kb from infected plants. Polyclonal and monoclonal antibodies produced against the virus and nucleic acid probes reacted with extracts from symptomatic plants indicating their involvement in disease etiology (Thomas & Dietzgen, 1991). Further studies by Burns, Harding, and Dale (1995) identified additional five circular ssDNA components separately encapsidated. The virus, BBTV, is currently the type member of the genus *Babuvirus*, family *Nanoviridae* (King, Adams, Lefkowitz, & Carstens, 2012). It contains at least six ssDNA components, each about 1.1 kbp encoding for a single open reading frame (ORF) in the virion sense strand designated as: DNA-R (rolling-circle replication initiation protein), -S (coat protein peptide of 19.6 kDa), -M (movement protein), -C (cell cycle link protein), -N (nuclear shuttle protein), and -U3 (a protein of unknown function) (Beetham, Harding, & Dale, 1999; Burns et al., 1995; Vetten et al., 2012). A few BBTV isolates, especially in the East Asian region, carry a satellite DNA component that encodes a protein homologous to Rep encoded in DNA-R (Horser, Karan, Harding, & Dale, 2001).

2.1.2 Symptoms and economic importance

BBTV induces characteristic discontinuous dark green flecks and streaks of variable length on the leaf sheath, midrib, leaf veins, and petioles (Fig. 2). New leaves emerging from the infected plants are narrower with wavy leaf lamina and yellow leaf margins (Nelson, 2004). Leaves produced are progressively shorter, narrow, and brittle in texture; these bunch together at the top and hence provide the name of the disease (Thomas, Iskra-Caruana, & Jones, 1994). Susceptible cultivars infected at a young stage and the suckers emerging from infected stools are severely stunted. Severely infected plants usually will not fruit, but if fruit is produced, the hands and fingers are likely to be distorted and twisted (Nelson, 2004). Occasionally, bracts of male flower buds turn to a leafy structure and exhibit dark green dots and streaks (Thomas et al., 1994). Emerging suckers from infected plants exhibit severe symptoms. Plants infected at a later stage do not normally show leaf symptoms, but dark green streaks can be seen on the tips of the bracts. Emerging suckers from such plants usually exhibit moderate symptoms or none. The incubation period from the time of virus

Figure 2 Severe bunching symptoms (A) and green streaks (B) symptoms in BBTV infected plant; yellow streak symptoms on leaves of the BSV infected plant (B and C); and BBrMV symptoms on petioles (D) and bracts (E). (See the color plate.)

inoculation to symptom expression varies between 19 and 125 days, depending on the stage of infection, cultivar, and weather (Allen, 1978; Hooks, Wright, Kabasawa, Manandhar, & Almeida, 2008). The shortest time for the diagnosis of BBTV using polymerase chain reaction (PCR) was 15 days after infection (Hooks et al., 2008). However, suckers emerging from the infected stools show symptoms from the time of emergence.

BBTD outbreaks in various countries have caused severe reductions in productivity. The first epidemic of BBTD recorded in Cavendish banana in Fiji reduced production from 788,000 bunches to 114,000 bunches at the end of the nineteenth century (Magee, 1953). A severe outbreak of BBTV in Australia in the 1920s destroyed approximately 90% of the crops in Queensland and New South Wales (Fist, 1970). One of the most significant BBTD epidemics was in Tamil Nadu, India, growing dessert banana cv. Virupakshi (Pome group, AAB), that reduced the production area from 18,000 to 2000 ha (Kesavamoorthy, 1980). The first BBTD epidemic in Sindh province in Pakistan reduced the production area from 60,000 to 26,000 ha (Khalid, Soomro, & Stover, 1993). New BBTD outbreaks 2007–2010 in Kodur, Andhra Pradesh, and Jalgaon, Maharashtra, India, caused an annual loss of production worth US$50 million (Selvarajan & Balasubramanian, 2014).

In Africa, BBTD is observed most commonly at elevations below 1300 m.a.s.l. (Kumar et al., 2011; Walangululu et al., 2010) and sparsely above 1700 m.a.s.l. in the hills in eastern DRC (Niyongere, Losenge, et al., 2011). The period of disease incubation is longer at higher altitudes (84 days at 2090 m.a.s.l.) compared with 21 days at 780 m.a.s.l. (Niyongere et al., 2013). Records of losses from BBTD outbreaks in Africa are lacking. In Malawi, a disease outbreak was first recorded in the mid-1990s; the estimated reduction in production was 50–95% at farm level and 30–50% at community level (Soko et al., 2009). Complete elimination of production was observed in Nkhatabay and Nkhotakota districts in the central region along Lake Malawi (Soko et al., 2009). BBTD has also been reported as a serious problem in Burundi, DRC, Gabon, and Rwanda and recently in Bénin, Nigeria, Congo Brazzaville, and the south of Cameroon. Although accurate estimates were lacking, a yield loss of 50–100% in severely infected plants of highly susceptible cultivars was reported from affected regions (Adegbola, Ayodeji, Awosusi, Atiri, & Kumar, 2013; Lokossou et al., 2012; Tushmereirwe & Bagabe, 1999). In general, mean incidence of BBTD in the affected regions across SSA varies from 5% to >70% and is known to result in variable production losses (Fouré & Manser, 1982; Kumar et al., 2011; Niyongere et al., 2013). Farmers in severely affected regions often abandon cultivation of the crop (Soko et al., 2009).

2.1.3 Transmission

BBTV is primarily disseminated through vegetative propagules, including the suckers, corms, and tissue-cultured plants (Drew, Moisander, &

Smith, 1989). The virus is vectored by a banana aphid, *P. nigronervosa* Coquerel, in which it is persistent and transmitted in a circulative manner (Anhalt & Almeida, 2008; Selvarajan, Balasubramanian, & Sathiamoorthy, 2006). Recently, another closely related species, *P. caladii* van der Goot, has been shown to transmit BBTV under experimental inoculation conditions but at a lower level of efficiency than *P. nigronervosa* (Watanabe, Greenwell, & Bressan, 2013).

Much of the attention on the transmission of BBTV by insects is focused on the banana aphid which is known to be highly specific to *Musa* species and present worldwide (Robson, Wright, & Almeida, 2006). Besides banana, host plants of *P. nigronervosa* include other members of the family *Musaceae*. On banana, aphids are frequently observed near the base of plants, followed by the newest unfurled leaf at the top (Robson et al., 2006). Studies in SSA indicated their distribution and abundance on *Musa* spp. are genotype dependent (Hanna et al., 2013). Aphids have a high rate of population growth at 25 °C compared with 20 or 30 °C (Robson, Wright, & Almedia, 2007). Allen (1987) estimated that the vector can spread virus from a primary source to a mean distance of 15.2 m in Australia.

In transmission experiments, the aphid was shown to transmit BBTV after the 4 h (minimum period) for acquisition access and 15 min of inoculation access (Hu, Wang, Sether, Xie, & Leonhardt, 1996). Further work by Anhalt and Almeida (2008) demonstrated a more efficient transmission of BBTV by adult aphids than by the third instar nymphs. Adults transmitted the virus more efficiently at 25 and 30 °C than at 20 °C. The minimum period for acquisition and inoculation for efficient transmission was in the range of 18–24 h. BBTV transmission by *P. nigronervosa* requires a latent period ranging from 20 to 28 h.

2.1.4 Geographic distribution and host range

BBTV is now known to occur in 36 countries; 14 are in Africa (Blomme et al., 2013; Kumar et al., 2011) and 22 in Asia and Oceania (Diekmann & Putter, 1996) (Fig. 1). BBTV recorded in Hawaii (Conant, 1992) was the only report of its occurrence in the New World. In Africa, occurrence of BBTD was first recorded from Egypt in 1901 (Fahmy, 1927), where there was an economically damaging spread of the disease in 1953 (Dale, 1987). BBTD was first discovered in SSA in the 1950s from DRC (Wardlaw, 1961) and later in Gabon, Burundi, Central African Republic, Equatorial Guinea, Rwanda, Malawi (Kenyon, Brown, & Khonje, 1997), Zambia (Gondwe, Mwenebanda, Natha, & Mutale,

2007), Angola (Kumar et al., 2008; Pillay, Blomme, Rodrigues, & Ferreira, 2005), and Cameroon (Oben et al., 2009). The first outbreak of BBTD in West Africa was recorded in the south-eastern part of the Republic of Benin (Lokossou et al., 2012) and spread from there into Nigeria (Adegbola et al., 2013). In Mozambique, the disease was seen in a banana nursery block in western Mozambique (Gondwe et al., 2007), but diseased plants were destroyed and new infections were not reported. An outbreak of BBTD was reported in Eritrea in 1964 (Saverio, 1964), but the current status is not known.

BBTV is known to infect natural and synthetic hybrids of *Musa paradisiaca*, abaca (*M. textilis*) (Manila hemp) (Sharman, Thomas, Skabo, & Holton, 2008) and *Ensete ventricosum* (Selvarajan & Balasubramanian, 2013). Searches for BBTV were negative in plants belonging to the species *Alpinia*, *Heliconia*, *Canna*, and *Strelitzia*, often found growing in the *Musa* production zones (Geering & Thomas, 1997). One report on BBTV detection in *Colocasia esculenta* in India (Ram & Summanwar, 1984) was not proved unequivocally (Geering & Thomas, 1997; Hu et al., 1996).

2.1.5 BBTV diversity

Various BBTV isolates characterized so far around the world have >85% homology (Banerjee et al., 2014). Although ABTV is also known to cause symptoms similar to those of BBTV in *Musa* spp., it is less prevalent and recognized so far only in the Philippines and Malaysia, mainly infecting abaca (Sharman et al., 2008). In general, the genetic diversity of BBTV isolates within the countries is very low [see studies in India (Selvarajan et al., 2010; Vishnoi, Raj, & Prasad, 2009), Pakistan (Amin, Qazi, Mansoor, Ilyas, & Briddon, 2008), Africa (Adegbola et al., 2013; Kumar et al., 2011), and Oceania (Stainton et al., 2012)]. However, in India, relatively greater diversity for BBTV was observed in the north-eastern region (Banerjee et al., 2014), including the identification of a new Babuvirus—*Cardamom bushy dwarf virus* (CBDV)—in cardamom (Mandal, Shilpi, Barman, Mandal, & Varma, 2013) (Fig. 3).

Based on the phylogenetic relationships among the DNA-R component sequences, various BBTV isolates were grouped into two different lineages: (i) the Pacific-Indian Oceans (PIO) group (formerly South Pacific group) comprising isolates in Africa, Australia, Hawaii, south Asia, Myanmar, and Tonga; and (ii) the South-East Asian (SEA) group (formerly Asian group) comprising isolates from China, Indonesia, Japan, the Philippines, Taiwan, and Vietnam (Banerjee et al., 2014; Karan, Harding, & Dale,

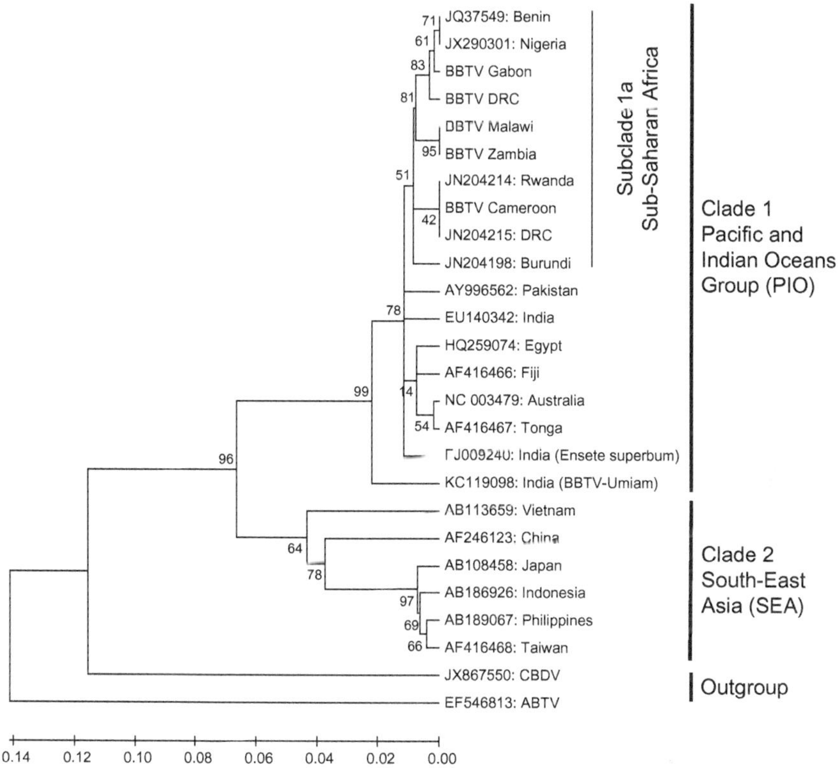

Figure 3 The evolutionary history of BBTV inferred using Neighbor-Joining method based on the ClustalW alignment of the 240 base pairs of Banana bunchy top virus (BBTV) DNA-R master replication-protein of various BBTV isolates. Country of origin of the virus isolates and corresponding NCBI GenBank accession numbers are listed. Bootstrap values (1000 replications) are shown as percentages at the branch points. *Abaca bunchy top virus* (ABTV) and Cardamom bushy dwarf virus (CBDV) are included as outgroup species using the Neighbor-Joining method. Phylogenetic analyses were conducted in MEGA4 (Tamura, Dudley, Nei, & Kumar, 2007).

1994; Stainton et al., 2012; Yu et al., 2012) (Fig. 3). Isolates from each country from these two groups are clustering together, indicating geographic separation. However, a BBTV isolate from Taiwan that has affinities with the PIO group is an exception (GenBank Accession no. EF095164).

Viral genome-based studies are also providing clues to the origin and pattern of BBTV spread. For instance, the DNA-R based sequence analysis of BBTV isolates from SSA, clustered together forming a unique sub-lineage within the PIO group, suggests a common origin for the isolates in SSA (Kumar et al., 2011) (Fig. 3). Historical data and farmers' knowledge of disease occurrence suggest the first occurrence was near Yangambi in the Kisangani region, DRC (Wardlaw, 1961). Subsequent reports were from regions and countries adjoining DRC (Central African Republic, Congo Brazzaville, Gabon, Equatorial Guinea, Rwanda, Burundi, northern Angola, and northern Zambia) indicating a strong potential for virus dispersal through infected planting material along with human movement (Blomme et al., 2013; Kumar et al., 2011). The recent spread of BBTV into Bénin and Nigeria appears to be through the introduction of infected planting material from the Central African region (Adegbola et al., 2013; Lokossou et al., 2012). However, the isolate from Egypt, first reported in 1901, was aligned more closely with isolates from Australia and Fiji than with those from SSA (Fig. 3). These accounts support the possibility that BBTV was introduced into Africa in two separate events, one in North Africa in Egypt and the second in SSA, which contributed to pandemic there (Dale, 1987; Fouré & Manser, 1982; Kumar et al., 2011; Wardlaw, 1961) The pattern of spread in Hawaii is similar to that in SSA. Almedia, Bennett, Anhalt, Tsai, and Grady (2009), using molecular data, suggested that the inter-island spread of BBTV in Hawaii was driven by the human displacement of infected planting material. All this evidence underscores the need for appropriate containment procedures to prevent spread through planting material.

2.1.6 BBTV diagnostics

First detection methods for BBTV were based on enzyme-linked immunosorbent assay (ELISA) using monoclonal and polyclonal antibodies (Thomas & Dietzgen, 1991; Wu & Su, 1990b), and are available commercially (www.agdia.com). Various formats of ELISA (triple antibody sandwich ELISA, plate-trapped antigen ELISA, and double antibody sandwich ELISA) have been established for the reliable detection of the virus in field-grown plants, tissue culture plants, and aphids (Geering & Thomas, 1996; Selvarajan, Balasubramanian, Dayakar, Sathiamoorthy, & Ahlawat, 2010; Thomas & Dietzgen, 1991; Wu & Su, 1990b). Virus in the inoculated plants can be detected after 12–25 days depending on the genotype and stage of infection (Hooks et al., 2008) in any part of the infected plant; however, the most sensitive detection is in samples from the mid-rib portion of the youngest leaf. Methods based on nucleic acid spot hybridization (NASH) using DNA probes have been applied for the sensitive detection of BBTV (Hafner, Harding, & Dale, 1997; Harding et al., 1991; Selvarajan & Balasubramanian, 2008; Xie & Hu, 1995).

Methods based on the PCR have become an important technique for virus detection in plants and vectors, because of their superior sensitivity and flexibility (Galal, 2007; Hu et al., 1996; Thiribhuvanamala, Doraiswamy, & Ganapathy, 2005; Xie & Hu, 1995). DNA primers have been described for the amplification of BBTV, components 1–6, and virus-associated satellite DNAs. Differential primers to distinguish PIO and SEA isolates have also designed (Burns et al., 1995; Mansoor et al., 2005; Sharman, Thomas, & Dietzgen, 2000; Stainton et al., 2012). A simple virus release protocol from tissues without homogenization has been used in PCR as a rapid method for virus detection (Thomson & Dietzgen, 1995). Real-time PCR assays with TaqMan™ probes have been developed the quantitative detection of viral DNA segments in both plants and aphid tissues (Bressan & Watanabe, 2011; Chen & Hu, 2013). Recently, isothermal DNA amplification methods, such as Loop-mediated Isothermal Amplification (LAMP) (Peng, Fan, & Huang, 2012) and Rolling Circle Amplification (RCA) have been developed (Stainton et al., 2012). LAMP products can be detected using either conventional agarose gel electrophoresis or by visual observation of turbidity/color changes (Peng, Zhang, et al., 2012). The latter option is convenient for quick diagnosis in the field or in areas where molecular laboratory facilities are not available. RCA is gaining popularity as research tool for amplification of full-length BBTV DNA segments (Stainton et al., 2012). Recombinase Polymerase Amplification is another isothermal technique being developed for the detection of BBTV (M. Amato, personal communication). This method can be performed at temperatures between 37–42 °C, and is a fast emerging, simple DNA amplification technique for rapid and sensitive detection outside a laboratory (Piepenburg, Williams, Stemple, & Armes, 2006).

2.1.7 Options for BBTV control

2.1.7.1 Integrated disease control by exclusion, eradication, and use of virus-free plants

Banana cultivars fully resistant to BBTV are not available. However, some with the B genome (AAB and ABB) are tolerant or express symptoms more slowly than those with the A genome (AA and AAA), such as the Cavendish subgroup (Espino, Magnaye, Johns, & Juanillo, 1993; Jose, 1981; Ngatat et al., 2013). Various *Musa* clones vary in their degree of susceptibility, even among cultivars with only an A genome composition (e.g., Gros Michel) (Hooks, Manandhar, Perez, Wang, & Almeida, 2009; Hooks et al., 2008, Magee, 1948). Tolerant clones have been utilized in the BBTD endemic areas, such as in Fiji (Magee, 1948), to recovery banana production. Current practices of BBTV management in many countries are based on the integration of several approaches, such as the eradication of infected plants by regular roguing, use of virus-free plants, and exclusion (quarantine) measures to prevent disease spread into other regions (Bouhida & Lockhart, 1990).

The availability of virus-free stocks is one of the major limitations in affected areas. *In vitro* methods have been established to generate virus-free planting material through meristem-tip culture combined with heat therapy (Lassois, Lepoivre, Swennen, van den Houwe, & Panis, 2013; Thomas, Smith, Kessling, & Hamill, 1995). These virus-free plants are then used as mother stocks for the mass propagation of virus-free planting material (Su, Hwang, Lee, & Chao, 2007). This approach, backed with certification systems, is now accepted for disease control in many countries in Asia, also in Australia and Hawaii. In some countries, such as India, recurring outbreaks in farmers' fields led to the formulation and strict enforcement of certification systems and commercial production units have been accredited to produce certified tissue cultured plants. More than 22 million TC plants were certified in 2013–2014 alone (R. Selvarajan, personal communication). In Australia, only pathogen-free stocks generated by the Queensland Banana Accredited Nursery (QBAN) are allowed to be used as foundation stock in the TC industry (QPPR, 2002). Presently, virus-free TC plants are widely used to manage BBTV in Taiwan, the Philippines, and India (Molina et al., 2009; Selvarajan, Balasubramanian, Sheeba, Raj Mohan, & Mustaffa, 2011; Su et al., 2007). However, in many SSA countries there are no guidelines or certification schemes for producing high quality planting material, thus increasing the risk of pathogen spread from the unregulated movement of uncertified planting material from one area to another. A very recent study in south-western DRC by Mukwa et al. (2014) attributed the high prevalence of BBTV in the region to the multiple introductions of infected planting materials in Bas Congo province, 1990–2002.

New plantations established in an endemic region are exposed to a high risk of infection. Regular monitoring has been recommended to identify and rogue infected plants to reduce virus spread as soon as symptoms appear (Allen, 1987; Hooks et al., 2008; Magnaye, 1994; Smith, Holt, Kenyon, & Foot, 1998). Allen (1987) also recommended uprooting apparently symptomless plants within 5 m radius of symptomatic plants as a precautionary principle. This however has been shown to result in the elimination of a considerable number of uninfected plants for every infected plant uprooted.

Injecting herbicides, such as fernoxone or 2,4-D, together with systemic insecticide sprays has been recommended to kill the infected plants and prevent the spread of aphids from the treated plants (Regupathy, Subramanian, & Naganathan, 1983; Robson et al., 2007). These methods are labor intensive, costly for many farmers, and of questionable effectiveness in smallholder farming systems where all farmers may not adopt common practices.

So far, intensive eradication of BBTV has been implemented only in Australia. These efforts have resulted in marked reductions in the prevalence of the virus, although outright eradication was not achieved (Allen, 1987; Cook et al., 2012). Using Monte Carlo simulation, Cook et al. (2012) estimated that the implementation of exclusion measures in Australia prevents annual losses of Aus$15.9–27.0 million for the banana industry. To have a similar program in developing countries where the crop is grown dominantly by smallholder farmers could be highly challenging. Recently, under CGIAR Roots Tubers and Banana Program, IITA, Bioversity International, CIRAD, and national programs in SSA partnered to implement a new initiative to contain BBTD (www.rtb.cgiar.org). This pilot phase, from 2014 to 2016, implemented in Bénin, Burundi, Cameroon, Congo Brazzaville, DRC, Malawi, and Nigeria, aims to pilot eradication of infected plants and recover production using healthy planting material in selected sites.

2.1.7.2 Host resistance

Use of disease resistant cultivars offers the most convenient and effective solution for the control of plant diseases. A diverse range of banana and plantain cultivars have been evaluated in Asia and Africa (Espino et al., 1993; Hooks, Manandhar, et al., 2009; Hooks et al., 2008; Jose, 1981, Magee, 1948; Niyongere, Ateka, Losenge, Lepoint, & Blomme, 2011). All these studies highlighted the lack of high levels of resistance to BBTD infection but tolerance was found, typified by a low or slow rate of infection. In field evaluation trials from 2004 to 2006 in Malawi, all 22 cultivars, including banana, plantain, and hybrids, were found to be susceptible to BBTV although with differences in symptom severity (Mwenebanda, Mwamlima, Msosa, & Banda, 2007). In an on-going trial in a BBTV endemic location in southern Cameroon, 16 *Musa* genotypes expressed varied symptoms (Ngatat et al., 2013). Williams (AAA) and the hybrid plantain PITA 23 (AAB) were the most susceptible with >90% infection. Local plantain landraces (AAB), several hybrid plantain (AAB and AAAB), a cooking banana (AAB), and Grande Nain (AAA) were moderately susceptible with a range of infection from about 30–60% infection. Gros Michel (AAA) and

Pisang Awak (ABB) were found to be least susceptible with infection less than 20%.

To overcome the current limitations of low resistance to BBTV in the *Musa* germplasm and breeding challenges for edible *Musa* spp., transgenic approaches based on the pathogen-derived resistance (PDR) strategies are being explored. Strategies using BBTV DNA-R gene or satellite DNA (DNA-S4) resulted partial resistance to BBTV (Tsao, 2008). More recent efforts using RNA interference (RNAi) have resulted in the development of clones with near immunity to infection (Borth et al., 2011; Elayabalan, Selvarajan, & Kumar, 2013). Shekhawat, Ganapathi, and Hadapad (2012) used intron-hairpin-RNA transcripts corresponding to the viral master replication initiation protein (Rep) to generate the BBTV-resistant transgenic banana cv. Rasthali. Transgenic plants challenged and inoculated with viruliferous aphids tested negative to infection 6 months after inoculation. Although field evaluation of these transgenic clones is still due, this approach has shown great potential to establish resistant clones and might prove useful to induce PDR in several popular *Musa* cultivars. Recently, IITA and Queensland University of Technology (QUT) have initiated an RNAi-based transgenic program to develop banana and plantain resistant to BBTV and aphids for Africa (L. Tripathi, personal communication).

Mutation breeding using gamma-irradiation has also been explored to induce genotyping and phenotyping variation for BBTV resistance. In the Philippines, mutation breeding through the gamma-irradiation of shoot-tip explants of cv. Lakatan (AAA) resulted in lines with resistance against BBTV (Damasco et al., 2006; Dizon et al., 2012). Of the 7366 regenerated plants screened, 29 lines were found to be highly resistant to the virus. Resistance in these lines was found to be due to the aphids' non-preference to colonize mutant banana lines and reproduce. Virus multiplication was reduced as indicated by low titer on mutant lines (Abustan, 2012).

Although significant progress has been achieved in the past decade to improve the levels of resistance in popular cultivars through transgenic approaches, none of these clones are available for commercial cultivation.

2.1.7.3 Vector control

Options for control of the banana aphid have so far received little attention. Pesticides for banana aphid control have been used mainly as a pre-emptive measure to prevent their spread from withering plants during eradication programs (Hooks, Fukuda, et al., 2009). It is unlikely that vector control using pesticides will gain popularity because of prohibitive costs and concerns with residual effects on health and environment. Biological control is so far not available as no natural biocontrol agents have been identified (R. Hanna et al., unpublished).

2.2. Banana streak disease

2.2.1 Disease discovery and biology

BSV causes chlorotic streak disease and is known to be the most widely distributed virus infecting banana and plantain around the world. The disease was first observed in the Nieky Valley on the Ivory Coast in 1958 (Lockhart & Jones, 2000) and later, in 1964, severe BSV chlorosis disease was reported in the Gros Michel triploid *M. acuminata* (AAA) banana cultivar. Lockhart (1986) purified bacilliform particles from field-grown Dwarf Cavendish banana (AAA) in southern Morocco, confirming the viral etiology of the disease. The data currently available on the disease show a complex of distinct BSVs, each causing the same disease.

BSVs are pararetroviruses belonging to the genus *Badnavirus*, family *Caulimoviridae*. The virus particles are bacilliform-shaped (120–150 × 30 nm), double-stranded noncovalently closed circular DNA (dsDNA) genome approximately 7.2–7.8 kb long that uses a virus-encoded reverse transcriptase (RT) to replicate. Harper and Hull (1998) were the first to describe the structure of the virus genome and named it *Banana streak Obino l'Ewai virus* (BSOLV). The virus genome has three consecutive ORFs on one strand (King et al., 2012). ORF1 and ORF2 potentially encode two small proteins of unknown function of 20.8 and 14.5 kDa. ORF3 is a large polyprotein of 220 kDa encoding at least four proteins, encompassing a putative cell-to-cell movement protein, a coat protein (analogous to retroviral GAG), an aspartic protease, and a viral replicase consisting of RT and RNase H domains (Harper & Hull, 1998; King et al., 2012). This polyprotein is cleaved into functional units by the aspartic protease once it has been fully translated. In contrast to retroviruses, BSV does not encode integrase, neither does it require integration into the host genome to replicate.

Two infectious forms of BSV exist: (i) the episomal form resulting from cells/plant infection following transmission by mealybugs and (ii) endogenous forms which are endogenous viral sequences of BSV (eBSV) integrated within the banana B genome (*M. balbisiana*). Physical stresses have been reported to induce *de novo* viral particles (episomal form) from eBSV (Cote et al., 2010; Lheureux, Carreel, Jenny, Lockhart, & Iskra-Caruana, 2003), possibly through intra-strand homologous recombination (HR)

(Chabannes & Iskra-Caruana, 2013; Iskra-Caruana, Baurens, Gayral, & Chabannes, 2010). Both episomal virus and infectious particles from eBSV give rise to systemic plant infection (Harper, Ganesh, Thottappilly, & Hull, 1999, Harper, Hart, Moult, & Hull, 2004; Iskra-Caruana et al., 2010). BSV particles from both origins can be transmitted by mealybugs (Dahal et al., 2000; Kubiriba, Legg, Tushemereirwe, & Adipala, 2001; Lockhart & Autrey, 1988)

2.2.2 Symptoms

BSVs infect different species of *Musa* and the natural and synthetic hybrids. Most isolates produce discontinuous yellow dots/streaks that turn necrotic on the leaves, and also pseudo-stem splitting (Fig. 2). Generally, symptoms are erratically distributed on the plant and not shown on all leaves. Some isolates immediately produce stunting with severe necrotic streaks resulting in cigar leaf necrosis and leading quickly to the death of the plant (Thangavelu, Selvarajan, & Singh, 2000). Symptomatic and symptomless stages alternate in infected plants but virus can be detected at all stages (Harper et al., 2002; Lassoudière, 1974; Lockhart & Jones, 2000).

Lassoudière (1974) recorded yield losses ranging up to 90% on Poyo (AAA, Cavendish subgroup) producing few if any exportable bunches. However, Daniells, Geering, Bryde, and Thomas (2001) observed only a mild effect (about 10% loss) on Cavendish subgroup cv. Williams triploid *M. acuminata* (AAA) under good growth conditions for banana with only a delay of bunch emergence and maturation.

2.2.3 Transmission and geographic distribution

BSV is not mechanically transmitted and field spread occurs by semi-persistent mealybug-mediated transmission and by the use of infected planting material, such as suckers. Field observations suggest that virus spread is slow (Daniells et al., 2001) with no difference in isolate transmission observed between the two main vector species *Planococcus citri* and *Pseudococcus* spp. (Dahal et al., 2000; Kubiriba et al., 2001; Matile-Ferrero & Williams, 1995). Other species reported to transmit BSV are *Dysmicoccus* spp. in West Africa and South America, *Planococcus musa* in Nigeria, *Ferrisia virgata* (striped mealybug) in India (Selvarajan et al., 2006), *D. brevipes* and *P. ficus* (Meyer, Kasdorf, Nel, & Pietersen, 2008) and *Paracoccus burnerae* (Muturi, Wachira, Karanja, Wambulwa, & Macharia, 2013) in South Africa. Transmission assays using *P. burnerae* demonstrated the inability of the vector to acquire and transmit the virus during hot conditions (24–30 °C).

However, under cool conditions (9–20 °C), a minimum of 6 h of feeding time was necessary for *P. burnerae* instars to become viruliferous (Meyer et al., 2008).

BSV is now reported to occur in over 43 countries of Africa, Asia, Australia, Europe, Oceania, and tropical America (Diekmann & Putter, 1996) (Fig. 1). However, the epidemiology of BSV remains surprisingly unclear and the role of eBSV still needs to be clarified. Iskra-Caruana, Chabannes, Duroy, and Muller (2014) proposed an evolutionary history based on both field studies and molecular epidemiology to explain the current situation, providing insights into the ecology and evolution of BSVs and *Musa* sp.

2.2.4 Virus diversity

The genus *Badnavirus* is both the most complex and the most diversified genus within the family *Caulimoviridae*, with at least three major clades (Harper et al., 2004, 2005; King et al., 2012). A final phylogeny of BSVs has been established to clarify whether partial sequences distributed over the three main clades of the genus *Badnavirus* correspond to episomal viruses with or without an endogenous counterpart (Gayral & Iskra-Caruana, 2009; Iskra-Caruana, Chabannes, Duroy, et al., 2014; Iskra-Caruana, Duroy, Chabannes, & Muller, 2014) (Fig. 4). Clades 1 and 3 are dedicated to BSV and Clade 2 gathers all *Musa* endogenous badnavirus sequences, with no episomal counterpart reported so far Chabannes et al., in preparation. Clade 1 also groups the four BSV species having an eBSV counterpart in the B genome (BSOLV, BSGFV, BSIMV, and BSMYV); Clade 3 groups only BSV species of Uganda (Fig. 4).

2.2.5 Diagnostics

Detection of BSV was first achieved using serological approaches although detection is complex due to the occurrence of a wide degree of diversity among virus isolates, some of which are serologically unrelated (Lockhart & Olszewski, 1993; Ndowora, 1998). For this purpose, a polyvalent polyclonal antiserum against BSV and *Sugarcane bacilliform virus* spp. has been developed (B.E.L. Lockhart, unpublished). It is capable of detecting most known BSV isolates but fails to detect some endemic BSV species (BSUIV, BSUJV, BSULV, BSUMV) from Uganda belonging to Clade 3 (Chabannes et al., unpublished). Meyer (2005) developed two other polyvalent BSV polyclonal antisera from 20 BSV isolates present in South Africa; these gave no background with plants but one failed to detect BSMYV

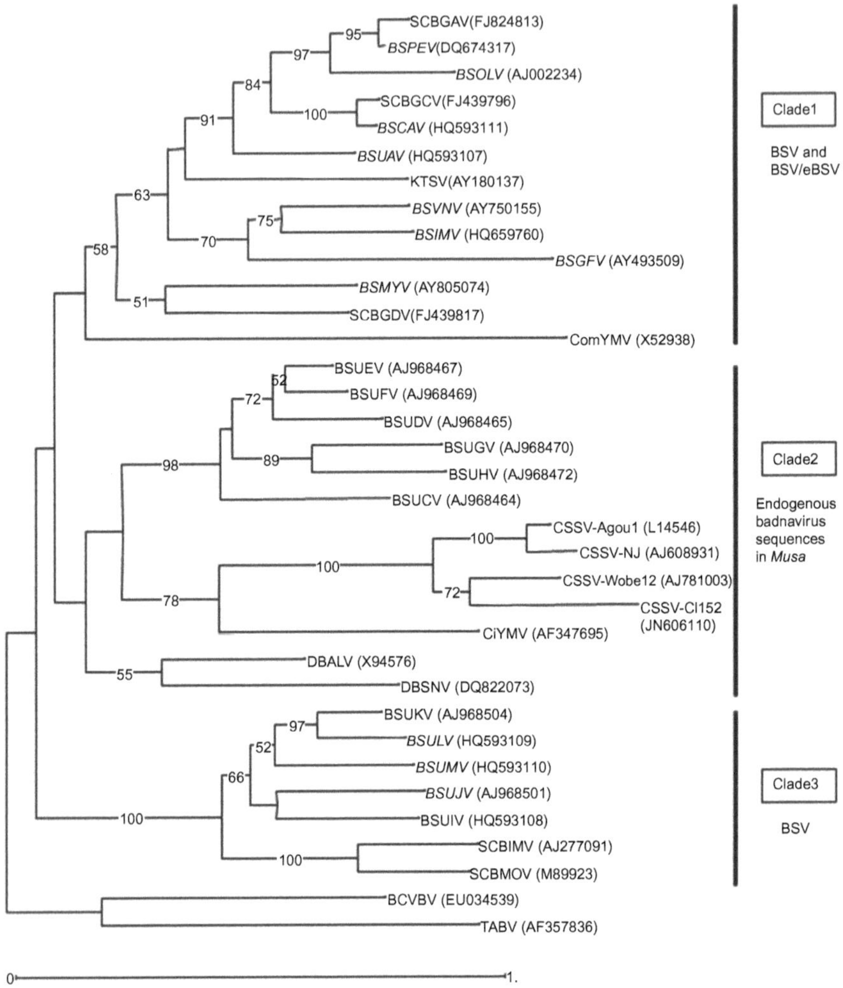

Figure 4 Maximum likelihood phylogeny of badnavirus sequences based on alignment of a 540-bp fragment of the RT/RNase H viral region. Bootstrap values of 500 replicates are given when >50%. *Taro bacilliform virus* (TABV) and *Bougainvillea spectabilis chlorotic vein-banding virus* (BCVBV) are given as outgroups. Viral sequences isolated from *Musa* are in bold. BSV species where a full-length sequence is available are in italic. The scale bar shows the number of substitutions per base. The GenBank accession numbers of sequences are given in parenthesis. *Cacao swollen shoot virus*-CSSV-Agou1, *Commelina yellow mottle virus*-ComYMV, *Citrus yellow mosaic virus*-CiYMV, *Dioscorea bacilliform AL virus*-DBALV, *Dioscorea bacilliform SN virus*-DBSNV, *Kalanchoe top-spotting virus*-KTSV, *Sugarcane bacilliform MO virus*-SCBMOV, *Sugarcane bacilliform IM virus*-SCBIMV, *Sugarcane bacilliform Guadeloupe A virus*-SCBGAV, *Sugarcane bacilliform Guadeloupe C virus*-SCBGCV, and *Sugarcane bacilliform Guadeloupe D virus*-SCBGDV. *Adapted from Iskra-Caruana, Duroy, (2014) and Iskra-Caruana, Chabannes, et al. (2014).* (See the color plate.)

(M.L. Iskra-Caruana, unpublished). Agindotan, Thottappilly, Uwaifo, and Winter (2003) reported high titered monoclonal antibodies for the detection of all isolates of BSV. Chen, Rao, Ruan, Liu, and Li (2013) reported the production of antibodies to the recombinant coat protein domain of BSV-Guangdon isolate which was effective at 1:600 to 1:6400 (v/v) dilutions in detecting BSV in infected plants. Serological detection of BSV can be achieved by different techniques, such as immunosorbent electron microscopy (ISEM) (Ndowora, 1998); DAS-ELISA (Meyer, 2005; Thottappilly, Dahal, & Lockhart, 1998) or Immuno-capture PCR (IC-PCR) (Harper, Osuji, Heslop-Harrison, & Hull, 1999; Le Provost, Iskra-Caruana, Acina, & Teycheney, 2006). Detailed and updated protocols for these three techniques can be found in the *Musa*Net Technical Guidelines (2014) for the safe movement of *Musa* germplasm. IC-PCR is more sensitive than immuno-electron microscopy (IEM) for detecting typical BSV; IEM sensitivity was proved to be similar to TAS-ELISA by sap dilution end point analyses (Agindotan et al., 2006). ISEM requires a partially purified plant extract and an electron microscope but is sensitive because BSV particles can be detected even in asymptomatic leaf tissue (B.E.L. Lockhart, unpublished). ELISA kits are now commercially available for BSV (www.sediag.com). Although both ELISA and ISEM techniques can detect most BSVs, species identification is not possible.

Diagnosis by PCR does not distinguish between infections (episomal) and eBSV. In addition, all *Musa* spp. carry partial or "dead" (truncated or mutated) integrated badnaviral sequences in their genomes that are incapable of generating infectious particles. Thus, PCR tests on any total genomic DNA extracted from a plant using moderate degenerated BSV primers may result in BSV-positive reactions. IC-PCR allows the detection of episomal particles after the binding of BSV to the polyvalent polyclonal antiserum and also identifies viral species using specific BSVs primers. This molecular approach is now the most commonly used diagnostic test for BSV. However, the presence of eBSV in *Musa* genomes could still interfere with PCR-episomal detection and leads to false positive results. To overcome this problem, Le Provost et al. (2006) used *Musa* sequence tagged microsatellite site primers developed by Lagoda et al. (1998) to detect genomic DNA contamination. These primers were used in multiplex with specific BSV species primers. Alternatively, a DNase I treatment step can be performed after the immune-capture stage to remove any plant DNA contamination (Chabannes et al., in preparation).

More recently, two new molecular techniques have been proposed as an alternative for the detection of BSV particles: the LAMP assay (Peng, Fan, et al., 2012) and RCA (James, Geijskes, Dale, & Harding, 2010, 2011). Both methods have the advantage of amplifying the target DNA without a thermal cycling instrument. RT multiplex PCR has been applied for the detection of BSV to avoid detection of eBSV sequences (Liu et al., 2012; Selvarajan et al., 2011). However, exclusive detection of episomal BSV by this method is difficult because eBSV sequences are known to be transcribed resulting in RNA transcripts that can be detected by RT-PCR leading to wrong assumptions.

2.2.6 Control

The main control method to limit BSV infection is by the production and multiplication of healthy banana plants. As mealybugs are slow moving; the spread of the disease is therefore easy to restrict by removing infected plants. Mass propagation is achieved either by *in vitro* plantlet multiplication, as used mainly for dessert banana and commercial plantations, or by the use of suckers. Vegetative multiplication from infected plants can significantly increase BSV incidence in the field. For instance, high BSV incidence in Uganda seems to be the result of the mass propagation of plantlets from symptomless BSV-infected plants. However, in vitro multiplication remains the most appropriate way for genotypes free of eBSVs to supply large quantities of safe *Musa* planting material without pathogens, such as bacteria, fungi, and nematodes. Only a few reports have been published regarding therapy procedures to eliminate BSV from infected plant tissues and the regeneration of BSV-free planting material. Helliot et al. (2002) developed a cryopreservation followed by apical meristem culture that significantly reduces virus titers. The same authors also demonstrated the use of antiviral, compounds adefovir, tenofovir, and 9-(2-phosphonomethoxyethyl)-2,6-diaminopurine (PMEDAP), to eradicate episomal forms (Helliot et al., 2003).

Efforts to breed cultivars resistant to streak disease are limited. Many A-genome *Musa* cultivars were found to be susceptible to BSV infection, whereas several BB genome cultivars are known to be resistant. Interestingly, natural interspecific hybrids showed high tolerance to BSV expression from eBSV (M.-L. Iskra-Caruana, unpublished). Dahal et al. (2000) evaluated 36 genotypes with diverse genetic backgrounds, including 14 tetraploid plantain (TMPx) and banana (TMBx) hybrids developed by IITA, and hybrids from Fundacio´ n Hondureo‘ a de Investigacio´ n Agri‘cola

(FHIA), for their response to BSV expression in Nigeria. Some of the evaluated hybrids (e.g., TMPx 548-9, TMPx 2637-49, TMPx 7002-1, and FHIA 21) despite being infected gave near normal yields and were regarded as disease tolerant. Developing transgenic BSV resistance has so far not succeeded (Tripathi, 2003).

The main problem, encountered for over 20 years, concerns the presence of infectious eBSV for at least three BSV species (BSOLV, BSIMV, and BSGFV) in all *M. balbisiana* genomes (Chabannes et al., 2013; Duroy, 2012; Gayral et al., 2008, 2010) that are able to release infectious particles spontaneously in interspecific contexts (AB, AAB, or AAAB) when hybrids are exposed to stresses such as wounding, heat shock, or *in vitro* culture. Therefore, eBSV became the main constraint to controlling BSV and for safe exchange of *Musa* germplasm. Recently, three infectious eBSV species were characterized in the seedy diploid *M. balbisiana* genitor Pisang Klutuk Wulung (PKW) (Chabannes et al., 2013; Gayral et al., 2008). This information was used to develop molecular markers to distinguish both eBSV (haplotypes) and alleles (infectious vs. non-infectious for BSOLV and BSGFV) (Fig. 5). These eBSV diagnostic markers have been used already to genotype the Bioversity International *M. balbisiana* germplasm collection. They have also greatly helped CIRAD breeders to produce by self-pollination and doubled haploid lines, *M. balbisiana* progenitors devoid of infectious eBSOLV and eBSGFV by segregation of eBSOLV and eBSGFV alleles (Pichaut et al., 2013; Umber et al., in preparation). Introduction of these improved diploid *M. balbisiana* plants which harbor only noninfectious eBSVs to the *Musa* breeding program can abolish the risk linked to the presence of eBSVs. Very recently, using such markers, Noumbissié (2014) demonstrated the possibility of producing an eBSV-free (i.e., without infectious and non-infectious eBSV) triploid hybrid (AAB) using a conventional ($4\times/2\times$) breeding strategy. These molecular tools, which are free for use by the worldwide *Musa* community, should in the next few years help to control the dissemination of BSV through the distribution of newly created interspecific banana hybrids.

2.3. Banana bract mosaic

2.3.1 Distribution and biology

Banana bract mosaic disease, caused by the BBrMV, was first reported in 1979 in the Philippines (Magnaye & Espino, 1990; Rodoni, Ahlawat, Varma, Dale, & Harding, 1997). Occurrence of the virus was discovered

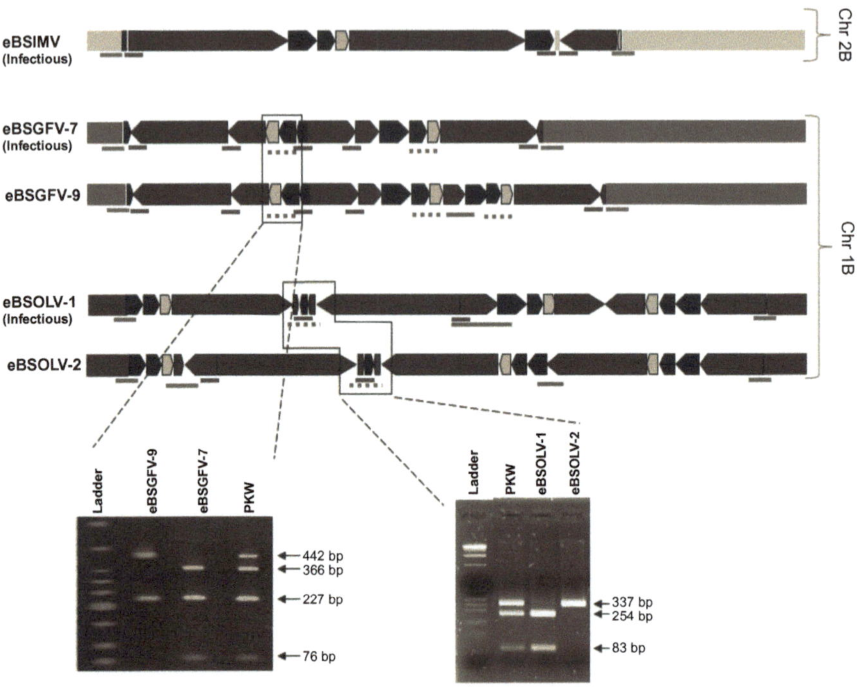

Figure 5 eBSV genotyping markers for eBSIMV, eBSGFV, and eBSOLV in Pisang Klutuk Wulung (PKW). Banana genomic sequences of chromosome 1 and 2 are in green. Dark blue, light blue, red, and black arrowed boxes represent ORF1, ORF2, ORF3, and the intergenic region (IG) of the BSV genome. Arrows indicate the orientation of the viral fragment. PCR fragments are indicated by color lines. Orange, purple, and pink, lines correspond to *Musa*-eBSV junctions, internal eBSV rearrangements, and eBSV allele-specific markers. Two kinds of markers have been developed to specifically discriminate alleles of each eBSV species: pink -full lines represent classical PCR markers whereas pink -dotted lines represent dCAPS markers. Agarose gel images below correspond to eBSGFV (left) and eBSOLV (right) allelic patterns obtained with dCAPS markers. PKW exhibits both alleles for each eBSV species. (See the color plate.)

in a few other countries in Asia and the South Pacific, including India, Samoa, Sri Lanka, Thailand, and Vietnam (Diekmann & Putter, 1996; Rodoni et al., 1997; Rodoni, Dale, & Harding, 1999). In Latin America, BBrMV occurrence was first reported in Colombia (Alarcon et al., 2006, cited in Quito-Avila et al., 2013). Recently, occurrence of this virus in a commercial farm was recorded in Ecuador (Quito-Avila et al., 2013). In Hawaii (USA), BBrMV was detected in ornamental ginger plants (*Alpinia purpurata*) but not in *Musa* (Wang, Sether, Melzer, Borth, & Hu, 2010).

BBrMV causes characteristic spindle-shaped, purplish streaks on bracts, pseudostems, midribs, peduncles, and even fruits (Rodoni et al., 1997; Selvarajan & Jeyabaskaran, 2006; Thomas, Geering, Gambley, Kessling, & White, 1997). Bunches from infected plants unusually contain a long or very short peduncle and in some cultivars, such as Nendran, the leaves appear as "traveler's palm" plant (Balakrishnan, Gokulapalan, & Paul, 1996). Necrotic streaks on fruits, leaves, pseudostems, and midribs have also been recorded (Selvarajan & Jeyabaskaran, 2006). Yield losses of between 30% and 70% have been recorded in India and the Philippines (Cherian, Menon, Suma, Nair, & Sudheesh, 2002; Magnaye & Espino, 1990; Thangavelu et al., 2000).

2.3.2 Host range and transmission

The main host of BBrMV is *Musa* spp., including abaca, widely grown in the Philippines (Sharman, Gambley, Oloteo, Abgona, & Thomas, 2000). Other natural hosts include small cardamom in India (Siljo, Bhat, Biju, & Venugopal, 2011) and flowering ginger, *A. purpurata*, a popular cut flower and tropical landscape plant in Hawaii (Wang et al., 2010). BBrMV is primarily transmitted through infected planting material. The virus is transmitted nonpersistently by several aphid species viz., *P. nigronervosa*, *Rhopalosiphum maidis*, *Aphis gossypii* (Magnaye & Espino, 1990; Munez, 1992), and *A. craccivora* (Selvarajan et al., 2006).

2.3.3 Virus diversity

BBrMV particles are flexuous and filamentous of ca. 750 × 11 nm with a single-stranded positive-sense RNA as its genome (Bateson & Dale, 1995; Thomas et al., 1997). The RNA genome is 9711 nucleotide long excluding the 3′-terminal poly A tail and contains a typical large ORF of 9378 nucleotides coding for a poly protein of 3125 amino acids, with 9 protease cleavage sites, potentially yielding 10 matured functional proteins that have motifs conserved among homologous proteins of other potyviruses (Balasubramanian & Selvarajan, 2012, Ha et al., 2008; Rodoni et al., 1997). The whole genome of BBrMV-TRY (India) and BBrMV-PHI (the Philippines) had 94% nucleotide sequence identity and 88–98% amino acid sequence identities (Balasubramanian & Selvarajan, 2012). The coat protein-encoding gene and 3′-untranslated region of BBrMV isolates from India, the Philippines, Samoa, and Vietnam, are 95.4–99% identical (Rodoni et al., 1999). The genetic diversity and recombination analysis in the CP gene of 49 isolates of BBrMV revealed a greater variation among them and two of the isolates from Tamil Nadu were distinct with 18–21% divergence at nucleotide level and 12–20% divergence at amino acid level (Balasubramanian & Selvarajan, 2014). Eleven putative recombinants were also detected; whereas the partial gene sequences from Hawaii and Ecuador are 99% identical, compared with the corresponding fragment of a BBrMV isolate from the Philippines (Quito-Avila et al., 2013; Wang et al., 2010).

2.3.4 Diagnosis

A range of diagnostic methods based on ELISA (Espino, Exconde, Zipagan, & Espino, 1990, Espino, Exconde, Zipagan, Maroon, & Espino, 1989; Thomas et al., 1997), RT-PCR (Bateson & Dale, 1995; Ha et al., 2008; Kiranmai et al., 2005; Thomas et al., 1997, Exconde, Zipagan, Maroon, & Espino,), and IC-RT-PCR has been used for the detection of BBrMV singly (Iskra-Caruana, Galzi, & Laboureau, 2008) or in combination with other banana infecting viruses in a multiplex diagnostic assay (Liu et al., 2012; Selvarajan, Balasubramanian, Dayakar, Sathaimoorthy, & Ahlawat, 2004; Sharman, Thomas, et al., 2000).

2.3.5 Control

Sources of resistance have not been reported for BBrMV. Standard control measures include the removal and destruction of affected plants along with the rhizome and the use of virus-free sources as planting material. Application of fertilizer was reported to mitigate the symptoms of BBrMV in certain commercial banana cultivars (Selvarajan et al., 2009). Recently, abaca cvs. Tinawagan Pula and Tangongon with putative resistance to BBrMV were developed through *in vitro* mutagenesis by gamma-irradiation (Dizon et al., 2012). Less than 2% of over 3000 plants irradiated were found to be virus negative even 3 months after inoculation. These promising lines have yet to be field tested or evaluated for other agronomic traits. Nonetheless, this approach demonstrated the promise of mutagenesis in developing virus resistant cultivars.

3. MINOR VIRUS DISEASES

3.1. Abaca bunchy top

Abaca bunchy top is caused by ABTV, a distinct member in the genus *Babuvirus*, family *Nanoviridae* (Sharman et al., 2008). The virus has been recorded from Malaysia and the Philippines (Ocfemia, 1930). ABTV causes vein clearing flecks and narrow, brittle leaves with chlorotic upturned margins in abaca; in Cavendish banana, symptoms are indistinguishable from those of BBTV (Diekmann & Putter, 1996; Su, Tsao, Wu, & Hung, 2003). The virus is transmitted through vegetative planting material and tissue cultured plants derived from infected sources. There are no reports on vector transmission. However, since the virus is a member of the genus *Babuvirus*, it is conceivable that the banana aphid, *P. nigronervosa*, may able to transmit ABTV in a persistent and circulative manner similar to BBTV. PCR assays have been established for ABTV detection (Sharman et al., 2008). Economic significance of this virus is not known. Since virus symptoms are similar to those of BBTV infection in banana, it may have a similar impact.

3.2. Abaca mosaic

Abaca mosaic is caused by SCMV-Ab, a distinct strain of SCMV (genus *Potyvirus*) (Eloja & Tinsley, 1963; Thomas et al., 1997). This virus has so far been recognized only in the Philippines. It induces spindle-shaped yellow chlorotic streaks on leaves, petioles, and midribs. The disease affects fiber yield and quality, and is a significant constraint to abaca production in the Philippines. Natural hosts recognized so far include *M. textilis*, *Maranta arundinacea*, and *Canna indica*. The virus is transmitted by vegetative propagation and tissue culture, also by aphids (mainly *R. maidis* and *A. gossypii*) in a nonpersistent manner (Diekmann & Putter, 1996). ELISA and RT-PCR assays have been established for virus detection (Gambley, Thomas, Magnaye, & Herradura, 2004). There are no specific recommendations for disease control, other than the use of clean planting material.

3.3. Banana mosaic

Banana mosaic, also known as infectious chlorosis, caused by CMV is one of the common viral diseases affecting banana and plantain worldwide. It causes variable symptoms from mild chlorosis to severe chlorotic streaks that are similar to the symptoms of BSV infection. Symptoms are known to fluctuate during the growing season depending on the temperature and rainfall. The virus has a wide host range, infecting over 900 species in almost every country with extensive agriculture. Most isolates of CMV from banana (CMV-B) have been identified to belong to subgroup I. Most common CMV strains do not produce severe symptoms or cause significant damage on banana. However, some severe strains cause severe necrosis in pseudostems (heart-rot) and even plant death (Bouhida & Lockhart, 1990).

The virus is transmitted through planting material and nonpersistently by several aphids, such as *A. gossypii*, *A. craccivora*, *R. maidis*, *R. prunifolium*, and *Myzus persicae* (Rao, 1980). CMV can be detected by several methods, such as mechanical inoculation to herbaceous plants, ELISA, NASH, RT-PCR methods, and electron microscopy (Hu, Li, Barry, Wang, & Jordan, 1995; Kiranmai, Sreenivasulu, & Nayudu, 1996). Kouassi, Wendy, Boonham, and Smith (2010) developed a quantitative real-time PCR (qRT-PCR) for screening banana planting material for CMV infection in Côte d'Ivoire. A RT-LAMP assay has also been established for CMV detection in banana (Peng, Shi, Xia, Huang, & Fan, 2012). Disease control is primarily through use of clean planting material. *In vitro* methods for the elimination of CMV from tissues are in common use (Helliot et al., 2002).

3.4. Banana mild mosaic

The disease is caused by BanMMV, an unassigned virus in the *Betaflexiviridae* (King et al., 2012). The virus has flexuous filamentous particles of about 580 nm in length, with a coat protein of ca. 26.8 kDa, and a single-stranded RNA genome (Gambley & Thomas, 2001). The virus often occurs in symptomless plants of *Musa* spp. and apparently has a worldwide distribution. It is often detected in mixed infection with BSV and BBrMV (Iskra-Caruana et al., 2008) and is reported to cause necrotic streaks in the presence of CMV infection in Guadeloupe (Iskra & Galzi, 1998). The virus spreads through the vegetative propagation of planting materials. Although a vector has not been identified, horizontal transmission through some unknown mechanism has been conjectured, based on the high heterogeneity of the viral genome and the temporal increase in virus incidence within a field (Teycheney, Laboureau, Iskra-Caruana, & Candresse, 2005). Polyclonal antisera to BanMMV have been developed which are used for virus detection by IC-RT-PCR and ISEM (Teycheney, Acina, Lockhart, & Candresse, 2007; Teycheney, Laboureau, et al., 2005). The virus is controlled mainly through the production and distribution of clean planting material. Meristem cultures and chemotherapy, especially with ribavirin, has been shown to eliminate the virus from TC plants (Busogoro, Vandermolen, Masquelier, & Jijakli, 2006)

3.5. Banana virus X

BVX is an unassigned member of the family *Betaflexiviridae* and has been detected mostly in symptomless *Musa* plants. The occurrence of the virus was first reported from Guadeloupe and it is known to be widespread in *Musa* germplasm collections, thus suggesting wide distribution (Teycheney, Marais, Svanella-Dumas, Dulucq, & Candresse, 2005). A nested PCR-assay has been established for virus detection (Teycheney et al., 2007). BVX is spread along with the planting material and no insect vector has been identified.

4. CONCLUSIONS

Banana and plantain are high priority crops in the developing countries because of their contribution to dietary energy, nutrition, and income for the millions of resource-poor farmers who grow over 85% of the world's banana. During the last decade, banana and plantain production around the world increased by 27% (FAOStat, 2012), indicating the high demand. The fruit are particularly valued in resource-poor agriculture because they yield, irrespective of the seasons. Viral, bacterial, and fungal pathogens and nematodes pose a particular concern as they can be moved through planting materials between fields and across borders. Virus disease not only causes yield reductions but is also a major constraint to the exchange of germplasm.

So far, eight viruses have been reported to infect *Musa* spp. Out of these, BBTV and BBrMV and the species complex, BSVs, are significant threats to banana and plantain. Although reports are from a few countries, it is widely believed that BSVs are distributed worldwide, especially in areas where banana and plantain with B genome composition are grown. All these viruses are spread around the world mainly through germplasm. Occurrence of BBTV has expanded markedly during the past 15 years, particularly in SSA, because many countries do not adhere to safe procedures for germplasm exchange.

Recommendations given in the Technical Guidelines for the Safe Movement of *Musa* Germplasm published by the FAO and the International Board for Plant Genetic Resources (IBPGR) in 1996 (Diekmann & Putter, 1996) provide useful information on best practices for the exchange of germplasm and the procedures for disease indexing available at that time. The International Transit Centre (ITC), established in 1985 at the Katholieke Universiteit Leuven in Belgium, has been serving as intermediary to accept germplasm of unknown health status to generate virus-free planting material for international distribution after checking at one of the two virus indexing centers based at QDPI (Australia) and at CIRAD (France). Despite the existence of such a facility, because of poor awareness, germplasm continues to be moved across porous borders, especially in SSA, as part of the traditional exchanges or for sale in local markets.

Many countries in Asia, Oceania, and Latin America have devised the large-scale production of virus-free planting material as a measure to contain virus spread and also to prevent yield losses in endemic areas. Although data on the effectiveness of this model in reducing disease impact are limited to Australia, similar measures are slowly gaining popularity because of the widespread epidemics of banana bacterial wilt disease in East Africa.

Efforts are also being made to (i) enhance capacity for the production and supply of virus-free planting material of farmer-preferred cultivars in BBTD-affected countries; (ii) increase vigilance in regions bordering BBTD-affected areas for pre-emptive action to be taken against the introduction of infected planting materials; (iii) demonstrate eradication programs and their benefits; (iv) improve monitoring capacity; and (v) deploy rapid eradication responses in mitigating the spread of the disease. Efforts on the development of host plant resistance through PDR are currently receiving high priority because of its potential to incorporate virus resistance in any cultivar.

REFERENCES

Abustan, S. A.-F. A. (2012). *IPB-UPLB rallies against banana disease, gets PCAARRD support.* http://www.pcaarrd.dost.gov.ph/home/ssentinel/index.php?option=com_content&view=article&id=2066%3Aipb-uplb-rallies-against-banana-disease-gets-pcaarrd-support&Itemid=41 (accessed on 24 Mar 14).

Adegbola, R. A., Ayodeji, O., Awosusi, O. O., Atiri, G. I., & Kumar, P. L. (2013). First report of banana bunchy top disease caused by *banana bunchy top virus* in banana and plantain (*Musa* spp.) in Nigeria. *Plant Disease, 97*, 290.

Agindotan, B. O., Thottappilly, G., Uwaifo, A., & Winter, S. (2003). Production of monoclonal and polyclonal antibodies against a Nigerian isolate of *banana streak virus. African Journal of Biotechnology, 2*, 171–178.

Agindotan, B. O., Winter, S., Lesemann, D., Uwaifo, A., Mignouna, J., Hughes,, J. d'A., et al. (2006). Diversity of banana streak-inducing viruses in Nigeria and Ghana: Twice as many sources detected by immunoelectron microscopy (IEM) than by TAS-ELISA or IC-PCR. *African Journal of Biotechnology, 5*, 1194–1203.

Allen, R. N. (1978). Spread of bunchy top disease in established banana plantations. *Australian Journal of Agricultural Research, 29*, 1223–1233.

Allen, R. N. (1987). Further studies on epidemiological factors influencing control of banana bunchy top disease and evaluation of control measures by computer simulation. *Australian Journal of Agricultural Research, 38*, 373–382.

Almedia, R. P., Bennett, G. M., Anhalt, M. D., Tsai, C.-W., & Grady, P. O. (2009). Spread of an introduced vector-borne banana virus in Hawaii. *Molecular Ecology, 18*, 136–146.

Amin, I., Qazi, J., Mansoor, S., Ilyas, M., & Briddon, R. W. (2008). Molecular characterization of *banana bunchy top virus* (BBTV) from Pakistan. *Virus Genes, 36*, 191–198.

Anhalt, M. D., & Almeida, R. P. P. (2008). Effect of temperature, vector life stage, and plant access period on transmission of banana bunchy top virus to banana. *Phytopathology, 98*, 743–748.

Balakrishnan, S., Gokulapalan, C., & Paul, S. (1996). A widespread banana malady in Kerala, India. *Infomusa, 5*, 28–29.

Balasubramanian, V., & Selvarajan, R. (2012). Complete genome sequence of a banana bract mosaic virus isolate infecting the French plantain cv. Nendran in India. *Archives of Virology, 157*, 397–400.

Balasubramanian, V., & Selvarajan, R. (2014). Genetic diversity and recombination analysis in the coat protein gene of banana bract mosaic virus. *Virus Genes, 48*, 509–517. http://dx.doi.org/10.1007/s11262-014-1056-x.

Banerjee, A., Roy, S., Beherea, G. T., Roy, S. S., Dutta, S. K., & Ngachana, S. V. (2014). Identification and characterization of a distinct banana bunchy top virus isolate of Pacific-Indian Oceans group from North-East India. *Virus Research, 183*, 41–49.

Bateson, M. F., & Dale, J. L. (1995). *Banana bract mosaic virus*: Characterization using potyvirus specific degenerate PCR primers. *Archives of Virology, 140*, 515–527.

Beetham, P. R., Harding, R. M., & Dale, J. L. (1999). *Banana bunchy top virus* DNA-2 to 6 are monocistronic. *Archives of Virology, 144*, 89–105.

Blomme, G., Ploetz, R., Jones, D., De Langhe, E., Price, N., Gold, C., et al. (2013). A historical overview of the appearance and spread of *Musa* pests and pathogens on the African continent: Highlighting the importance of clean *Musa* planting materials and quarantine measures. *The Annals of Applied Biology, 162*, 4–26.

Borth, W., Perez, E., Cheah, K., Chen, Y., Xie, W. S., Gaskill, D., et al. (2011). Transgenic banana plants resistant to banana bunchy top virus infection. *Acta Horticulturae, 897*, 449–457.

Bouhida, M., & Lockhart, B. E. L. (1990). Increase in importance of cucumber mosaic virus infection in greenhouse grown bananas in Morocco. *Phytopathology, 80*, 981.

Bressan, A., & Watanabe, S. (2011). Immunofluorescence localization of banana bunchy top virus (family Nanoviridae) within the aphid vector, *Pentalonia nigronervosa*, suggests a virus tropism distinct from aphid-transmitted luteoviruses. *Virus Research, 155*, 520–525.

Burns, T. M., Harding, R. M., & Dale, J. L. (1995). The genome organization of banana bunchy top virus: Analysis of six ssDNA components. *The Journal of General Virology, 76*, 1471–1482.

Busogoro, J.-P., Vandermolen, M., Masquelier, L., & Jijakli, H. (2006). Development of a chemotherapy protocol to sanitise banana genotypes infected by banana mild mosaic virus (BanMMV). In *XVII Reuniao Internacional da Associação para a Cooperação nas Pesquisas sobre Banana no Caribe e na América Tropical (ACORBAT)* (pp. 752–756), Brazil: Joinville-Sanla Catarina.

Chabannes, M., Baurens, F. C., Duroy, P.-O., Bocs, S., Vernerey, M.-S., Rodier-Goud, M., et al. (2013). Three infectious viral species lying in wait in the banana genome. *Journal of Virology, 87*, 8624–8637.

Chabannes, M., & Iskra-Caruana, M.-L. (2013). Endogenous pararetroviruses—A reservoir of virus infection in plants. *Current Opinion in Virology, 3*, 615–620.

Chen, Y., & Hu, X. (2013). High-throughput detection of banana bunchy top virus in banana plants and aphids using real-time TaqMan®PCR. *Journal of Virological Methods, 193*, 177–183.

Chen, X., Rao, X.-Q., Ruan, X. L., Liu, F.-X., & Li, H.-P. (2013). Prokaryotic expression and antiserum preparation of functional domain of *banana streak virus* coat protein gene. *Acta Horticulturae Sinica, 40*, 2401–2408.

Cherian, A. K., Menon, R., Suma, A., Nair, S., & Sudheesh, M. V. (2002). Impact of banana bract mosaic diseases on the yield of commercial banana varieties of Kerala. In *Global conference on banana and plantain, Bangalore, 28–31 October, 2002*, Abstract pp. 155.

Conant, P. (1992). Banana bunchy top disease, a new threat to banana cultivation in Hawaii. *Proceedings of the Hawaiian Entomological Society, 31*, 91–95.

Cook, D. C., Liu, S., Edwards, J., Villalta, O. N., Aurambout, J.-P., Kriticos, D. J., et al. (2012). Predicting the benefits of banana bunchy top virus exclusion from commercial plantations in Australia. *PLoS One*, 7, e42391.

Cote, F. X., Galzi, S., Folliot, M., Lamagnere, Y., Teycheney, P. Y., & Iskra-Caruana, M.-L. (2010). Micropropagation by tissue culture triggers differential expression of infectious endogenous *banana streak virus* sequences (eBSV) present in the B genome of natural and synthetic interspecific banana plantains. *Molecular Plant Pathology, 11*, 137–144.

Dahal, G., Ortiz, R., Tenkouano, A., Hughes, J. d'A., Thottappilly, G., Vuylsteke, D., et al. (2000). Relationship between natural occurrence of *banana streak badnavirus* and symptom expression, relative concentration of viral antigen, and yield characteristics of some micropropagated *Musa* spp. *Plant Pathology, 49*, 68–79.

Dale, J. L. (1987). Banana bunchy top: An economically important tropical plant virus disease. *Advances in Virus Research, 33*, 301–325.

Damasco, O. P., Estrella, J. B., Caymo, L. C., Dizon, T. O., Rabara, R. C., de la Cruz, F. S., Jr., et al. (2006). *Banana bunchy top virus* (BBTV) resistance in cultivar 'Lakatan' developed by gamma irradiation of shoot tips. *Philippine Journal of Crop Science, 31*, 21–34.

Daniells, J. W., Geering, A. D. W., Bryde, N. J., & Thomas, J. E. (2001). The effect of *banana streak virus* on the growth and yield of dessert bananas in tropical Australia. *The Annals of Applied Biology, 139*, 51–60.

De Langhe, E., & De Maret, P. (1999). Tracking the banana: Its significance in early agriculture. In C. Gosden & J. Hather (Eds.), *The prehistory of food: Appetites for change* (pp. 377–396). New York, USA and London, UK: Routledge.

Diekmann, M., & Putter, C. A. J. (1996). FAO/IPGRI technical guidelines for the safe movement of germplasm. In C. A. J. Putter (Ed.), *Musa* (2nd ed., p. 28). Rome: Food and Agriculture Organization of the United Nations/International Plant Genetic Resources Institute.

Dizon, O. T., Damasco, O. P., Lobina, I. T., Pinili, M. S., Lalusin, A. G., & Natsuaki, K. T. (2012). Induction of putative resistant lines of abaca (Musa textiles Nee) to banana bunchy top virus and banana bract mosaic virus through in vitro mutagenesis. *Journal of ISSAAS, 18*, 87–99.

Drew, R. A., Moisander, J. A., & Smith, M. K. (1989). The transmission of banana bunchy-top virus in micropropagated bananas. *Plant Cell Tissue and Organ Culture, 16*, 187–193.

Duroy, P.-O. (2012). *Quels sont les enjeux au cours de l'évolution du bananier (Musa sp.) qui ont conduit au maintien de séquences virales de Banana streak virus dans son génome?* Thèse université Montpellier II, Ecole doctorale SIBAGHE, 260.

Elayabalan, S., Selvarajan, R., & Kumar, K. K. (2013). Development of *Agrobacterium*-mediated transformation of highly valued hill banana cultivar Virupakshi (AAB) for resistance to BBTV disease. *World Journal of Microbiology and Biotechnology, 29*, 589–596.

Eloja, A. L., & Tinsley, T. W. (1963). Abaca mosaic virus and its relationship to sugarcane mosaic. *The Annals of Applied Biology, 51*, 253–258.

Espino, T. M., Exconde, S. B., Zipagan, F. B., & Espino, R. R. C. (1990). Banana bract mosaic, a new disease of banana II. Isolation and purification for monoclonal antibody production. *The Philippine Agriculture, 73*, 61–68.

Espino, T. M., Exconde, S. B., Zipagan, F. B., Maroon, M. C. M., & Espino, R. R. C. (1989). Production of monoclonal antibodies for diagnosis of *banana bunchy top virus*. *The Philippine Agriculture*, 72, 473–485.

Espino, R. C., Magnaye, L. V., Johns, A. P., & Juanillo, C. (1993). Evaluation of Philippine banana cultivars for resistance to bunchy-top and fusarium wilt. In R. V. Valmayor, S. C. Hwang, R. Ploetz, S. C. Lee, & N. V. Roa (Eds.), *Proceedings: International symposium on recent developments in banana cultivation technology, 14–18 December 1992, Chiuju,*

Pingtung, Taiwan (pp. 89–102), Los Banos, Philippines: Taiwan Banana Research Institute, INIBAP/ASPNET.

Fahmy, T. (1927). Plant diseases of Egypt. Minerals and Agriculture in Egypt. *Bulletin, 30.*

FAOStat. (2012). *Banana exports by region 2010-2012.* www.fao.org/economic/est/est-commodities/bananas/banana-expors/en/ (accessed on 20 Mar 14).

FAOStat. (2014). *FAO production statistics for banana and plantain 2012.* http://faostat.fao.org/site/567/DesktopDefault.aspx?PageID=567#ancor (accessed on 20 Mar 14).

Fist, S. (1970). The history of plant pathology in Australia. *Annual Review of Phytopathology, 8*, 13–36.

Fouré, E., & Manser, P. D. (1982). Note surl 'apparition au Gabon d'unegrâv emaladie virale des bananiers et plantains: le bunchy top. *Fruits, 37*, 409–414.

Galal, A. M. (2007). Use of polymerase chain reaction for detecting *banana bunchy top nano virus. Biotechnology, 6*, 53–56.

Gambley, C. F., & Thomas, J. E. (2001). Molecular characterisation of banana mild mosaic virus, a new filamentous virus in *Musa* spp. *Archives of Virology, 146*, 1369–1379.

Gambley, C. F., Thomas, J. E., Magnaye, L. V., & Herradura, L. (2004). Abacá mosaic virus: A distinct strain of sugarcane mosaic virus. *Australasian Plant Pathology, 33*, 475–484.

Gayral, P., & Iskra-Caruana, M.-L. (2009). Phylogeny of *banana streak virus* reveals recent and repetitive endogenization in the genome of its banana host (*Musa* sp.). *Journal of Molecular Evolution, 69*, 65–80.

Gayral, P., Blondin, L., Guidolin, O., Carreel, F., Hippolyte I., Perrier, X., et al. (2010). Evolution of endogenous sequences of *Banana streak virus*: what can we learn from banana (*Musa sp.*) evolution? *Journal of Virology*, 84, 7346–7359.

Gayral, P., Noa, C., Lescot, M., Lheureux, F., Lockhart, B., Matsumoto, T., et al. (2008). A single *banana streak virus* integration event in the banana genome as the origin of infectious endogenous pararetrovirus. *Journal of Virology, 82*, 6697–6710.

Geering, A. D. W., & Thomas, J. E. (1996). A comparison of four serological tests for the detection of *banana bunchy top virus* in banana. *Australian Journal of Agricultural Research, 47*, 403–412.

Geering, A. D. W., & Thomas, J. E. (1997). Search for alternative hosts of *banana bunchy top virus* in Australia. *Australasian Plant Pathology, 26*, 250–254.

Gondwe, W. T., Mweenebanda, B. M. T., Natha, E., & Mutale, P. (2007). Banana bunchy top disease in Mozambique and Zambia. *Infomusa, 16*, 38–39.

Ha, C., Coombs, S., Revil, P. A., Harding, R. M., Vu, M., & Dale, J. L. (2008). Design and application of two novel degenerate primer pairs for the detection and complete genomic characterization of potyviruses. *Archives of Virology, 153*, 25–36.

Hafner, G. J., Harding, R. M., & Dale, J. L. (1997). A DNA primer associated with *banana bunchy top virus. The Journal of General Virology, 78*, 1795–1799.

Hanna, R., Ngatat, S., Ndjab, M. R., Kuate, A. F., Doumtsop, A. R. F., & Kumar, P. L. (2013). Ecology of the banana aphid *Pentalonia nigronervose*, the vector of *banana bunchy top virus*. In *Program and book of abstracts, 12th international symposium on plant virus epidemiology* (p. 45), Arusha, Tanzania: IITA.

Harding, R. M., Burns, T. M., & Dale, J. L. (1991). Virus-like particles associated with banana bunchy top disease contains small single stranded DNA. *The Journal of General Virology, 72*, 225–230.

Harper, G., Ganesh, D., Thottappilly, G., & Hull, R. (1999). Detection of episomal *banana streak virus* by IC-PCR. *Journal of Virological Methods, 79*, 1–8.

Harper, G., Hart, D., Moult, S., & Hull, R. (2002). Detection of Banana streak virus in field samples of bananas from Uganda. *Annals of Applied Biology, 141*, 247–257.

Harper, G., Hart, D., Moult, S., & Hull, R. (2004). *Banana streak virus* is very diverse in Uganda. *Virus Research, 100*, 51–56.

Harper, G., Hart, D., Moult, S., Hull, R., Geering, A., & Thomas, J. (2005). The diversity of *Banana streak virus* in Uganda. *Archives of Virology, 150*, 2407–2420.

Harper, G., & Hull, R. (1998). Cloning and sequence analysis of *banana streak virus* DNA. *Virus Genes, 17*, 271–278.

Harper, G., Osuji, J. O., Heslop-Harrison, J. S., & Hull, R. (1999). Integration of *banana streak badnavirus* into the *Musa* genome: Molecular and cytogenetic evidence. *Virology, 255*, 207–213.

Helliot, B., Panis, B., Frison, E., De Clercq, E., Swennen, R., Lepoivre, P., et al. (2003). The acyclic nucleoside phosphanate analogues, adefovir, tenofovir and PMEDAP, efficiently eliminate *banana streak virus* from banana (*Musa* spp.). *Antiviral Research, 59*, 121–126.

Helliot, B., Panis, B., Poumay, Y., Swennen, R., Lepoivre, P., & Frison, E. (2002). Cryopreservation for the elimination of cucumber mosaic and banana streak viruses from banana (*Musa* spp.). *Plant Cell Reports, 20*, 1117–1122.

Heslop-Harrison, J. S., & Schwarzacher, T. (2007). Domestication, genomics and the future for banana. *Annals of Botany, 100*, 1073–1084.

Hooks, C. R. R., Fukuda, S., Perez, E. A., Manandhar, R., Wang, K.-H., Wright, M. G., et al. (2009). Aphid transmission of banana bunchy top virus to bananas after treatment with a bananacide. *Journal of Economic Entomology, 102*, 493–499.

Hooks, C. R. R., Manandhar, R., Perez, E. P., Wang, K.-H., & Almeida, R. P. P. (2009). Comparative susceptibility of two banana cultivars to *banana bunchy top virus* under laboratory and field environments. *Journal of Economic Entomology, 102*, 897–904.

Hooks, C. R. R., Wright, M. G., Kabasawa, D. S., Manandhar, R., & Almeida, R. P. P. (2008). Effect of banana bunchy top virus infection on morphology and growth characteristics of banana. *The Annals of Applied Biology, 153*, 1–9.

Horser, C. L., Karan, M., Harding, R. M., & Dale, J. L. (2001). Additional Rep-encoding DNAs associated with *banana bunchy top virus*. *Archives of Virology, 146*, 71–86.

Hu, J. S., Li, H. P., Barry, K., Wang, M., & Jordan, R. (1995). Comparison of dot blot, ELISA, and RT-PCR assays for detection of two *cucumber mosaic virus* isolates infecting banana in Hawaii. *Plant Disease, 79*, 902–906.

Hu, J. S., Wang, M., Sether, D., Xie, W., & Leonhardt, K. W. (1996). Use of polymerase chain reaction (PCR) to study transmission of *banana bunchy top virus* by the banana aphid (*Pentalonia nigronervosa*). *The Annals of Applied Biology, 128*, 55–64.

Iskra, M. L., & Galzi, S. (1998). Identification of uncharacterised filamentous viral particles on banana plants. *Acta Horticulturae, 490*, 323–335.

Iskra, M. L., Garnier, M., & Bove, J. M. (1989). Purification of banana bunchy top virus. *Fruits, 44*, 63–66.

Iskra-Caruana, M. L., Baurens, F. C., Gayral, P., & Chabannes, M. (2010). A four-partner plant-virus interaction: Enemies can also come from within. *Molecular Plant-Microbe Interactions, 23*, 1394–1402.

Iskra-Caruana, M.-L., Chabannes, M., Duroy, P.-O., & Muller, M. (2014). A possible scenario for the evolution of *banana streak virus* in banana. *Virus Research, 186*, 155–162.

Iskra-Caruana, M.-L., Duroy, P.-O., Chabannes, M., & Muller, M. (2014). The common evolutionary history of badnaviruses and banana. *Infection, Genetics and Evolution, 21*, 83–89.

Iskra-Caruana, M.-L., Galzi, S., & Laboureau, N. (2008). A reliable IC One-step RT-PCR method for the detection of BBrMV to ensure safe exchange of Musa germplasm. *Journal of Virological Methods, 153*, 223–231.

James, J. A., Geijskes, R. J., Dale, J. L., & Harding, R. M. (2010). Development of a novel rolling-circle amplification technique to detect *banana streak virus* which also discriminates between integrated and episomal virus sequences. *Plant Disease, 95*, 57–62.

James, A. P., Geijskes, R. J., Dale, J. L., & Harding, R. M. (2011). Molecular characterisation of six badnavirus species associated with leaf streak disease of banana in East Africa. *The Annals of Applied Biology, 158*, 346–353.

Jones, D. R. (2002). Risk of spread of banana diseases in international trade and germplasm exchange. In *Proceedings XV ACORBAT meeting 2002*, Colombia: Cartagena de Indias.

Jones, D. R. (2013). Emerging banana diseases—New threats from old problems. In *Proceedings: XX Reunião Internacional da Associação para a Cooperação em Pesquisa e Desenvolvimento Integral das Musáceas (Bananas e Plátanos), 9–13 September 2013, Fortaleza, CE.*

Jose, P. C. (1981). Reaction of different varieties of banana against bunchy top disease. *Agricultural Research Journal of Kerala, 9*, 96–97.

Karan, M., Harding, R. M., & Dale, J. L. (1994). Evidence for two groups of banana bunchy top virus isolates. *The Journal of General Virology, 75*, 3541–3546.

Kenyon, L., Brown, M., & Khonje, P. (1997). First report of *banana bunchy top virus* in Malawi. *Plant Disease, 81*, 1096.

Kesavamoorthy, R. C. (1980). Radical changes in ecosystem in the Pulney hills. In C. R. Muthukrishnan, Abdul Chaser, & J. B. M. Md (Eds.), *Proc. 13th national seminar on banana production technology* (pp. 23–28). Coimbatore: TNAU.

Khalid, S., Soomro, M. H., & Stover, R. H. (1993). First report of *banana bunchy top virus* in Pakistan. *Plant Disease, 77*, 101.

King, A. M. Q., Adams, M. J., Lefkowitz, E. J., & Carstens, E. B. (Eds.), (2012). *Virus taxonomy: Ninth report of the international committee on taxonomy of viruses*. NY, USA: Elsevier Academic Press.

Kiranmai, G., Kumar, P. L., Hema, M., Venkatramana, M., Kirshna-Prasadji, Mandheva-Rao, J., et al. (2005). Partial characterization of a potyvirus causing bract mosaic of banana in Andhra Pradesh. *Indian Journal of Virology, 16*, 7–11.

Kiranmai, G., Sreenivasulu, P., & Nayudu, M. V. (1996). Comparison of three different tests for detection of cucumber mosaic cucumovirus in banana (*Musa paradisica*). *Current Science, 71*, 764–767.

Kouassi, N. K., Wendy, M., Boonham, N., & Smith, J. (2010). Development of a diagnostic protocol for *Cucumber mosaic virus* for screening banana (*Musa* spp.) planting material in Ivory Coast. In T. Dubois, et al. (Eds.), *Proc. IC on banana and plantain in Africa*, Acta Hort. 879, ISHS 2010.

Kubiriba, J., Legg, J. P., Tushemereirwe, W., & Adipala, E. (2001). Vector transmission of *banana streak virus* in the screenhouse in Uganda. *The Annals of Applied Biology, 139*, 37–43.

Kumar, P. L., Ayodele, M., Oben, T. T., Mahungu, N. M., Beed, F., Coyne, D., et al. (2008). First report of banana bunchy top virus in banana and plantain (*Musa* spp.) in Angola. *Plant Pathology, 58*, 402.

Kumar, P. L., Hanna, R., Alabi, O. J., Soko, M. M., Oben, T. T., Vangu, G. H., et al. (2011). Banana bunchy top virus in sub-Saharan Africa: Investigations on virus distribution and diversity. *Virus Research, 159*, 171–182.

Lagoda, P. J. L., Noyer, J.-L., Dambier, D., Baurens, F.-C., Grapin, A., & Lanaud, C. (1998). Sequence tagged microsatellite site (STMS) markers in *Musaceae*. *Molecular Ecology, 7*, 657–666.

Lassois, L., Lepoivre, P., Swennen, R., van den Houwe, I., & Panis, B. (2013). Thermotherapy, chemotherapy, and meristem culture in banana. In M. Lambardi, A. E. Ozudogru, & S. M. Jain (Eds.), *Protocols for micropropagation of selected economically-important horticultural plants* (pp. 419–433). New York: Springer Science + Business Media.

Lassoudière, A. (1974). La mosaïque dite "à tirets" du bananier Poyo en Côte d'Ivoire. *Fruits, 29*, 349–357.

Le Provost, G., Iskra-Caruana, M. L., Acina, I. N., & Teycheney, P. Y. (2006). Improved detection of episomal banana streak viruses by multiplex immunocapture PCR. *Journal of Virological Methods, 137*, 7–13.

Lheureux, F., Carreel, F., Jenny, C., Lockhart, B. E. L., & Iskra-Caruana, M.-L. (2003). Identification of genetic markers linked to banana streak disease expression in inter-specific *Musa* hybrids. *Theoretical and Applied Genetics, 106*, 594–598.

Liu, F., Feng, L., Chen, X., Han, Y., Li, W., Xu, W., et al. (2012). Simultaneous detection of four banana viruses by multiplex PCR. *Journal of Phytopathology, 160*, 622–627.

Lockhart, B. E. L. (1986). Purification and serology of a bacilliform virus associated with a streak disease of banana. *Phytopathology, 76*, 995–999.

Lockhart, B. E. L., & Autrey, L. J. C. (1988). Occurence in sugarcane of a bacilliform virus related serologically to *banana streak virus*. *Plant Disease, 72*, 230–233.

Lockhart, B. E. L., & Jones, D. R. (2000). Banana mosaic. In D. R. Jones (Ed.), *Diseases of banana, abaca and enset* (pp. 256–263). Wallingford, UK: CAB International.

Lockhart, B. E. L., & Olszewski, N. E. (1993). Serological and genomic heterogeneity of *banana streak badnavirus*: Implications for virus detection in *Musa* germplasm. In J. Ganry (Ed.), *Breeding banana and plantain for resistance to diseases and pests* (pp. 105–113). Montpellier, France: CIRAD/INIBAP.

Lokossou, B., Gnanvossou, D., Ayodeji, O., Akplogan, F., Safiore, A., Migan, D. Z., et al. (2012). Occurrence of *banana bunchy top virus* in banana and plantain in (*Musa* sp.) in Benin. *New Disease Reports, 25*, 13.

Magee, C. J. P. (1927). Investigations of the bunchy top disease of bananas. *Bulletin of the Council for Scientific and Industrial Research (Australia), 30*, 64.

Magee, C. J. P. (1940). Transmission studies on the banana bunchy top virus. *Journal of the Australian Institute of Agricultural Science, 6*, 109–110.

Magee, C. J. P. (1948). Transmission of banana bunchy top to banana varieties. *The Journal of the Australian Institute of Agricultural Science, 14*, 18–24.

Magee, C. J. P. (1953). Some aspects of bunchy top disease of banana and other Musa spp. *Journal and Proceedings of the Royal Society of New South Wales, 87*, 1–18.

Magnaye, L. X. (1994). Virus diseases of banana and current studies to eliminate the virus by tissue culture. In N. G. Tangonan (Ed.), *Towards making pest and disease management relevant to Big and small banana growers Proceedings of the 1st PPS-SMD national symposium on pests and diseases in the Philippines, 23–24 April 1993* (pp. 38–43). Davao City, Philippines: Phytopathological Society Inc., Southern Mindanao Division.

Magnaye, L. V., & Espino, R. R. C. (1990). Note: Banana bract mosaic, a new disease of banana I. Symptomatology. *The Philippine Agriculturist, 73*, 55–59.

Mandal, B., Shilpi, S., Barman, A. R., Mandal, S., & Varma, A. (2013). Nine novel DNA components associated with the foorkey disease of large cardamom: evidence of a distinct babuvirus species in Nanoviridae. *Virus Research, 178*(2), 297–305.

Mansoor, S., Qazi, J., Amin, L., Khatri, A., Khan, I. A., Raza, S., et al. (2005). A PCR based method, with internal control for the detection of *banana bunchy top virus* in banana. *Molecular Biotechnology, 30*, 167–170.

Mathews, R. E. F. (1982). Classification and nomenclature of viruses. *Intervirology, 17*, 1–199.

Matile-Ferrero, D., & Williams, D. J. (1995). Recent outbreaks of mealybugs on plantain (*Musa* spp.) in Nigeria including a new record for Africa and a description of a new species of *Planococcus* (Ferris) (Homoptera: *Pseudococcidae*). *Bulletin de la Societe Entomologique de France, 100*, 445–449.

Meyer, J. B. (2005). *Banana streak badnavirus (BSV) in South Africa: Incidence, transmission, and the development of an antibody-based detection system* (p. 179). Pretoria, South Africa: University of Pretoria. *http://upetd.up.ac.za/thesis/available/etd-02092007-171659/*.

Meyer, J. B., Kasdorf, G. G. F., Nel, L. H., & Pietersen, G. (2008). Transmission of activated episomal *banana streak OL (badna)virus* (BSOLV) to cv. Williams banana (*Musa* sp.) by three mealybug species. *Plant Disease, 92*, 1158–1163.

Molina, A. B., Sinohin, V. G. O., de la Cueva, F. M., Esguerra, A. V., Crucido, S. S., Vida, E., et al. (2009). Managing *banana bunchy top virus* in smallholdings in the Philippines. *Acta Horticulturae, 828*, 383–387.

Morton, J. (1987). *Banana. Fruits of warm climates* (pp. 29–46). Miami, FL: Julia F. Morton.

Mukwa, L. F. T., Muengula, M., Zinga, I., Kalonji, A., Iskra-Caruana, M.-L., & Bragard, C. (2014). Occurrence and distribution of *banana bunchy top virus* related agro-ecosystem in south western, Democratic Republic of Congo. *American Journal of Plant Sciences, 5*, 647–658.

Munez, A. R. (1992). Symptomatology, transmission and purification of banana bract mosaic virus (BBMV) in 'giant cavendish' banana. *Faculty of graduate school* (pp. 1–57). Los Baños: University of Philippines.

Muturi, S. M., Wachira, F. N., Karanja, L. S., Wambulwa, M. C., & Macharia, E. (2013). *Paracoccus burnerae* (Homoptera; Planococcidae) as a vector of *banana streak virus*. *Journal of Experimental Biology and Agricultural Sciences, 1*, 405–414.

Mwenebanda, B. M. L., Mwamlima, L. H., Msosa, T., & Banda, D. L. N. (2007). Evaluation of FHIA hybrids for resistance to banana bunchy top virus and yield potential in Malawi *Program and abstracts, 10th international plant virus epidemiology symposium: Controlling epidemics of emerging and established plant virus diseases—The way forward* (p. 161). Hyderabad, India: ICRISAT.

Ndowora, T. C. (1998). Banana streak virus*: Development of an immunoenzymatic assay for detection and characterisation of sequences that are integrated into the genome of the host* Musa *sp.* Thesis (p. 83). Saint Paul: University of Minnesota.

Nelson, S. C. (2004). Banana bunchy top: Detailed signs and symptom. *Cooperative extension service college of tropical agriculture and human resources* (p. 22). Mānoa: University of Hawai'i.

Ngatat, S., Hanna, R., Kumar, P. L., Gray, S. M., Cilia, M., & Fontem, A. (2013). Long term evaluation of the susceptibility of 16 *Musa* genotypes to banana bunchy top disease in Cameroon, Central Africa. *Phytopathology, 103*(Suppl. 6), S103.

Niyongere, C., Ateka, E. A., Losenge, T., Lepoint, P., & Blomme, G. (2011). Screening Musa genotypes for banana bunchy top disease resistance in Burundi. *Acta Horticulturae, 897*, 439–447.

Niyongere, C., Losenge, T., Ateka, E. A., Nkezabahizi, D., Blomme, G., & Lepoint, P. (2011). Occurrence and distribution of banana bunchy top disease in the Great Lakes region of Africa. *Tree and Forestry Science and Biotechnology, 6*, 102–107.

Niyongere, C., Losenge, T., Ateka, E. M., Ntukamazina, N., Ndayiragije, P., Simbare, A., et al. (2013). Understanding banana bunchy top disease epidemiology in Burundi for an enhanced and integrated management approach. *Plant Pathology, 62*, 562–570.

Noumbissié, T. G. (2014). *Ségrégation des Chromosomes dans un croisement interspécifique de bananiers (AAABxAA) et redistribution des séquences du* Banana streak virus *intégrées au génome B.* Thèse Université de l'école doctorale SIBAGHE Montpellier II, SupAgro Montpellier, 200 pp.

Nweke, F. I., Njoku, J. E., & Wilson, G. F. (1988). Productivity and limitations of plantain (*Musa spp.* Cv.AAB) production in compound gardens in South East Nigeria. *Fruits, 43*, 161–166.

Oben, T. T., Hanna, R., Ngeve, J., Alabi, O. J., Naidu, R. A., & Kumar, P. L. (2009). Occurrence of *banana bunchy top virus* in Cameroon. *Plant Disease, 93*, 1076.

Ocfemia, G. O. (1930). Bunchy-top of abaca or Manila hemp I. A study of the cause of the disease and its method of transmission. *American Journal of Botany, 17*, 1–18.

Ortiz, R. (2013). Conventional banana and plantain breeding. *Acta Horticulturae, 986*, 77–194.

Ortiz, R., & Swennen, R. (2014). From crossbreeding to biotechnology-facilitated improvement of banana and plantain. *Biotechnology Advances, 32*, 158–169.

Peng, J., Fan, Z., & Huang, J. (2012). Rapid detection of *banana streak virus* by loop-mediated isothermal amplification assay in South China. *Journal of Phytopathology, 160*, 248–250.

Peng, J., Shi, M., Xia, Z., Huang, J., & Fan, Z. (2012). Detection of cucumber mosaic virus isolates from banana by one-step reverse transcription loop-mediated isothermal amplification. *Archives of Virology, 157*, 2213–2217.

Peng, J., Zhang, J., Xia, Z., Li, Y., Huang, J., & Fan, Z. (2012). Rapid and sensitive detection of *banana bunchy top virus* by loop-mediated isothermal amplification. *Journal of Virological Methods, 185*, 254–258.

Perrier, X., De Langhe, E., Donohue, M., Lentfer, C., Vrydaghs, L., Bakry, F., et al. (2011). Multidisciplinary perspectives on banana (*Musa* spp.) domestication. *Proceedings of the National Academy of Sciences of the United States of America, 108*, 11311–11318.

Pichaut, J. P., Umber, M., Laboureau, N., Farinas, B., Chabannes, M., Duroy, P. O., et al. (2013). Towards the end of the BSV constraint for breeding banana interspecific hybrids. In *ACORBAT Brasil 2013. XX Reunion ACORBAT, Fortaleza, Brazil, 2013/09/09-13, ACORBAT, 2013.*

Piepenburg, O., Williams, C. H., Stemple, D. L., & Armes, N. A. (2006). DNA detection using recombination proteins. *PLoS Biology, 4*, e204. http://dx.doi.org/10.1371/journal.pbio.0040204.

Pillay, M., Blomme, G., Rodrigues, E., & Ferreira, A. L. (2005). Presence of *banana bunchy top* virus in Angola. *Infomusa, 14*, 44–45.

Pollefeys, P., Sharrock, S., & Arnaud, E. (2004). *Preliminary analysis of the literature on the distribution of wild Musa species using MGIS and DIVA-GIS* (p. 68). Montpellier, France: INIBAP. http://www.musalit.org/seeMore.php?id=8616 (accessed on 18 Jan 2014).

QPPR (2002). *Queensland plant protection (banana pest quarantine) regulation 1999.* https://www.legislation.qld.gov.au/LEGISLTN/REPEALED/P/PlantProBanPesQuarR99_01_.pdf (accessed on 16 Jan 2014).

Quito-Avila, D. F., Ibarra, M. A., Alvarez, R. A., Ratti, M. F., Espinoza, L., Cevallos-Cevallos, J. M., et al. (2013). First report of *banana bract mosaic virus* in 'cavendish' banana in Ecuador. *Plant Disease, 97*, 1003.

Ram, R. D., & Summanwar, A. S. (1984). *Colocasia esculenta* (L) Schott. A reservoir of bunchy top disease of banana. *Current Science, 53*, 145–146.

Rao, D. G. (1980). Studies on a new strain of banana mosaic virus in South India. In C. R. Muthukrishnan & J. B. M. AbdulKhader (Eds.), *Proc. national seminar on banana production technology* (pp. 155–159). Coimbatore, India: Tamil Nadu Agricultural University.

Regupathy, A., Subramanian, K. S., & Naganathan, T. G. (1983). Evaluation of certain aphicides in the containment of banana bunchy top. *Pesticide, 17*, 35–36.

Robson, J. D., Wright, M. G., & Almedia, R. P. P. (2007). Biology of *Pentalonia nigronervosa* (Hemiptera, Aphididae) on banana using different rearing methods. *Environmental Entomology, 36*, 46–52.

Robson, J. D., Wright, M. G., & Almeida, R. P. P. (2006). Within-plant distribution and binomial sampling plan of *Pentalonia nigronervosa* (Hemiptera, Aphididae) on banana. *Journal of Economic Entomology, 99*, 2185–2190.

Rodoni, B. C., Ahlawat, Y. S., Varma, A., Dale, J. L., & Harding, R. M. (1997). Identification and characterization of banana bract mosaic virus in India. *Plant Disease, 81*, 669–672.

Rodoni, B. C., Dale, J. L., & Harding, R. M. (1999). Characterization and expression of the coat protein-coding region of banana bract mosaic potyvirus, development of diagnostic assays and detection of the virus in banana plants from five countries in Southeast Asia. *Archives of Virology, 144*, 1725–1737.

Saverio, B. (1964). *Banana cultivation in Eritrea and its problems* (p. 56). Milan, Italy: Edagricole.

Selvarajan, R., & Balasubramanian, V. (2008). Banana viruses. In Govind P. Rao, Arben Myrta, & Kai-shu Ling (Eds.), *Characterization, diagnosis and management of plant viruses: 2.* (pp. 109–124). Texas, USA: Studium press LLC.

Selvarajan, R., & Balasubramanian, V. (2013). Natural occurrence of *banana bunchy top virus* in *Ensete superbum* in India. *Indian Journal of Virology, 24*, 97–98.

Selvarajan, R., & Balasubramanian, V. (2014). Host interaction host–virus interactions in banana-infecting viruses. In R. K. Gaur, Thomas Hohn, & Pradeep Sharma (Eds.), *Plant virus–host interaction molecular approaches and viral evolution* (pp. 57–78). MA, USA: Elsevier Academic Press.

Selvarajan, R., Balasubramanian, V., Dayakar, S., Sathaimoorthy, S., & Ahlawat, Y. S. (2004). Simultaneous detection of Indian isolates of banana bunchy top virus, banana streak virus and banana bract mosaic virus by multiplex-PCR technique. In B. Bandyopadhyay, S. K. Sarkar, D. K. Misra, & F. K. Bauri (Eds.), *Banana research in India Proceedings of national seminar on banana industry—Present scenario and future strategies held at the Bidhan Chandra Krishi Viswavidyala, FTC, Kalyani, West Bengal, 11 to 13 June, 2004* (pp. 237–240).

Selvarajan, R., Balasubramanian, V., Dayakar, S., Sathiamoorthy, S., & Ahlawat, Y. S. (2010). Evaluation of immunological and molecular techniques for the detection of different isolates of *banana bunchy top virus* in India. *Indian Phytopathology, 63*, 333.

Selvarajan, R., Balasubramanian, V., Jeyabaskaran, K. J., Pandey, S. D., & Mustaffa, M. M. (2009). Management of banana bract mosaic disease through higher dose of fertilizer application in banana cv. Ney Poovan (AB). *Indian Journal of Horticulture, 66*, 301–305.

Selvarajan, R., Balasubramanian, V., & Sathiamoorthy, S. (2006). Vector transmission of banana bract mosaic and *banana streak viruses* in India. In *Abstracts of XVI annual convention and international symposium on management of vector-borne viruses, ICRISAT, 7–10th February, 2006* (p. 110).

Selvarajan, R., Balasubramanian, V., Sheeba, M. M., Raj Mohan, R., & Mustaffa, M. M. (2011). Virus-indexing technology for production of quality banana planting material: A boon to the tissue-culture industry and banana growers in India. In I. Van den Bergh, et al. (Eds.), *Acta Horticulturae: 897. Proc. Int'l ISHS-ProMusa symp. on global perspectives on Asian challenges* (pp. 463–469).

Selvarajan, R., & Jeyabaskaran, K. J. (2006). Effect of *banana bract mosaic virus* (BBrMV) on growth and yield of cultivar nendran (plantain, AAB). *Indian Phytopathology, 59*, 496–500.

Selvarajan, R., Mary Sheeba, M., Balasubramanian, V., Rajmohan, R., Lakshmi Dhevi, N., & Sasireka, T. (2010). Molecular characterization of geographically different banana bunchy top virus isolates in India. *Indian Journal of Virology, 21*, 110–116.

Sharman, M., Gambley, C. F., Oloteo, E. O., Abgona, R. V. J., & Thomas, J. E. (2000). First record of natural infection of abaca (*Musa textilis*) with banana bract mosaic potyvirus in the Philippines. *Australasian Plant Pathology, 29*, 69.

Sharman, M., Thomas, J. E., & Dietzgen, R. G. (2000). Development of a multiplex immunocapture PCR with colorimetric detection for viruses of banana. *Journal of Virological Methods, 89*, 75–88.

Sharman, M., Thomas, J. E., Skabo, S., & Holton, T. A. (2008). Abaca bunchy top virus—A new member of the genus *Babuvirus* (family *Nanoviridae*). *Archives of Virology, 153*, 135–147.

Shekhawat, U. K. S., Ganapathi, T. R., & Hadapad, A. B. (2012). Transgenic banana plants expressing siRNAs targeted against viral replication initiation gene display high-level resistance to banana bunchy top virus infection. *The Journal of General Virology, 93*, 1804–1813.

Siljo, A., Bhat, A. I., Biju, C. N., & Venugopal, M. N. (2011). Occurrence of *banana bract mosaic virus* on cardamom. *Phytoparasitica, 40*, 77.

Simmonds, H. W. (1931). Noxious weeds and their control in Fiji. II biological control. *Fiji Agricultural Journal, 4*, 29.

Smith, M. C., Holt, J., Kenyon, L., & Foot, C. (1998). Quantitative epidemiology of banana bunchy Top virus disease and its control. *Plant Pathology, 47*, 177–187.

Soko, M. M., Dale, J., Kumar, P. L., James, A., Izquierdo, L., Fiaboa, K., et al. (2009). *Banana bunchy top disease survey reports, previous control efforts and the way forward* (p. 200). Limbe, Malawi: Bvumbwe Agricultural Research Station, P.O. Box 5748.

Stainton, D., Kraberger, S., Walters, M., Wiltshire, E. J., Rosario, K., Halafihi, M., et al. (2012). Evidence of inter-component recombination, intra-component recombination and reassortment in banana bunchy top virus. *The Journal of General Virology, 93*, 1103–1119.

Su, H. J., Hwang, A. S., Lee, S. Y., & Chao, C. P. (2007). Conservation, disease indexing and utilization of pathogen free citrus and banana genetic resources in Taiwan. In *International training workshop on the conservation and utilization of Tropical/ Subtropical plant genetic resources* (pp. 1–24).

Su, H.-J., Tsao, L.-Y., Wu, M.-L., & Hung, T.-H. (2003). Biological and molecular categorization of strains of banana bunchy top virus. *Journal of Phytopathology, 151*, 290–296.

Tamura, K., Dudley, J., Nei, M., & Kumar, S. (2007). MEGA4: Molecular evolutionary genetics analysis (MEGA) software version 4.0. *Molecular Biology and Evolution, 24*, 1596–1599.

Teycheney, P. Y., Acina, I., Lockhart, B. E. L., & Candresse, T. (2007). Detection of banana mild mosaic virus and banana virus X by polyvalent degenerate oligonucleotide RT-PCR (PDO-RT-PCR). *Journal of Virological Methods, 142*, 41–49.

Teycheney, P.-Y., Laboureau, N., Iskra-Caruana, M.-L., & Candresse, T. (2005). High genetic variability and evidence for plant-to-plant transfer of banana mild mosaic virus. *The Journal of General Virology, 86*, 3179–3187.

Teycheney, P.-Y., Marais, A., Svanella-Dumas, L., Dulucq, M.-J., & Candresse, T. (2005). Molecular characterization of banana virus X (BVX), a novel member of the *Flexiviridae* family. *Archives of Virology, 150*, 1715–1727.

Thangavelu, R., Selvarajan, R., & Singh, H. P. (2000). Status of *banana streak virus* and *banana bract mosaic virus* diseases in India. In H. P. Singh, & K. L. Chadha (Eds.), *Banana: Improvement, production and utilization. Proceedings of the conference on challenges for banana production and utilization in 21st century* (pp. 364–376). Trichy, India: AIPUB, NRCB.

Thiribhuvanamala, G., Doraiswamy, Sabitha, & Ganapathy, T. (2005). Detection of *banana bunchy top virus* in the aphid vector using double antibody sandwich (DAS) ELISA. *Indian Journal of Virology, 16*, 12–14.

Thomas, J. E., & Dietzgen, R. G. (1991). Purification, characterization and serological detection of virus-like particles associated with *banana bunchy top virus* in Australia. *The Journal of General Virology, 72*, 217–224.

Thomas, J. E., Geering, A. D. W., Gambley, C. F., Kessling, A. F., & White, M. (1997). Purification, properties and diagnosis of banana bract mosaic potyvirus and its distinction from abaca mosaic potyvirus. *Phytopathology, 87*, 698–705.

Thomas, J. E., Iskra-Caruana, M. L., & Jones, D. R. (1994). *Banana bunchy top disease*. Musa *disease fact sheet No. 4*. Montpellier, France: INIBAP, pp. 2.

Thomas, J. E., Smith, M. K., Kessling, A. F., & Hamill, S. D. (1995). Inconsistent transmission of banana bunchy top virus in micro-propagated bananas and its implications for germplasm screening. *Australian Journal of Agricultural Research, 46*, 663–671.

Thomson, D., & Dietzgen, R. G. (1995). Detection of DNA and RNA plant viruses by PCR and RT-PCR using a rapid virus release protocol without tissue homogenization. *Journal of Virological Methods, 54*, 85–95.

Thottappilly, G., Dahal, G., & Lockhart, B. E. L. (1998). Studies on a Nigerian isolate of banana streak badnavirus. Purification and enzyme-linked immunoassay. *The Annals of Applied Biology, 132*, 253–261.

Tripathi, L. (2003). Genetic engineering for improvement of *Musa* production in Africa. *African Journal of Biotechnology, 12*, 503–508.

Tsao, T. T.-H. (2008). *Towards the development of transgenic banana bunchy top virus (BBTV)-resistant banana plants: Interference with replication.* PhD thesis, Brisbane: Queensland University of Technology. http://eprints.qut.edu.au/17031/ (accessed on 24 Mar 2014).

Tushmereirwe, W. K., & Bagabe, M. (1999). Review of disease distribution and pest status in Africa. In *Review of IPM research activities—Pathology* (pp. 139–147).

Vetten, H. J., Dale, J. L., Grigoras, I., Gronenborn, B., Harding, R., Randles, J. W., et al. (2012). Family Nanoviridae. In *Virus taxonomy: Ninth report of the international committee on taxonomy of viruses* (pp. 395–404).

Vishnoi, R., Raj, S. K., & Prasad, V. (2009). Molecular characterization of an Indian isolate of banana bunchy top virus based on six genomic DNA components. *Virus Genes, 38*, 334–344.

Walangululu, M. J., Matara, M. R., Bahati, L., Niyongere, C., Lepoivre, P., & Blomme, G. (2010). Assessing the spread and seasonal influence of fruit peel disease and banana bunchy top disease in South Kivu, Eastern DR-Congo. *Tree and Forestry Science and Biotechnology, 4*, 98–104.

Wang, I.-C., Sether, D. M., Melzer, M. J., Borth, W. B., & Hu, J. S. (2010). First report of *banana bract mosaic virus* in flowering ginger in Hawaii. *Plant Disease, 94*(7), 921.

Wardlaw, C. W. (1961). Mosaic, infectious chlorosis and other virus diseases. *Banana diseases, including plantains and abaca* (pp. 116–145). London: Longmans.

Watanabe, S., Greenwell, A. M., & Bressan, A. (2013). Localization, concentration, and transmission efficiency of *banana bunchy top virus* in four asexual lineages of *Pentalonia* aphids. *Viruses, 5*, 758–775. http://dx.doi.org/10.3390/v5020758.

Wu, R. Y., & Su, H. J. (1990a). Purification and characterization of banana bunchy top virus. *Journal of Phytopathology, 128*, 153–160.

Wu, R. Y., & Su, H. J. (1990b). Production of monoclonal antibodies against *banana bunchy top virus* and their use in enzyme-linked immunosorbent assay. *Journal of Phytopathology, 128*, 203–208.

Xie, W. S., & Hu, J. S. (1995). Molecular cloning, sequence analysis and detection of *banana bunchy top virus* in Hawaii. *Phytopathology, 85*, 339–347.

Yu, N. T., Zhang, Y. L., Feng, T. C., Wang, J. H., Kulye, M., Yang, W. J., et al. (2012). Cloning and sequence analysis of two banana bunchy top virus genomes in Hainan. *Virus Genes, 44*, 488–494.

CHAPTER EIGHT

Control of Virus Diseases of Berry Crops

Robert R. Martin*,[1], Ioannis E. Tzanetakis†
*USDA-ARS Horticultural Crops Research Unit, Corvallis, Oregon, USA
†Department of Plant Pathology, Division of Agriculture, University of Arkansas, Fayetteville, Arkansas, USA
[1]Corresponding author: e-mail address: bob.martin@ars.usda.gov

Contents

Abstract

Virus control in berry crops starts with the development of plants free of targeted pathogens, usually viruses, viroids, phytoplasmas, and systemic bacteria, through a combination of testing and therapy. These then become the top-tier plants in certification programs and are the source from which all certified plants are produced, usually after multiple cycles of propagation. In certification schemes, efforts are made to produce plants free of the targeted pathogens to provide plants of high health status to berry growers. This is achieved using a systems approach to manage virus vectors. Once planted in fruit production fields, virus control shifts to disease control where efforts are focused on controlling viruses or virus complexes that result in disease. In fruiting fields, infection with a virus that does not cause disease is of little concern to growers. Virus control is based on the use of resistance and tolerance, vector management, and isolation.

Advances in Virus Research, Volume 91
ISSN 0065-3527
http://dx.doi.org/10.1016/bs.aivir.2014.10.003

1. INTRODUCTION

The term berry (small fruit) refers primarily to the genera *Fragaria* (strawberry), *Rubus* (blackberry, raspberry, and their hybrids), *Vaccinium* (blueberry and cranberry), *Ribes* (currants and gooseberry), and *Sambucus* (elderberry). Traditionally, berry crops have been collected from the wild at the dawn of the human species and only recently they have become agricultural crops, especially after the development of the modern strawberry (*F.* x *ananassa*). In a short period of time and because of their ever growing popularity with consumers, berry crops have been developed to grow across the globe from subtropical to subarctic environments. The major expansion in production and the environments where these crops are grown, along with the rapid changes in the genotypes grown commercially, have resulted in a very diverse virosome in berry crops (Martin, MacFarlane, et al., 2013; Martin, Peres, & Whidden, 2013; Martin, Polashock, & Tzanetakis, 2012; Martin & Tzanetakis, 2006). Those changes and the development of new technologies allowing for the rapid discovery of new viruses have led to a substantial increase in the number of known berry viruses, which has more than doubled in the last 20 years. The number of new discoveries still increases with more than eight berry viruses identified in each of the last 3 years (Martin et al., 2012; Martin, MacFarlane, et al., 2013; Martin, Peres, et al., 2013; Martin et al., unpublished; Tzanetakis et al., unpublished). This communication aims to provide a general overview of how basic knowledge on berry viruses can be used for efficient virus control during the propagation process and field establishment.

There are more than 80 species in 30 virus genera known to infect the major berry crops (here we will only present the ones communicated in peer-reviewed publications as of May 2014). Most viruses identified before the 1990s are well-studied at the molecular and biological level, whereas, newly identified viruses are characterized primarily at the molecular level. This has led to knowledge gaps that need to be addressed. For the purpose of this communication, we will provide control strategies for those less-studied viruses based on their taxonomic placement and the accumulated biological knowledge on closely related viruses. Based on molecular data, the best guesses of potential vectors need to be considered, ranging from whiteflies to eriophyid mites to fungi. The number of vector taxa in a particular area may be such that control of all vectors is unfeasible. In general terms, areas with tradition of berry production have a set group of viruses of concern. In most cases, there is the know-how on viruses/vectors and their control, whereas new production areas tend to have a wider array of viruses, many of which have only recently been discovered. A list of viruses infecting each of the major berry crops, their mode of transmission and their geographic range (where known) are presented in individual tables. Aphids are major virus vectors for all berries crops. They are abundant at the traditional production areas in the temperate regions around the world and just a few years ago they were considered the only vector of concern for most production areas. In most crops, the prominent aphid species colonizing the crop are also the major vectors, able to transmit an array of viruses that tend to act synergistically to cause disease as in the case of strawberry decline or raspberry mosaic (Martin & Tzanetakis, 2006; Quito-Avila, Lightle, & Martin, 2014). Notwithstanding, there are opportunistic feeders that may transmit without colonizing plants (*Myzus ornatus*; Lowery, Bernardy, Deyoung, & French, 2008). When an aphid species is crop-specific, control can be achieved through chemical sprays or even more efficiently, in the case of strawberry with crop-free times, an applicable practice for nurseries that are not neighboring commercial fields.

Nematode-transmitted viruses can cause significant losses and even plant death (Martin, MacFarlane, et al., 2013; Martin, Peres, et al., 2013; Martin et al., 2012; Martin & Tzanetakis, 2006). The major viruses infecting berry crops belong to the genus *Nepovirus* and the unassigned *Strawberry latent ringspot virus*. One of the major caveats with members of this group is their wide host range that extends to several common weeds present in berry fields. As nematode movement is restricted to less than 2 m/year, they have been successfully controlled with the use of potent nematicides such as methyl bromide. With the phase-out of the more efficient nematicides, this virus group may reemerge; especially, in areas where the nematode vectors are endemic.

Pollen and seed-transmitted viruses (PSTVs) present the most challenging group when it comes to control because the only viable control method is avoidance. In many cases, PSTVs cause significant losses in single or mixed infections whereas several of them infect multiple berry crops (MacDonald, Martin, & Bristow, 1991; Martin, MacFarlane, et al., 2013; Martin, Peres, et al., 2013; Pallas et al., 2012) Some of the PSTVs do not have an active vector and can move passively by wind currents and pollinating arthropods such as bees, thrips (Sdoodee & Teakle, 1987), and mites, whereas others may have insect or nematode vectors, in which case control of their primary vectors can significantly reduce spread.

Whiteflies have only been identified recently as virus vectors in berry crops although there have been diseases, first reported in the 1950s, that have now been proven to be caused by whitefly-transmitted viruses (Tzanetakis et al., 2004; Tzanetakis, Wintermantel, & Martin, 2003). The emergence of this virus group as a major limiting factor to production is primarily due to the expansion of berry production to the subtropics and the extension of the geographic range of whiteflies to the temperate regions around the world. Notwithstanding, the number of whitefly-transmitted viruses is still limited and only includes species that infect strawberry and blackberry. In all cases, those viruses do not cause high-impact symptoms in single infections but act synergistically with other viruses in mixed infections to cause detrimental symptoms that can even lead to plant death (Martin & Tzanetakis, 2013; Susaimuthu, Tzanetakis, Gergerich, Kim, & Martin, 2008; Susaimuthu, Tzanetakis, Gergerich, & Martin, 2008). This should be a concern for certification programs and for nursery systems where quite often disease evaluation is based on visual observations. With many viruses of the berry crops, plants infected with one or two viruses may be asymptomatic in nursery plants and pass visual inspections, but when moved to production fields and infected with additional viruses they can decline rapidly (Martin & Tzanetakis, 2013).

Virus-vectoring eriophyid mites were discovered recently in berry crops although some of the diseases they are associated with have been reported since the beginning of the twentieth century (Jones, Gordon, & Jennings, 1984). The lack of knowledge on the biology of these vectors and the absence of systemic miticides make vector control challenging. To date, most berry mite-transmitted viruses belong to the genus *Emaravirus* (Hassan, Keller, Martin, Sabanadzovic, & Tzanetakis, 2013; McGavin, Mitchell, Cock, Wright, & MacFarlane, 2012), viruses that appear to be localized at the mite feeding sites, but there are indications that several new berry-infecting systemic RNA viruses are also mite-transmitted and involved in important diseases, primarily as part of virus complexes (Sabanadzovic, Abou Ghanem-Sabanadzovic, & Tzanetakis, 2011). A major exception is *Black currant reversion virus* (BRV), which taxonomically belongs to the genus *Nepovirus* but is transmitted by the black currant gall mite (Susi, 2004). BRV exemplifies the need for experimental verification of a virus vector as predictions based on molecular data and taxonomy may be deceiving.

Leafhoppers are another group recently identified as potential virus vectors for berry crops. To date, there are at least four blackberry viruses in the genus *Marafivirus* (Sabanadzovic & Abou Ghanem-Sabanadzovic, 2009; Sabanadzovic, Ghanem-Sabanadzovic, & Gorbalenya, 2009; Sabanadzovic et al., unpublished), and other viruses in this genus are known to be transmitted by leafhoppers. A study in the southeastern United States identified about 50 leafhopper species in production fields (Johnson et al., unpublished), indicative of the complex virus/vector interactions that need to be elucidated before the development of meaningful control strategies.

The only thrips-transmitted virus known to infect berry crops is *Impatiens necrotic spot virus* (INSV), detected in blackberry (Tzanetakis, Guzmán-Baeny, VanEsbroeck, Fernandez, & Martin, 2009). There is no detailed work performed on transmission of the virus in blackberry that would elucidate the efficiency of transmission or the thrips species that are the primary vectors of the virus in blackberry. However, the excessive number of thrips found in several fields and the low numbers of INSV-infected plants indicate rather inefficient transmission probably because of the thrips composition in blackberry fields, or the presence of diverse flora in many blackberry fields in the southeastern United States.

A new insect group was recently added in the list of berry virus vectors. After the discovery of *Blackberry vein banding associated virus* (BVBaV; Thekke-Veetil et al., 2013), a member of the genus *Ampelovirus*, experiments were performed with mealybugs that colonize plants and successfully demonstrated transmission (Sabanadzovic et al., unpublished). As is the case with several of the berry viruses, BVBaV does not appear to cause symptoms in single infections but is often found in mixed virus infections in declining plants.

The last virus group with a single representative in the list of berry viruses to date is that transmitted by fungi. Blueberry mosaic is a disease that was first described about 60 years ago and only recently was a virus associated with the disease, *Blueberry mosaic associated virus* (BlMaV). BlMaV belongs to the genus *Ophiovirus* and has been detected in all plants with typical disease symptoms (Thekke Veetil, Ho, Keller, Martin, & Tzanetakis, 2014). Transmission experiments are underway, and the virus is hypothesized to be transmitted by members of the genus *Olpidium*, as is the case with other members of the genus.

2. VIRUS CONTROL DURING PLANT PROPAGATION

All berry crops are clonally propagated. Breeders perform crosses and evaluate progeny annually until they identify and select plants with the desired attributes. This process normally takes three to seven or more years depending on the crop. The selection process is usually performed in the same fields with little or no rotation leading to the establishment of viruses that circulate within the system year after year. This may the underlying reason for the tolerance present in the majority of modern berry cultivars to most single virus infections as susceptible progeny do not progress to the next selection level. However, plants in a breeding program that are grown for multiple seasons in the field in close proximity to field plots of the same crop of multiple ages are likely to be infected with one or several viruses during the selection process. This is where virus testing combined with virus clean-up programs are used to ensure that plants of new cultivars are free of targeted viruses and virus-like pathogens. These "clean" plants then enter into certification schemes to ensure the propagation of elite material with full potential for productivity and longevity in commercial fields.

3. DETECTION

Virus detection is the first step in the quest for plants that are free of targeted viruses. For the majority of the virus species that infect berry crops, there are rapid, sensitive laboratory tests available. Most viruses can be tested using molecular (polymerase chain reaction (PCR) and variants of PCR, loop-mediated isothermal amplification (LAMP), rolling circle amplification for DNA viruses, etc.), biochemical (nucleic acid hybridization, dsRNA isolation), or immunological (enzyme-linked immunosorbent assay (ELISA) in multiple formats, immunoblots, etc.) tests. In general terms, molecular tests are most widely used for newly discovered viruses, whereas immunological tests are applied to viruses where there are good-quality antibodies available. Still there are viruses or virus-like agents without laboratory tests available. In those cases, grafting onto indicators, an approach developed before the development of any of the aforementioned laboratory-based tests, is used to test for the presence of virus. Indicator plants are usually older cultivars or clones, which exhibit a range of symptoms when infected with one or more viruses. The selection of the indicators was done at a time when a subset of the viruses that infect berry viruses were known to exist and several of them are actually clones infected with an asymptomatic virus that develops symptoms when grafted with material infected with additional viruses. *Fragaria vesca* clone EMC, is infected with *Strawberry crinkle virus* (SCV) and has been used widely as a highly sensitive indicator for other strawberry viruses (Frazier, 1953). Many of the recently discovered berry viruses are asymptomatic in the established indicators used for berry crops (Susaimuthu et al., 2007). Probably, one of the reasons that those viruses remained unidentified until the development of generic techniques that allow virus identification even in plants that lack any obvious symptoms.

Today, we have tools available that allow for virus detection and identification without any previous knowledge of the molecular, biochemical, or immunological properties of the virus. The development of macro- and microarrays with several hundreds or thousands of probes allows for the hybridization and subsequent identification of viruses with only marginal homology to known taxa (Agindotan & Perry, 2007; Thompson et al., 2014; Wang et al., 2002). Arrays present a powerful tool, allowing for broad spectrum detection of known and unknown viruses, but they may miss viruses that do not show homology with established taxa, and thus not hybridize to sequence-specific probes. This issue can be overcome with the use of next-generation sequencing as a tool for virus detection and identification (Barba, Czosnek, & Hadidi, 2014). Material can be processed using an un-bias approach where nucleic acids are amplified and analyzed for detection of known viruses and discovery of new viruses. Even when a virus does not have significant homology to established taxa, bioinformatics and structural analyses of putative proteins can lead to the discovery of previously unknown agents (Ho & Tzanetakis, 2014). This new technology is powerful and in use for the delivery—for the first time—of virus-free plants. It has been shown to be effective in multiple cases, but its implementation as the sole detection platform will require careful evaluation to ensure that it is at least as good as other methods currently in use. It will be some time before national regulatory agencies, the Regional Plant Protection Organizations, and the International Plant Protection Organization agree that these new technologies are acceptable to meet the phytosanitary testing requirements for intra- and international trade.

4. CERTIFICATION SCHEMES

Certification programs for berry crops aim to safeguard plant material during the vegetative propagation of millions of plants from a single plant to provide the best-quality planting stock to growers that have full potential for sustained production over many years. From a nursery's or fruit grower's standpoint, economics are the driving force to ensure viability and profitability. Thus, the nursery owner and fruit grower have a vested interest in certification standards that are scientifically sound in terms of disease control, biologically doable in terms of producing high-quality plants, and economical in terms of producing high-quality plants at a reasonable cost per plant. Here, we outline state-of-the-art certification standard for berry crops based on ongoing work on *Rubus*, blueberry and strawberry certification schemes in the United States.

All cultivars start as a cross of two parents with desirable properties. Hundreds or thousands of seedlings go through rigorous evaluations before an individual is selected as a potential new cultivar. Because of the long evaluation process, many, if not all plants are infected with one or more viruses, by the time they are selected. During the last few years, with the development of new powerful technologies, the processes to obtain plants clean from all targeted pathogens have changed. Traditionally, material has been grafted onto virus-susceptible indicators. Today, in addition to grafting, advanced selections are also evaluated using molecular and serological assays and very recently next-generation sequencing (Barba et al., 2014). Data are fed into a bioinformatics pipeline that identifies all known virus sequences (detection) or even discover new viruses (Ho & Tzanetakis, 2014). After heat/chemo/cryo-therapy and meristem tissue culture (covered elsewhere in this communication), the regenerated plant is tested again for all targeted pathogens. The term "targeted" is important when accessing the economic feasibility of virus-clean-up. Several viruses may not be pathogenic, e.g., cryptic or amalgaviruses, which may be ubiquitous in specific cultivars, species, or genera of plants. These viruses lack a movement protein and appear to be present in every cell of the plant, are seed transmitted at a rate of 100%, and are not graft transmissible. These viruses have not been eliminated by any type of therapy, and their presence is not known to cause any pathology either when present alone or to have an interaction with other viruses that infect the plants. These viruses should not be included in quarantine or certification programs, thus the use of the term targeted viruses or pathogens.

When free of all targeted pathogens, the material is designated as Generation 1, or G1, and becomes the top-tier plant in a certification scheme and is the ultimate source for each plant of this cultivar. The elite material (G1) is maintained in designated vector-free facilities (screenhouses, screened greenhouses) that are free of any other plant species. The G1 plants are maintained in pots and not allowed to be in direct contact with soil to avoid soil-borne vector movement, and the plants are deflowered to minimize the risk of infection by pollen-borne viruses. Plants are tested regularly for all targeted pathogens to assure their health status.

Propagation material (cuttings, stolons, etc.) from G1 plants can be used to increase the number of plants within the same facility, but if moved off-site the material is automatically designated as Generation 2 (G2). They can be reassigned the designation G1 after being tested following establishment in the new location and found free of targeted pathogens. This material can be maintained in tissue culture, in screenhouses/screened greenhouses, or the field as long as appropriate phytosanitary conditions as defined in a certification standard are met (soil barriers to eliminate nematode movement, weed-free areas and deflowering where possible to minimize infection by pollen-borne viruses, etc.). All materials other than that in tissue culture are tested regularly (normally every 3 years), as defined in the certification standard for all major circulating viruses in this particular region ("canary" viruses). If found free of targeted pathogens, the G2 plants may continue serving as material for the next-propagation steps, whereas if the testing regime timeline is lapsed, they are eliminated from the propagation scheme. The requirements for G2 plants grown in the field are more stringent and require a higher level or frequency of testing for the "canary" viruses. Infection of G2 material leads to plant discard.

Propagation material obtained from G2 plants is designated as Generation 3, or G3, and they are normally grown in the field. The general rules applied in field-grown G2 plants also apply for the G3 plants including testing. Given the higher number of plants at this level, it is usually not feasible/economic to test all plants for "canary" viruses. For this reason, testing using a hypergeometric distribution model (ISPM No. 31) that would detect a 1% infection with 95% confidence is recommended, but may vary depending on the certification standard. Depending on the size of the block tested, if infections exceed the threshold level of 1% the block can be subdivided to smaller blocks, all of which need to be tested based on the same threshold level to identify and remove infection hotspots.

Generation 4, or G4, is the last certification level and includes all plants that are propagated for sale to fruit growers, certified plants. All rules applicable for G3 plants are also relevant for G4s, but the virus infection threshold is relaxed to identify major problems in the system that may lead to a virus epidemic. In the United States, blueberry and *Rubus* certification schemes aim to identify infection rates of more than 5% with 95% confidence. If infection exceeds the G4 threshold, subplots can be sampled to identify blocks that meet the standard. The plants from infection hotspots will not be given a certification tag and cannot be sold as certified plants.

5. GENERATING AND TESTING G1 PLANTS

The key to developing and implementing an effective certification program is to start with plants of the highest possible health status. In the United States, "Clean Plant Centers" for berries, grapes, tree fruits, and citrus have evolved since the mid-twentieth century (Gergerich et al., 2015). The role of these centers has been to produce and maintain the G1 plants. For many years, these centers were funded through a combination of fees, funding from grower organizations, and grants. In 2008, the National Clean Plant Network (NCPN) was established as part of the Farm Bill, which has led to more stable baseline funding for these programs with the goal of developing and maintaining G1 plants. Additionally, the NCPN is to provide an avenue for the introduction of cultivars into the United States that includes testing, and clean-up if necessary, with the aim of reducing the introduction of pathogens that could threaten domestic production or natural environments. There are currently two funded Centers that focus on producing the G1 plants for berry crops in the United States. The G1 collections are often maintained by state or federal agencies or some private/government arrangement. Additionally, some large, private, vertically integrated berry companies produce and maintain their own proprietary G1 plants, though the testing needs to be done or confirmed by a third party. As outlined above, certification schemes for berries crops start with a single plant that is fully tested and found free of targeted pathogens, which usually includes all viruses known to infect the crop, as well as phytoplasmas and some systemic bacteria. For the berry crops, these G1 plants are maintained in protected culture, either screenhouses or screened greenhouses.

Plants that enter into G1 collections come from multiple sources including new cultivars developed in breeding programs, cultivars in use commercially that have not been tested for all viruses of the crop, heritage cultivars that are from various collections, cultivars entering from another country. Before any plant receives G1 status, it must go through extensive testing to ensure that it is free of viruses and other systemic pathogens identified in the certification standard (Fig. 1). For materials coming from a foreign or domestic clean plant program, the plants are immediately put into the virus testing program. First, all laboratory tests are run and if all negative, the bioassays are performed. If the bioassays are all negative, the plants are ready for release. This may require submission of the test results to the regulatory body responsible for plant importation into the country to get final approval for release of the plants from foreign sources for commercial production.

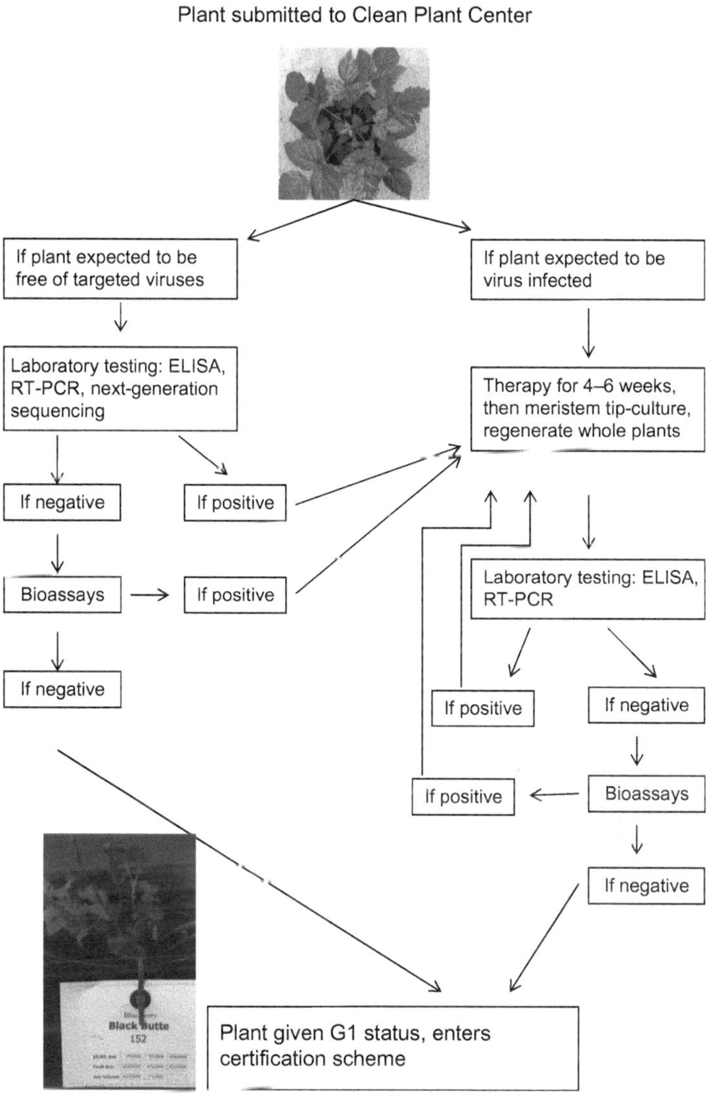

Figure 1 Path of a plant submitted to a Clean Plant Center from submission to obtaining G1 status. (See the color plate.)

Based on experience with viruses and plants from the various regions within the United States, domestic materials coming from breeding programs can follow one of several paths when received by the Clean Plant Center. For example, due to disease pressure, blackberry plants from the breeding program in Arkansas have a very high chance of being infected and enter directly into the clean-up process on arrival. This is also true for raspberry plants coming from the breeding program in Washington State. In contrast, blackberry and raspberry plants from the breeding program in Oregon have a greater than 50% chance of being free of targeted pathogens and enter into the virus testing program when they arrive at the clean plant facility in Corvallis, Oregon. In the case of strawberries, all plants received from breeding programs are subjected to therapy and meristem-tip culture when they arrive since the disease pressure is high in most growing areas in North America for at least several of the common strawberry viruses. Having information on the prevalence of viruses by region that infect a crop helps inform decision making on what type of action to implement when new material arrives at a Clean Plant Center. This information is also used in developing the list of "canary viruses" that are the primary focus of certification programs at the G2, G3, and G4 levels. As an example, canary viruses for blueberry certification in the Pacific Northwest of the United States would be different from those in the southeastern United States.

Plants that enter into the virus clean-up process are established in large pots or in tissue culture. Then these plants (potted or tissue culture) are subjected to therapy (heat-, chemo-, or cryo-) for a period of 3–8 weeks. Then meristems, generally 0.5 mm or less are removed, grown in tissue culture until rooted, and finally potted and maintained in a screened greenhouse. At this point, plants are removed from tissue culture and potted, and a few leaves are removed for virus testing by ELISA, RT-PCR, or PCR. If any of these tests are positive, the plants are subjected to a second cycle of therapy and the process repeated. If the laboratory tests are all negative, the plants are grown until large enough for biological indexing. At this time, the plants are grafted onto indicators and retested by ELISA, RT-PCR, or PCR. If all tests are negative, the plants enter the G1 collection; if any positive tests are obtained, the clean-up cycle is repeated. Each cycle from beginning of therapy to complete testing of plants in bioassays takes about 2 years for *Rubus*. Therefore, if a positive test is obtained, the clean-up process can take 4 or more years. For this reason, we work with breeders to obtain their top selections 2 years before they are ready to name and release them as cultivars, with the goal of having G1 plants available at the time of the cultivar release. This means that we often get three to five plants into the clean-up process for each cultivar that is released. If the breeder makes a decision to drop one of the advanced selections, they work with us so that we can drop it from the clean-up process. With this cooperation, the breeders are able to release new cultivars without waiting to have a G1 source available. From the time of submission to the Clean Plant Center until all testing is completed and plants are found free of targeted pathogens, they are considered candidate G1 plants. Once they have indexed negative in the full suite of tests required in the certification standard, they become G1 plants. Figure 1 shows the path of a plant entering a Clean Plant Center until it obtains G1 status in the certification scheme.

There is a common misconception by growers and many researchers that the use of tissue culture for plant propagation will result in plants free of viruses. *This is not true.* Tissue culture for plant propagation generally uses large tissue pieces, and plants are not subjected to any therapy treatment before hand. It would be very rare for plants to escape virus infection through routine tissue culture multiplication.

6. VIRUS CONTROL IN BERRY CROPS

For berry crops, as with any of the vegetatively propagated crops, there are two strategies for virus control depending on what product is being produced. For plant production, which includes all nursery systems, the objective is to control all targeted viruses in the crop to ensure production of plants with highest health status possible. Control of viruses at this level ensures that trade in berry plants does not result in movement of viruses to new locations, which can lead to increased vector management costs and thus production costs should a new virus be introduced. It has been shown that starting with clean plants results in better establishment and production than virus-infected plants (Quito-Avila, Lightle, & Martin, 2014). Also, virus-infected plants in a nursery setting may appear healthy, but digging, shipping, and replanting puts the plants under significant additional stresses that can lead to poor establishment and fruit production from these virus-infected plants (Martin, MacFarlane, et al., 2013; Martin, Peres, et al., 2013). The second type of virus control is used by fruit growers, where disease control is the objective rather than virus control. In these situations, latent viruses or mild viruses may have minimal impact on fruit production. Also, with the berry crops, most virus diseases are caused by mixed infections, thus, rather than controlling all viruses it is much more efficient to control one or two viruses that are critical for disease development. In cases where fields are established with plants carrying a symptomless virus, the addition of a second virus could lead to reduced yields or quality in the fruiting fields, thus, the importance of nurseries managing viruses rather than virus diseases.

7. VIRUS CONTROL IN NURSERIES

The viruses that have been reported to infect each of the major berry crops are shown in Tables 1–5: strawberry (Table 1); raspberry and blackberries (Table 2); blueberry and cranberry (Table 3); currants and gooseberry (Table 4); and elderberry (Table 5). The virus genera, means of transmission, detection methods, and distribution at a very large scale are also shown in these tables. Virus distribution in many cases is listed as N/A, which means the testing has not been done at a large scale and the presence or absence of the virus is not known. Also, note that the distribution is on a very large (continental) scale and for making management decisions a finer scale is much more useful. As an example, the presence of aphid-transmitted strawberry viruses in Africa and North America does not provide enough detail to make management decisions in Morocco or Florida, respectively. Information on virus distribution on a smaller scale is needed to make management decisions for a nursery or fruit grower. As an example, Raspberry leaf curl virus has been reported in eastern North America but not in west of the Rocky Mountains (Stace-Smith & Converse, 1987). Thus, management for this virus in the western United States is not a concern other than testing germplasm that comes from areas where the virus occurs, whereas, if managing this virus on the basis of its occurrence in North America, there would be resources and chemicals used for controlling this virus in the western United States There are similar cases in blueberry and blackberry, *Blueberry necrotic ring blotch virus* and most of the viruses involved in blackberry yellow vein disease occur in the southeastern United States but have not been detected in the western parts of the country. The better information available on the distribution of the viruses and their vectors, the more informed decisions growers can make on disease management.

Virus control in nurseries is based on Best Management Practices (BMPs) combined with testing. BMPs are predicated on knowing the viruses and their vectors that occur in the area where the nursery is located. If nurseries are receiving G1 plants that are free of targeted pathogens, then the nursery's objective is to prevent infection of these plants during the plant multiplication cycles. If done by conventional propagation this process may take 3–5 years, during which the plants are grown in a field setting and exposed to virus vectors and possible virus infection. In cases where tissue culture propagation is used, thousands or millions of plants can be produced in tissue culture without exposure to vectors. These tissue culture plants can then be

Table 1 List of viruses known to infect strawberry worldwide, including acronym, genus, means of transmission, laboratory tests available for detection, and occurrence of each virus on a continental scale

Virus name	Acronym	Genus	Transmission	Laboratory detection	Regional occurrence					
					NA	SA	Europe	Africa	Asia	Australia/NZ
Apple mosaic	ApMV	*Ilarvirus*	Pollen/seed	ELISA, RT-PCR[a]	Yes	Yes	Yes	Yes	Yes	Yes
Arabis mosaic	ArMV	*Nepovirus*	Nematodes/semi-persistent[b,c]	ELISA, RT-PCR	Yes	Yes	Yes	Yes	Yes	Yes
Beet pseudo-yellows	BPYV	*Crinivirus*	Whiteflies/semi-persistent	ELISA, RT-PCR	Yes	Yes	Yes	Yes	Yes	Yes
Cucumber mosaic	CMV	*Cucumovirus*	Aphids/nonpersistent	ELISA, RT-PCR	Yes	Yes	Yes	Yes	Yes	Yes
Fragaria chiloensis cryptic	FCCV	*Alphacryptovirus?*	Pollen/seed[d]	RT-PCR	No	Yes	N/A	N/A	N/A	N/A
Fragaria chiloensis latent	FClLV	*Ilarvirus*	Pollen/seed	ELISA, RT-PCR	Yes	Yes	N/A	N/A	N/A	N/A
Raspberry ringspot	RpRSV	*Nepovirus*	Nematodes/semi-persistent	ELISA, RT-PCR	No	No	Yes	N/A	Yes	No
Strawberry chlorotic fleck	SCFaV	*Closterovirus*	Aphids/semi-persistent	RT-PCR	Yes	N/A	N/A	N/A	N/A	N/A
Strawberry crinkle	SCV	*Cytorhabdovirus*	Aphids/persistent	RT-PCR	Yes	Yes	Yes	N/A	Yes	Yes

Continued

Table 1 List of viruses known to infect strawberry worldwide, including acronym, genus, means of transmission, laboratory tests available for detection, and occurrence of each virus on a continental scale—cont'd

					Regional occurrence					
Virus name	**Acronym**	**Genus**	**Transmission**	**Laboratory detection**	**NA**	**SA**	**Europe**	**Africa**	**Asia**	**Australia/NZ**
Strawberry latent C	SLCV	*Nucleorhabdovirus*	Aphids/persistent?	N/A	Yes	N/A	N/A	N/A	N/A	N/A
Strawberry latent ringspot	SLRSV	Unassigned	Nematodes/semi-persistent	ELISA, RT-PCR	Yes	No	Yes	Yes	Yes	Yes
Strawberry leaf curl	StLCV	*Begomovirus*	Whiteflies/semi-persistent	ELISA, PCR	N/A	N/A	N/A	Yes	N/A	N/A
Strawberry mild yellow edge	SMYEV	*Potexvirus*	Aphids/persistent	ELISA. RT-PCR	Yes	Yes	Yes	Yes	Yes	Yes
Strawberry mottle	SMoV	*Sadwavirus*	Aphids/Semi-persistent	RT-PCR	Yes	Yes	Yes	N/A	Yes	Yes
Strawberry necrotic shock	SNSV	*Ilarvirus*	Pollen/seed	ELISA, RT-PCR	Yes	N/A	Yes	N/A	Yes	Yes
Strawberry pallidosis	SPaV	*Crinivirus*	Whiteflies/semi-persistent	RT-PCR	Yes	Yes	N/A	N/A	N/A	Yes
Strawberry pseudo mild yellow edge	SPMYEV	*Carlavirus*	Aphids/nonpersistent	ELISA?	Yes	N/A	N/A	N/A	Yes	N/A
Strawberry vein banding	SVBV	*Caulimovirus*	Aphids/semi-persistent	PCR	Yes	Yes	Yes	Yes	Yes	Yes

Tobacco necrosis D	TNVD	*Necrovirus*	Olpidium/?	ELISA/RT-PCR	Yes	Yes	Yes	Yes	Yes	Yes
Tobacco streak virus	TSV	*Ilarvirus*	Pollen/seed[d]	ELISA, RT-PCR	Yes	Yes	Yes[e]	Yes	Yes	Yes[e]
Tomato black ring	TBRV	*Nepovirus*	Nematodes/semi-persistent	ELISA, RT-PCR	No	N/A	Yes	N/A	Yes	N/A
Tomato ringspot	ToRSV	*Nepovirus*	Nematodes/semi-persistent	ELISA/RT-PCR	Yes	Yes	Yes	Yes	Yes	Yes

[a]Indicative of conventional and qRT-PCR.
[b]Pollen and seed transmitted.
[c]Also transmitted by pollen-feeding arthropods.
[d]Not confirmed in *Fragaria X ananassa*.
[e]May have been an isolate of SNSV.

Table 2 List of viruses known to infect raspberry, blackberry, and their hybrids worldwide, including acronym, genus, means of transmission, laboratory tests available for detection, and occurrence of each virus on a continental scale

					Regional occurrence					
Virus name	**Acronym**	**Genus**	**Transmission**	**Laboratory detection**	**NA**	**SA**	**Europe**	**Africa**	**Asia**	**Australia/ NZ**
Apple mosaic	ApMV	*Ilarvirus*	Pollen/seed[a]	ELISA, RT-PCR[b]	Yes	Yes	Yes	Yes	Yes	Yes
Arabis mosaic	ArMV	*Nepovirus*	Nematodes/semi-persistent[c,d]	ELISA, RT-PCR	Yes[e]	Yes[e]	Yes	Yes[e]	Yes[e]	Yes[e]
Beet pseudo-yellows	BPYV	*Crinivirus*	Whiteflies/semi-persistent	ELISA, RT-PCR	Yes	Yes[e]	Yes[e]	Yes[e]	Yes[e]	Yes[e]
Blackberry chlorotic ringspot	BCRV	*Ilarvirus*	Pollen/seed	RT-PCR	Yes	N/A	Yes	N/A	Yes	N/A
Blackberry vein banding	BVBaV	*Ampelovirus*	Mealybugs/semi-persistent?	RT-PCR	Yes	N/A	N/A	N/A	N/A	N/A
Blackberry virus E	BVE	*Unassigned*	Unknown (eriophyid mites?)	RT-PCR	Yes	N/A	N/A	N/A	N/A	N/A
Blackberry virus S	BVS	*Marafivirus*	Unknown (leafhoppers?)	RT-PCR	Yes	N/A	N/A	N/A	N/A	N/A
Blackberry virus Y	BVY	*Brambyvirus*	Unknown (eriophyid mites?)	RT-PCR	Yes	N/A	N/A	N/A	N/A	N/A
Blackberry yellow vein	BYVaV	*Crinivirus*	Whiteflies/semi-persistent	RT-PCR	Yes	N/A	N/A	N/A	N/A	N/A

Black raspberry necrosis	BRNV	*Sadwavirus*	Aphids/ nonpersistent	RT-PCR	Yes	N/A	Yes	N/A	N/A	N/A
Cherry leaf roll	CLRV	*Nepovirus*	Pollen/seed (nematodes?)[a]	ELISA, RT-PCR	Yes	Yes[e]	Yes	N/A	Yes[e]	Yes
Cherry rasp leaf	CRLV	*Cheravirus*	Pollen/seed (nematodes?)	ELISA, RT-PCR	Yes	N/A	N/A	N/A	N/A	N/A
Grapevine Syrah Virus 1	GSyV-1	*Marafivirus*	Unknown (leafhoppers?)	RT-PCR	Yes	Yes[e]	N/A	N/A	N/A	N/A
Impatiens necrotic spot	INSV	*Tospovirus*	Thrips/persistent	ELISA, RT-PCR	Yes	Yes[e]	Yes[e]	Yes[e]	Yes[e]	Yes[e]
Raspberry bushy dwarf	RBDV	*Ideovirus*	Pollen/seed	ELISA, RT-PCR	Yes	Yes	Yes	Yes	Yes	Yes
Raspberry latent	RpLV	*Unassigned*	Aphids/persistent	RT-PCR	Yes	N/A	N/A	N/A	N/A	N/A
Raspberry leaf curl	RLCV	*Unknown*	Unknown	Grafting	Yes	N/A	N/A	N/A	N/A	N/A
Raspberry leaf mottle	RLMV	*Closterovirus*	Aphids/semi-persistent	RT-PCR	Yes	N/A	Yes	N/A	N/A	N/A
Raspberry leaf blotch	RLBV	*Emaravirus*	Unknown (eriophyid mites?)	RT-PCR	N/A	N/A	Yes	N/A	N/A	N/A
Raspberry vein chlorosis	RVCV	*Rhabdovirus*	Aphids?	RT-PCR	Yes?	N/A	Yes	N/A	N/A	Yes
Raspberry ringspot	RpRSV	*Nepovirus*	Aphids/semi-persistent	ELISA, RT-PCR	No	No	Yes	N/A	Yes	No

Continued

Table 2 List of viruses known to infect raspberry, blackberry, and their hybrids worldwide, including acronym, genus, means of transmission, laboratory tests available for detection, and occurrence of each virus on a continental scale—cont'd

					Regional occurrence					
Virus name	**Acronym**	**Genus**	**Transmission**	**Laboratory detection**	**NA**	**SA**	**Europe**	**Africa**	**Asia**	**Australia/ NZ**
Rubus canadensis 1	RuCV-1	Unassigned	Unknown	RT-PCR	Yes	N/A	N/A	N/A	N/A	N/A
Rubus yellow net	RYNV	*Badnavirus*	Aphids/semi-persistent	PCR	Yes	N/A	Yes	N/A	N/A	N/A
Sowbane mosaic	SoMV	*Sobemovirus*	Pollen/seed	ELISA, RT-PCR	Yes	Yes[e]	Yes	Yes[e]	Yes[e]	Yes[e]
Strawberry latent ringspot	SLRSV	Unassigned	Nematodes/semi-persistent	ELISA, RT-PCR	Yes[e]	No	Yes	Yes[e]	Yes[e]	Yes[e]
Strawberry necrotic shock	SNSV	*Ilarvirus*	Pollen/seed	ELISA, RT-PCR	Yes	N/A	Yes?	N/A	Yes?	Yes?
Tomato black ring	TBRV	*Nepovirus*	Nematodes/semi-persistent[d]	ELISA, RT-PCR	No	N/A	Yes	N/A	Yes[e]	N/A
Tomato ringspot	ToRSV	*Nepovirus*	Nematodes/semi-persistent[d]	ELISA/RT-PCR	Yes	Yes[e]	Yes[e]	Yes[e]	Yes[e]	Yes[e]
Tobacco ringspot	TRSV	*Nepovirus*	Nematodes/semi-persistent	ELISA/RT-PCR	Yes	Yes[e]	Yes[e]	Yes[e]	Yes[e]	Yes[e]
Wineberry latent/ Blackberry calico	WLV/BlCV	Unassigned	Unknown		Yes	N/A	Yes	N/A	N/A	N/A
Blackberry leaf mottle	BLMaV	*Emaravirus*	Eriophyid mites?	RT-PCR	Yes	N/A	N/A	N/A	N/A	N/A

[a]Not confirmed in *Rubus*.
[b]Indicative of conventional and qRT-PCR.
[c]Pollen and seed transmitted.
[d]Also transmitted by pollen-feeding arthropods.
[e]Present but not reported in Rubus species.

Table 3 List of viruses known to infect blueberry and cranberry worldwide, including acronym, genus, means of transmission, laboratory tests available for detection, and occurrence of each virus on a continental scale

					Regional occurrence					
Virus name	Acronym	Genus	Transmission	Laboratory detection	NA	SA	Europe	Africa	Asia	Australia/ NZ
Blueberry shock	BlShV	*Ilarvirus*	Pollen/seed	ELISA, RT-PCR[a]	Yes	N/A	N/A	N/A	N/A	N/A
Blueberry latent	BlLV	*Amalgavirus*	Unknown	RT-PCR	Yes	N/A	N/A	N/A	Yes	N/A
Blueberry latent spherical	BLSV	*Nepovirus*	Nematodes/semi-persistent[b,c]/?	RT-PCR	N/A	N/A	N/A	N/A	Yes	N/A
Blueberry leaf mottle	BLMoV	*Nepovirus*	Nematodes/semi-persistent?[b,c]	ELISA, RT-PCR	Yes	N/A	N/A	N/A	N/A	N/A
Blueberry mosaic	BlMV	*Ophiovirus*	Olpidium/?	RT-PCR	Yes	Yes	Yes	Yes	Yes	Yes
Blueberry necrotic ring blotch	BNLBV	Unassigned	Unknown	RT-PCR	Yes	N/A	N/A	N/A	N/A	N/A
Blueberry red ringspot	BRRV	*Soymovirus*	Unknown	ELISA/PCR	Yes	N/A	Yes	N/A	Yes	Yes
Blueberry scorch	BlScV	*Carlavirus*	Aphids/nonpersistent	ELISA, RT-PCR	Yes	N/A	Yes	N/A	N/A	N/A
Blueberry shoestring	BlSSV	*Sobemovirus*	Aphids/nonpersistent	ELISA, RT-PCR	Yes	N/A	N/A	N/A	N/A	N/A

Continued

Table 3 List of viruses known to infect blueberry and cranberry worldwide, including acronym, genus, means of transmission, laboratory tests available for detection, and occurrence of each virus on a continental scale—cont'd

					Regional occurrence					
Virus name	Acronym	Genus	Transmission	Laboratory detection	NA	SA	Europe	Africa	Asia	Australia/NZ
Blueberry virus A	BVA	*Closterovirus*	Aphids/semi-persistent?	RT-PCR	Yes	N/A	N/A	N/A	Yes	Yes
Cherry leaf roll	CLRV	*Nepovirus*	Pollen/seed (nematodes?)	ELISA, RT-PCR	Yes[d]	Yes[d]	Yes[d]	N/A	Yes[d]	Yes
Peach rosette mosaic	PRMV	*Nepovirus*	Nematodes/semi-persistent[b,c]	ELISA, RT-PCR	Yes	N/A	Yes[d]	N/A	Yes[d]	N/A
Strawberry latent ringspot	SLRSV	Unassigned	Nematodes/semi-persistent[b,c]	ELISA, RT-PCR	Yes[d]	No	Yes[d]	Yes[d]	Yes[d]	Yes
Tobacco ringspot	TRSV	*Nepovirus*	Nematodes/semi-persistent[b,c]	ELISA/RT-PCR	Yes	Yes[d]	Yes[d]	Yes[d]	Yes[d]	Yes[d]
Tobacco streak	TSV	*Ilarvirus*	Pollen/seed	ELISA, RT-PCR	Yes	Yes[d]	Yes[d]	Yes[d]	Yes[d]	Yes[d]
Tomato ringspot	ToRSV	*Nepovirus*	Nematodes/semi-persistent[b,c]	ELISA/RT-PCR	Yes	Yes	Yes[d]	Yes[d]	Yes[d]	Yes[d]

[a]Indicative of conventional and qRT-PCR.
[b]Pollen and seed transmitted.
[c]Also transmitted by pollen-feeding arthropods.
[d]Present but not in blueberry or cranberry.

Table 4 List of viruses known to infect currant and gooseberry worldwide, including acronym, genus, means of transmission, laboratory tests available for detection, and occurrence of each virus on a continental scale

					Regional occurrence					
Virus name	**Acronym**	**Genus**	**Transmission**	**Laboratory detection**	**NA**	**SA**	**Europe**	**Africa**	**Asia**	**Australia/NZ**
Cucumber mosaic	CMV	*Cucumovirus*	Aphids/nonpersistent	ELISA, RT-PCR[a]	Yes[d]	Yes[d]	Yes	Yes[d]	Yes[d]	Yes[d]
Alfalfa mosaic	AlMV	*Alfamovirus*	Aphids/nonpersistent	ELISA, RT-PCR	Yes[d]	Yes[d]	Yes	Yes[d]	Yes[d]	Yes[d]
Arabis mosaic	ArMV	*Nepovirus*	Nematodes/semi-persistent[b,c]	ELISA, RT-PCR	Yes[d]	Yes[d]	Yes	Yes[d]	Yes[d]	Yes[d]
Black currant reversion	BRV	Nepovirus	Eriophyid mites/semi-persistent	ELISA, RT-PCR	No	N/A	Yes	N/A	Yes	Yes
Blackcurrant leafroll 1	BCLRaV-1	*Closterovirus*	Aphids/semi-persistent?	RT-PCR	N/A	N/A	Yes	N/A	N/A	N/A
Gooseberry vein banding	GVBaV	Badnavirus	Seed/pollen/aphids/semi-persistent	PCR	Yes	N/A	Yes	N/A	N/A	N/A
Raspberry ringspot	RpRSV	*Nepovirus*	Nematodes/semi-persistent	ELISA, RT-PCR	No	No	Yes	N/A	Yes[d]	No
Sowbane mosaic	SoMV	*Sobemovirus*	Pollen/seed	ELISA, RT-PCR	Yes[d]	Yes[d]	Yes	Yes[d]	Yes[d]	Yes[d]
Strawberry latent ringspot	SLRSV	Unassigned	Nematodes/semi-persistent	ELISA, RT-PCR	Yes[d]	No	Yes	Yes[d]	Yes[d]	Yes[d]

Continued

Table 4 List of viruses known to infect currant and gooseberry worldwide, including acronym, genus, means of transmission, laboratory tests available for detection, and occurrence of each virus on a continental scale—cont'd

					Regional occurrence					
Virus name	**Acronym**	**Genus**	**Transmission**	**Laboratory detection**	**NA**	**SA**	**Europe**	**Africa**	**Asia**	**Australia/ NZ**
Tobacco rattle	TRV	*Tobravirus*	Nematodes/semi-persistent	ELISA/ RT-PCR	Yes[d]	Yes[d]	Yes	Yes[d]	Yes[d]	Yes[d]
Tomato ringspot	ToRSV	*Nepovirus*	Nematodes/semi-persistent	ELISA/ RT-PCR	Yes[d]	Yes[d]	Yes	Yes[d]	Yes[d]	Yes[d]

[a]Indicative of conventional and qRT-PCR
[b]Pollen and seed transmitted.
[c]Also transmitted by pollen-feeding arthropods.
[d]Not confirmed in Ribes.

Table 5 List of viruses known to infect elderberry worldwide, including acronym, genus, means of transmission, laboratory tests available for detection, and occurrence of each virus on a continental scale

Virus name	Acronym	Genus	Transmission	Laboratory detection	Regional occurrence					
					NA	SA	Europe	Africa	Asia	Australia/NZ
Apple mosaic	ApMV	*Ilarvirus*	Pollen/seed	ELISA, RT-PCR[a]	Yes	Yes[d]	Yes	Yes[d]	Yes[d]	Yes[d]
Arabis mosaic	ArMV	*Nepovirus*	Nematodes/ semi-persistent[b,c]	ELISA, RT-PCR	Yes[d]	Yes[d]	Yes	Yes[d]	Yes[d]	Yes[d]
Blueberry scorch	BlScV	*Carlavirus*	Aphids/ nonpersistent	ELISA, RT-PCR	Yes[d]	N/A	Yes	N/A	N/A	N/A
Cherry leaf roll	CLRV	*Nepovirus*	Pollen/seed (nematodes?)	ELISA, RT-PCR	Yes[d]	Yes[d]	Yes	N/A	Yes[d]	Yes[d]
Cucumber mosaic	CMV	*Cucumovirus*	Aphids/ nonpersistent	ELISA, RT-PCR	Yes	Yes[d]	Yes	Yes[d]	Yes[d]	Yes[d]
Elderberry carlaviruses	ECVs	*Carlavirus*	Aphids/ nonpersistent?	RT-PCR	Yes	N/A	Yes	N/A	N/A	N/A
Elderberry latent	ELV	Carmovirus	Soil/pollen/seed	ELISA, RT-PCR	Yes	N/A	N/A	N/A	N/A	N/A
Strawberry latent ringspot	SLRSV	Unassigned	Nematodes/ semi-persistent	ELISA, RT-PCR	Yes[d]	No	Yes	Yes[d]	Yes[d]	Yes[d]
Tobacco ringspot	TRSV	*Nepovirus*	Nematodes/ semi-persistent	ELISA/ RT-PCR	Yes	Yes[d]	Yes[d]	Yes[d]	Yes[d]	Yes[d]

Continued

Table 5 List of viruses known to infect elderberry worldwide, including acronym, genus, means of transmission, laboratory tests available for detection, and occurrence of each virus on a continental scale—cont'd

					Regional occurrence					
Virus name	**Acronym**	**Genus**	**Transmission**	**Laboratory detection**	**NA**	**SA**	**Europe**	**Africa**	**Asia**	**Australia/NZ**
Tomato black ring	TBRV	*Nepovirus*	Nematodes/ semi-persistent	ELISA, RT-PCR	No	N/A	Yes	N/A	Yes	N/A
Tomato bushy stunt	TBSV	Tombusvirus	Soil/pollen/seed	ELISA, RT-PCR	Yes	Yes	Yes	Yes	Yes	No
Tomato ringspot	ToRSV	*Nepovirus*	Nematodes/ semi-persistent	ELISA/ RT-PCR	Yes	Yes	Yes	Yes	Yes	Yes

[a]Indicative of conventional and qRT-PCR.
[b]Pollen and seed transmitted.
[c]Also transmitted by pollen-feeding arthropods.
[d]Not confirmed in elderberry.

potted and acclimated in screenhouses with very minimal exposure to virus vectors. In most certification standards, plants are not allowed to flower or fruit in nurseries, to minimize the risk of infection by pollen-borne viruses. Thus, somewhere in the plant multiplication process, plants need to evaluated for trueness-to-type to ensure that an off-type was not propagated, this is true for conventional or tissue culture propagated nursery stock. In the United States, this is the responsibility of the nursery, not the Clean Plant Centers.

A golden rule for nursery production of berry crops, or for any crop, is to locate nurseries so that they are isolated from production fields of the same crop. As mentioned above, production fields are managed to control disease rather than controlling viruses. Therefore, if located nearby to a production field, nurseries could be under extreme virus pressure based on vectors in the area. This juxtaposition of nurseries and production fields, likely contributed to the high level of virus infection in nursery stock in strawberry in eastern Canada in 2012 (Martin, MacFarlane, et al., 2013; Martin, Peres, et al., 2013).

There are generally two to three cycles of propagation in nurseries, and each successive stage should require less stringent testing, thus, a G2 block would require more testing than a G3 block, and a G3 block would require more testing than a G4 block. G2 material is propagated to produce G3 material, and then G3 material is propagated to produce G4 material, which is sold to fruit growers. Therefore, an infection at the G2 level will result in its multiplication in the nursery system as well as provide a source of inoculum for further spread within the nursery. This is the rationale for having more stringent testing at the G2 level than G3, etc. There is debate whether the final stage of propagation in the nursery should be subjected to visual inspection or if there should be some level of testing. It is our contention that there should be a level of testing for the one or two highest risk viruses in the region. Based on hypergeometric sampling strategies, only 59 plants need to be tested to provide a 95% confidence that there is less than a 5% infection rate in the block (Anonymous, 2008). We would suggest that serological tests (ELISA) be developed for these high-risk or "canary" viruses to keep the cost of testing low, but effective. This level of testing will ensure that there has not been a complete breakdown in the system such as occurred recently in strawberry nurseries in Canada and California. As was shown in those cases, visual inspection was not adequate to detect the problem, which only became obvious once the plants were transplanted in fruit growers fields, and millions of dollars of crop were lost (Martin, MacFarlane, et al., 2013; Martin, Peres, et al., 2013; Martin & Tzanetakis, 2013).

The BMPs are based on the ecology and biology of the vectors of viruses that occur in the region, and are often defined in a certification standard. An explanation of why requirements are in the standard in terms of the vectors and the virus–vector relationships helps the nursery managers to appreciate the importance of following the BMPs, rather than just viewing them as arbitrary dictates from government regulators. For nematode-vectored viruses, cropping history, site selection, preplant nematode testing, and preplant soil treatment are addressed in the certification standard. Cropping history information is related to potential survival of viruses with nematode or fungal vectors. Therefore, the cropping history information often inquires about crops that can serve as hosts for these viruses and vectors rather than being limited to history of the nursery crop that is planned for the site. This is especially true for many of the nematode-transmitted viruses, which have fairly broad host ranges and thus, the potential to remain viable at a site for many years. As an example, a research plot used for *Tomato ringspot virus* research in strawberry in the 1980s had viable virus at the site in 2013 even though strawberries had not been grown at the site in more than 30 years (Martin, personnel observation). The virus has a broad host range and can perennate at a site for many years by infecting a range of broad leaf weeds, etc., or in this case, raspberries had also been grown at the site off and on during the 30 years. There are multiple methods to eliminate or greatly reduce nematode populations, including the use of soil fumigation and fallowing. Fallowing requires the absence of any plants that can serve as hosts for the nematode. Eliminating virus from a site can be accomplished in several ways in addition to soil fumigation, fallowing the location, or planting with a nonhost for the virus can be very effective (Pinkerton & Martin, 2005). The latter did not eliminate nematodes from the site, but virus control was equivalent to that obtained with fumigation with methyl bromide.

For aphid-transmitted viruses, isolation distances listed in a certification standard will be dependent on the type of virus/vector interaction and efficiency of aphid transmission, i.e., nonpersistent viruses will have a shorter isolation distance than persistently transmitted viruses that can be carried a long distance by the aphid vector. Another important consideration is host range of the aphid and virus, for example, the strawberry aphid has a very limited host range that includes *Fragaria* spp., *Potentilla* spp., and very occasionally Rosa spp. (Forbes & Chan, 1989), thus isolation from these hosts is critical rather than from all vegetation. Additionally, the four most important aphid-transmitted viruses of strawberry (Strawberry crinkle, Strawberry mild yellow edge, Strawberry mottle, and Strawberry veinbanding viruses) have limited host ranges and have only been reported in *Fragaria* spp. in nature, making most non-*Fragaria* vegetation near a nursery very low risk as virus reservoirs for these viruses. In blueberry, *Blueberry scorch virus* is transmitted very inefficiently (Lowery et al., 2008) where groups of 25 aphids only transmitted the virus to about 20% of the test plants using the most efficient of 8 aphid species tested. Additionally, this virus is transmitted in a non-persistent manner suggesting a relatively short isolation distance could greatly reduce the risk from this virus to nursery plants. In several cases, ilarviruses have been shown to be transmitted by thrips, thus if ilarviruses are present in the area of a nursery thrips transmission should be considered.

In most cases, pollen transmitted viruses should be of little concern to nursery planting stock, if the nursery plants are not allowed to flower. In the case of *Blueberry shock virus*, the horizontal transmission in field settings can be very rapid depending on genotype, thus even few flowers in a nursery setting could provide a conduit for transmission, if there are infected field or native plants nearby. Recent, work with *Raspberry bushy dwarf virus* (RBDV) demonstrated the ability of cross-family transmission via pollen through stigma infection, even though the pollen tube did not completely penetrate the stigma (Isogai, Yoshida, Nakanowatari, & Yoshikawa, 2014). This is likely a very rare event that would pose very little risk in a nursery setting with limited flowering of the plants. It may explain a mechanism for host jumping of viruses that are limited to pollen transmission.

Recently, several eriophyid mite-transmitted viruses, or viruses where the best guess vector is an eriophyid mite based on sequence analysis, have been identified in berry crops (Hassan et al., 2013; McGavin et al., 2012; Quito-Avila, Brannen, Cline, Harmon, & Martin, 2013). These vectors are extremely small and screening would not prevent them from entering a screenhouse or screened greenhouse. Thus, in areas where these viruses are present along with the vectors, production of nursery stock free of these viruses will be very difficult. Thrips-transmitted viruses pose a similar risk in that thrips-proof screening greatly limits any airflow and in warmer environments it would be very difficult to prevent extreme heat buildup in screenhouses. In such cases, it would be necessary to develop a pest management program to control the eriophyid mites and/or thrips.

8. BMPS, KNOWING THE HIGH-RISK VIRUSES

There are more than 80 viruses reported to infect berry crops (Martin, MacFarlane, et al., 2013; Martin, Peres, et al., 2013; Martin & Tzanetakis,

2006; Martin, Polashock, & Tzanetakis, 2012), but fortunately in any given area the number of viruses that need to be managed is much smaller. A key to managing viruses in nursery settings is to know what are the high-risk viruses in the region and where the nursery is located (Martin & Tzanetakis, 2013). As an example, for *Rubus* nurseries in the Pacific Northwest there are four viruses of concern given that site selection and preparation have addressed the issue with nematode-transmitted viruses. These viruses include *Black raspberry necrosis virus* (Secoviridae, unassigned genus), *Raspberry leaf mottle virus* (Closterovirus), *Raspberry latent virus* (Reoviridae, unassigned genus), and *Rubus yellow net virus* (Badnavirus), all of which are transmitted by the large raspberry aphid, *Amphorophora agathonica* (Converse, Stace-Smith, & Jones, 1987). Thus, controlling a single vector should be the main focus in managing these viruses. In addition, studies on flights of this vector have shown that it is not moving early in the season (approximately 950 days, base 50 from Jan 1), in northern Washington flights of this aphid start mid to late June, with a sharp peak through mid-July then very little aphid movement until late in the season (Lightle, Quito-Avila, Martin, & Lee, 2014). This suggests that vector control can be targeted, with most efforts aimed at the peak flight times. Additionally, for certifying agencies, monitoring of virus infection in *Rubus* nurseries in this region should focus on these four viruses for floricane type *Rubus* cultivars, since they do not flower in the first year canes that would be present in nurseries. For primocane fruiting cultivars, it will be necessary to consider the pollen-borne viruses: RBDV, *Strawberry necrotic shock virus*, *Apple mosaic virus*, and *Blackberry chlorotic ringspot virus*, since these cultivars flower in first year canes and will very likely have flowers in the nursery mid to late in the summer (Martin, MacFarlane, et al., 2013; Martin, Peres, et al., 2013).

9. VIRUS CONTROL IN COMMERCIAL FIELDS

Fruit growers are interested in disease control rather than virus control, which is fortunate since many of the viruses of berry crops are symptomless (Martin, MacFarlane, et al., 2013; Martin, Peres, et al., 2013; Martin et al., 2012; Martin & Tzanetakis, 2013) or produce symptoms but have a very minor impact on crop production when present in single infections (Martin, MacFarlane, et al., 2013; Martin, Peres, et al., 2013). *Blackberry calico virus* produces dramatic symptoms but has minimal impact on yield or quality (Converse, 1987), it spreads rapidly and thus control in a grower's field could not be justified. In many parts of the world, strawberry production is done with annual plantings, which means that if plants from the nursery are free of, or have a very low incidence of virus infection, growers can produce a profitable crop with little costs for virus control. In perennial strawberry production, vector control is often required. The other berry crops are all grown as longer term perennials and in the case of blueberries it may take 3–5 years after planting before the first crop is harvested, thus, vector control can be very important depending on vector populations and transmission efficiency. *Blueberry shoestring virus* in areas where there is an efficient vector and virus inoculum present is an example of a case where vector control is necessary to manage this disease in the field (Ramsdell, 1987).

9.1. Virus Resistance and Tolerance

There are few examples of virus resistance in berry crops, the best known and most widely used is resistance to RBDV, which is controlled by a single dominant gene. This source of resistance is used widely in breeding programs but appears to be closely linked with some negative horticultural traits, since breeders have had difficulty developing acceptable cultivars with RBDV resistance. A resistance breaking strain of RBDV (RBDV-RB) has been reported in Europe and Russia (Wilson, Knight, & Barbara, 1983) but has not been detected in other parts of world. RBDV-RB infects almost all raspberry cultivars that have been reported to be resistant to the virus (Knight & Barbara, 1999). The fact that there are cultivars that are resistant to the RBDV-RB strain suggests that there are multiple genes for resistance to RBDV that could be pyramided for a more durable resistance. In blueberry, "Bluecrop" is resistant to *Tomato ringspot virus* and can be used in areas where the virus and nematode vector are present. Since blueberry is a long-term crop, ideally >20 years, soil fumigation to control nematodes is not very effective since nematodes below the fumigant control profile in the soil will migrate up and infect plants before they are in full production.

Virus tolerance is very common in the cultivars of strawberry, raspberry, and blackberry. In these crops, very few viruses other than the nepoviruses cause symptoms in single infections. In some cases, such as "Totem" or "Puget Reliance" strawberries, plants are tolerant to mixed infections with *Strawberry veinbanding, Strawberry mottle*, and *Strawberry mild yellow edge viruses*. Prior to 2000, this tolerance had been very useful for production in Oregon, Washington, and British Columbia where aphid populations are very high most years, since these were the three common aphid-borne viruses in this region. Starting about 2000, the incidence of SCV increased rapidly in this area, and since then the virus tolerance has been much less effective. In annual strawberry production, the plants are in the fruiting fields for less than a year; thus if starting with "clean" plants, virus diseases often are not a problem since very few plants will have a disease causing complex at the end of the production cycle (Martin & Tzanetakis, 2013). In areas where the strawberry aphid (*Chaetosiphon* spp.) does not occur, viruses should be a minor problem for strawberry production. The four viruses that cause most damage in strawberry worldwide are transmitted by aphids in this genus, and if these viruses are not present the whitefly-transmitted criniviruses are asymptomatic. Raspberry mosaic disease is caused by a virus complex where many cultivars are tolerant to any single component of the complex, but mixed infections can lead to a serious disease (Martin, MacFarlane, et al., 2013; Martin, Peres, et al., 2013). The same is true with blackberry yellow vein disease, which can be caused by a number of different virus combinations, and the more viruses the more serious the disease (Martin, MacFarlane, et al., 2013; Martin, Peres, et al., 2013). For fruit growers, virus tolerance is a very important component in disease management.

When vector management is necessary, growers can target specific virus vectors to control part of a virus complex rather than trying to control all viruses. Identifying which vector to target should be based on the vector biology, i.e., is it dispersing throughout the growing season (whiteflies)? Is it a sedentary feeder most of the season as is the case with many aphids? In the case of the aphid-borne viruses in red raspberry, aphid flights occur over a short time period during the season (Lightle et al., 2014) making control easier than for many other types of aerial vectors. Thus, control measures should focus on the easiest vector(s) to manage to achieve disease control, rather than trying to control all vectors. A potential problem with tolerant cultivars is that they can serve as a source of inoculum that can spread to adjacent cultivars that lack virus tolerance.

9.2. Vector Resistance

Aphid resistance has been deployed very successfully in red raspberry that has resulted in resistance to the raspberry mosaic disease complex (Martin, MacFarlane, et al., 2013; Martin, Peres, et al., 2013). In North America, a single dominant gene has been used for over 40 years and provided high levels of resistance. In the last decade, there have been biotypes of the large raspberry aphid that has overcome this resistance (Dossett & Kempler, 2012). Aphid resistance to the European large raspberry aphid (*Amphorophora idaei*) has been less successful. Multiple dominant genes for resistance to this aphid have been deployed, Ag_1 from red raspberry and Ag_{10} from black raspberry, as well as several minor genes for resistance, and in all cases new biotypes of the aphid developed that overcame the resistance. Recently, aphid resistance has been identified in native populations of *Rubus occidentalis* from several locations in North America (Dossett & Finn, 2010), and the mechanism of the aphid resistance in black raspberry is different from that in red raspberry (Lightle, 2013). This suggests that it should be possible to pyramid the two types of aphid resistance to develop a more durable resistance that can be used to protect raspberries against aphid-borne viruses. It is not known if the mechanism of resistance from the Ag_{10} gene that came from black raspberry is the same as that of the resistance gene identified in black raspberry more recently in North America.

There are two sources of resistance to the blackcurrant gall mite (*Cecidophyopsis ribis*) that have been used successfully for mite control, which is a serious pest of black currants and also vectors BRV (Jones, Brennan, McGavin, & Lemmetty, 1998), the most damaging virus of blackcurrants. One gene, *P*, is from *Ribes nigrum* ssp. *sibiricum* (Anderson, 1971), and the second, *Ce* (Knight, Keep, Briggs, & Parker, 1974) is derived from gooseberry. These genes are used widely in black currant breeding programs, and efforts are underway using molecular markers to pyramid these genes to develop more durable resistance (Mazeikiene, Bendokas, Stanys, & Siksnianas, 2012).

9.3. High-risk Viruses and Mixed Infections

Effective virus disease control in fruit production fields requires knowledge of the viruses that are present and spreading in the area, which of these viruses lead to disease when they occur in mixed infections. These are the high-risk viruses for the area. How each of the high-risk viruses is vectored is also important in developing a virus disease control program. When the risk of virus disease is evaluated, cultivar selection should consider any vector resistance and virus resistance or tolerance that is available. In the Pacific Northwest, the four aphid-transmitted viruses are the high-risk strawberry viruses. In this region, virus disease control in strawberry cultivars that have tolerance to multiple aphid-transmitted viruses but exhibit disease when infected with three or four of the aphid-vectored viruses can be accomplished with vector control that targets the weakest links in the disease complex. In this case, SCV is the easiest virus to target since has a long latent period in the vector, is only transmitted by *Chaetoshipon* spp., and this aphid has a relatively narrow window of dispersal during the growing season (Sylvester, Richardson, & Frazier, 1974). Thus, one to three well-timed sprays during peak aphid flights can be quite effective at controlling this virus, and as a result virus disease in cultivars with tolerance to multiple aphid-vectored viruses.

In blackberries in the southeastern United States, there are several virus complexes that cause blackberry yellow vein disease, with a different complex observed in Arkansas, Mississippi, and North Carolina (Martin, MacFarlane, et al., 2013; Martin, Peres, et al., 2013; Susaimuthu, Tzanetakis, Gergerich, Kim, et al., 2008; Susaimuthu, Tzanetakis, Gergerich, & Martin, 2008; Thekke-Veetil et al., 2013). This disease is more complicated to control since within each complex there are multiple vectors that transmit viruses (Poudel, Ho, Laney, Khadgi and Tzanetakis, 2014; Poudel et al., 2013). However, the same principle applies: know the viruses in the area that contribute to the disease, know their vectors, and when they disperse and develop control measures based on the easiest virus(es) to control given the biology of the vector and virus/vector interactions. In all berry crops other than strawberry, the plantings are expected to be fruitful for 10 years or more. With these crops, there is a longer period of time for plantings to come into production, so replanting is much more expensive in terms of lost production. Thus, virus control is more critical since the plants need to be protected for multiple growing seasons before they are productive.

9.4. Coordinated Control Efforts

In some situations, the primary virus inoculum in a region is from the crop itself rather than from native vegetation. This is the case for the aphid-borne viruses in strawberry in production areas where there are few native strawberries. Even if there are native strawberries, the production fields are often the most important source of virus inoculum. In areas where strawberries are grown as perennials, with a cropping cycle of 3–4 years, the aphid-borne viruses are primarily circulating between the production fields. This is the case in the Pacific Northwest of the North America (Oregon, Washington, and British Columbia). Effective virus control in these situations is best achieved through a coordinated vector control program for a full cropping cycle. Thus, implementation of an area-wide aphid management program for 4 years to reduce the aphid and virus pressure in the area will greatly reduce virus inoculum. In this case, the viruses and vector are primarily limited to *Fragaria* species. Without a coordinated effort, the vectors and viruses will continue to rotate between fields and a high level of vector and virus pressure will be a never ending problem.

Controlling whitefly-transmitted viruses in strawberry in California is much more challenging than controlling the aphid-borne viruses, since the whitefly (*Trialeurodes vaporariorum*) has a broad host range and is an active flyer throughout the season (Tzanetakis, Martin, & Wintermantel, 2013; Wintermantel, 2004). Additionally, one of the criniviruses in strawberry (*Beet pseudo yellows virus*) has a broad host range and the strawberry production in many parts of California occurs in areas with very diverse agriculture. In this case, the best strategy is to focus efforts on the aphid-transmitted viruses simply because the vector and viruses are primarily limited to strawberry hosts. Additionally, in tests with mixed infections of the whitefly-transmitted criniviruses in strawberry there were not observed disease symptoms in the absence of any aphid-borne viruses in several cultivars tested including "Hood" one of the most sensitive strawberry cultivars in terms of virus disease.

Isolation from other production fields of the same crop is very effective at limiting virus disease pressure if one starts with "clean" plants. In the absence of virus, vector populations often are not a problem. If vector populations reach levels to be a pest problem in the absence of virus, control often only needs to be targeted to reduce populations at harvest time to minimize contaminants in the harvested fruit. In some cases, vectors can reach populations to become serious pests for crop production, and control is required regardless of the presence of virus. However, in most cases economic thresholds for vector control are much lower if managing for virus control than they are if managing for vector control in the absence of virus. As an example, strawberry fields located on Lulu Island along the coast in British Columbia had no virus symptoms after 4 years even though the vector populations were quite high. There is only one strawberry grower on the island, and the air currents are coming from Pacific Ocean. This was at the same time, when strawberry fields 40 km inland from Lulu Island were experiencing severe virus problems (Martin, unpublished). Isolation from other commercial fields can be very effective in control of pollen-borne viruses.

Effective control of virus diseases in berry crops is dependent on knowledge of the viruses, vectors, the virus/vector relationship, their distribution geographically, vector phenology, host resistance and tolerance to viruses and/or vectors. As researchers, we are trained to focus on the diseases and their management, but growers also need to consider all factors in growing, harvesting, and marketing their crops. In many cases, what we consider optimal disease management strategies may not fit well into fruit production systems. Labor availability may be more critical for when a specific management practice is implemented then when an aphid flight is occurs. The marketability of a specific cultivar may be much more important than whether it is virus or vector resistant. As researchers and extension specialists, we provide information to growers, but must realize that they need to consider the entire fruit production enterprise rather than just disease control to remain profitable. Understanding the constraints that fruit growers have to deal with will help us develop control measures that can better serve the industries.

REFERENCES

Agindotan, B., & Perry, K. L. (2007). Macroarray detection of plant RNA viruses using randomly primed and amplified complementary DNAs from infected plants. *Phytopathology*, *97*, 119–127.

Anderson, M. M. (1971). Resistance to gall mite (*Phytoptus ribis* Nal.) in the Eucoreosma section of ribes. *Euphytica*, *20*, 422–426.

Anonymous (2008). *ISPM No. 31, International standards for phytosanitary measures*. Rome, Italy: IPPC Secretariat.

Barba, M., Czosnek, H., & Hadidi, A. (2014). Historical perspective, development and applications of next-generation sequencing in plant virology. *Viruses*, *6*, 106–136.

Converse, R. H. (1987). Blackberry calico. In R. H. Converse (Ed.), *USDA-ARS Agric Handbook Number 631. Virus diseases of small fruits* (pp. 245–246). Washington, D.C.: U.S. Govt Printing Office.

Converse, R. H., Stace-Smith, R., & Jones, A. T. (1987). Raspberry mosaic. In R. H. Converse (Ed.), *USDA-ARS Agric Handbook Number 631. Virus diseases of small fruits* (pp. 245–246). Washington, D.C.: U.S: Govt Printing Office.

Dossett, M., & Finn, C. E. (2010). Identification of resistance to the large raspberry aphid in black raspberry. *Journal of American Society for Horticultural Science*, *135*, 438–444.

Dossett, M., & Kempler, C. (2012). Biotypic diversity and resistance to the raspberry aphid *Amphorophora agathonica* in Pacific Northwestern North America. *Journal of American Society for Horticultural Science*, *137*, 445–451.

Forbes, A. R., & Chan, C. K. (1989). Aphids of British Columbia. In *Technical Bulletin, 1989-IE* Vancouver, B.C., Research Branch, Agriculture Canada, 260 p.

Frazier, N. W. (1953). A latent virus of *Fragaria vesca*. *Plant Disease Reporter*, *37*, 606–608.

Gergerich, R. C., Welliver, R., Gettys, S., Osterbauer, N. K., Kamenidou, S., Martin, R. R., et al. (2015). Safeguarding fruit crops in the age of agricultural globalization. *Plant Disease*, http://dx.doi.org/10.1094/PDIS-07-14-0762-FE..

Hassan, M., Keller, K. E., Martin, R. R., Sabanadzovic, S., & Tzanetakis, I. E. (2013). A new emaravirus detected in blackberries affected by yellow vein disease. *Petria*, *22*, 357–361.

Ho, T., & Tzanetakis, I. E. (2014). Developing a virus detection and discovery pipeline using next generation sequencing. *Virology*, *471–473*, 54–60.

Isogai, M., Yoshida, T., Nakanowatari, C., & Yoshikawa, N. (2014). Penetration of pollen tubes with accumulated *Raspberry bushy dwarf virus* into stigmas is involved in initial infection of maternal tissue and horizontal transmission. *Virology*, *452–453*, 247–253.

Jones, A. T., Brennan, R. M., McGavin, W. J., & Lemmetty, A. (1998). Galling and reversion disease incidence in a range of blackcurrant genotypes, differing in resistance to the blackcurrant gall mite (*Cecidophyopsis ribis*) and blackcurrant reversion disease. *Annals of Applied Biology, 133*, 375–384.

Jones, A. T., Gordon, S. C., & Jennings, D. L. (1984). A leaf-blotch disorder of tayberry associated with the leaf and bud mite (*Phyllocoptes gracilis*) and some effects of three aphid-borne viruses. *Journal of Horticultural Science, 59*, 523–528.

Knight, V. H., & Barbara, D. J. (1999). A review of Raspberry bushy dwarf virus at the HRI-East Malling and the situation on a sample of commercial holdings in England in 19967 and 1996. *Acta Horticulturae, 505*, 263–271.

Knight, R. L., Keep, E., Briggs, J. B., & Parker, J. (1974). Transference of resistance to black currant gall mite *Cecidophyopsis ribis*, from gooseberry to black currant. *Annals of Applied Biology, 76*, 123–130.

Lightle, D. (2013). *Interactions of Amphorophora agathonica with raspberry viruses and resistant red and black raspberry.* Ph.D. Thesis, Oregon State University, 109 p.

Lightle, D., Quito-Avila, D. F., Martin, R. R., & Lee, J. C. (2014). Seasonal phenology of *Amphorophora agathonica* and spread of viruses in red raspberry in Washington. *Environmental Entomology, 43*, 467–473.

Lowery, T. D., Bernardy, M. G., Deyoung, R. M., & French, C. J. (2008). Identification of new aphid vector species of *Blueberry scorch virus*. *Journal of the Entomological Society of British Columbia, 105*, 27–33.

MacDonald, S. G., Martin, R. R., & Bristow, P. R. (1991). Characterization of an ilarvirus associated with a necrotic shock reaction in blueberry. *Phytopathology, 81*, 210–214.

Martin, R. R., MacFarlane, S., Sabanadzovic, S., Quito, D., Poudel, B., & Tzanetakis, I. E. (2013). Viruses and virus diseases of *Rubus*. *Plant Disease, 97*, 169–182.

Martin, R. R., Peres, N. A., & Whidden, A. (2013). Virus outbreak in several Nova Scotia strawberry nurseries affects fruit growers in the United States. *Phytopathology, 103*, S2.90–S2.91.

Martin, R. R., Polashock, J. J., & Tzanetakis, I. E. (2012). New and emerging viruses of blueberry and cranberry. *Viruses, 4*, 2831–2852.

Martin, R. R., & Tzanetakis, I. E. (2006). Characterization, detection and management of strawberry viruses. *Plant Disease, 90*, 384–396.

Martin, R. R., & Tzanetakis, I. E. (2013). High risk strawberry viruses by region in the United States and Canada: Implications for certification, nurseries and fruit production. *Plant Disease, 97*, 1358–1362.

Mazeikiene, I., Bendokas, V., Stanys, V., & Siksnianas, T. (2012). Molecular markers linked to resistance to the gall mite in blackcurrant. *Plant Breeding, 131*, 762–766.

McGavin, W. J., Mitchell, C., Cock, P. J. A., Wright, K. M., & MacFarlane, S. A. (2012). Raspberry leaf blotch virus, a putative new member of the genus Emaravirus, encodes a novel genomic RNA. *Journal of General Virology, 93*, 430–437.

Pallas, V., Aparicio, F., Herranz, M. C., Amari, K., Sanchez-Pina, M. A., Myrta, A., et al. (2012). Ilarviruses of *Prunus* spp.: A continued concern for fruit trees. *Phytopathology, 102*, 1108–1120.

Pinkerton, J. N., & Martin, R. R. (2005). Management of tomato ringspot virus in red raspberry with crop rotation. *International Journal Fruit Science, 5*, 55–67.

Poudel, B., Ho, T., Laney, A., Khadgi, A., & Tzanetakis, I. E. (2014). Epidemiology of *Blackberry chlorotic ringspot virus*. *Plant Disease, 98*, 547–550.

Poudel, B., Wintermantel, W. M., Cortez, A. A., Ho, T., Khadgi, A., & Tzanetakis, I. E. (2013). Epidemiology of *Blackberry yellow vein associated virus*. *Plant Disease, 97*, 1352–1357.

Quito-Avila, D. F., Brannen, P. M., Cline, W. O., Harmon, P. F., & Martin, R. R. (2013). Genetic characterization of *Blueberry necrotic ring blotch virus*, a novel RNA virus with unique genetic features. *Journal of General Virology, 94*, 1426–1434.

Quito-Avila, D. F., Lightle, D., & Martin, R. R. (2014). Effect of *Raspberry bushy dwarf virus*, *Raspberry leaf mottle virus*, and *Raspberry latent virus* on plant growth and fruit crumbliness in 'Meeker' red raspberry. *Plant Disease*, *98*, 176–183.

Ramsdell, D. C. (1987). Blueberry shoestring. In R. H. Converse (Ed.), *USDA-ARS Agric Handbook Number 631. Virus diseases of small fruits* (pp. 103–105). Washington, D.C.: U.S. Govt Printing Office.

Sabanadzovic, S., & Abou Ghanem-Sabanadzovic, N. (2009). Identification and molecular characterization of a marafivirus in Rubus spp. *Archives of Virology*, *154*, 1729–1735.

Sabanadzovic, S., Abou Ghanem-Sabanadzovic, N., & Tzanetakis, I. E. (2011). *Blackberry virus E*: An unusual flexivirus. *Archives of Virology*, *156*, 1665–1669.

Sabanadzovic, S., Ghanem-Sabanadzovic, N. A., & Gorbalenya, A. E. (2009). Permutation of the active site of putative RNA-dependent RNA polymerase in a newly identified species of plant alpha-like virus. *Virology*, *394*, 1–7.

Sdoodee, R., & Teakle, D. S. (1987). Transmission of *Tobacco streak virus* by *Thrips tabaci*: A new method of plant virus transmission. *Plant Pathology*, *36*, 377–380.

Stace-Smith, R., & Converse, R. H. (1987). Raspberry leaf curl. In R. H. Converse (Ed.), *USDA-ARS Agric Handbook Number 631. Virus diseases of small fruits* (pp. 187–190). Washington, D.C.: U.S. Govt Printing Office.

Susaimuthu, J., Gergerich, R. C., Bray, M. M., Clay, K. A., Clack, J. R., Tzanetakis, I. E., et al. (2007). The incidence and ecology of *Blackberry yellow vein associated virus*. *Plant Disease*, *91*, 809–813.

Susaimuthu, J., Tzanetakis, I. E., Gergerich, R. C., Kim, K. S., & Martin, R. R. (2008). Viral interactions lead to decline of blackberry plants. *Plant Disease*, *92*, 1288–1292.

Susaimuthu, J., Tzanetakis, I. E., Gergerich, R. C., & Martin, R. R. (2008). A member of a new genus in the *Potyviridae* infects *Rubus*. *Virus Research*, *131*, 145–151.

Susi, P. (2004). Black currant reversion virus, a mite-transmitted nepovirus. *Molecular Plant Pathology*, *5*, 167–173.

Sylvester, E. S., Richardson, J., & Frazier, N. W. (1974). Serial passage of *Strawberry crinkle virus* in the aphid *Chaetosiphon jacobi*. *Virology*, *59*, 301–306.

Thekke-Veetil, T., Aboughanem Sabanadzovic, N., Keller, K. E., Martin, R. R., Sabanadzovic, S., & Tzanetakis, I. E. (2013). Molecular characterization and population structure of *Blackberry vein banding associated virus*, new ampelovirus associated with yellow vein disease. *Virus Research*, *178*, 234–240.

Thekke-Veetil, T., Ho, T., Keller, K. E., Martin, R. R., & Tzanetakis, I. E. (2014). A new *Ophiovirus* is associated with blueberry mosaic disease. *Virus Research*, *189*, 92–96.

Thompson, J. R., Fuchs, M., McLane, H., Celebi-Toprak, F., Fischer, K. F., Potter, J. L., et al. (2014). Profiling viral infections in grapevine using a randomly primed reverse transcription-polymerase chain reaction/macroarray multiplex platform. *Phytopathology*, *104*, 211–219.

Tzanetakis, I. E., Guzmán-Baeny, T. L., VanEsbroeck, Z. P., Fernandez, G. E., & Martin, R. R. (2009). First report of *Impatiens necrotic spot virus* in blackberry Southeastern United States. *Plant Disease*, *93*, 432.

Tzanetakis, I. E., Halgren, A. B., Keller, K. E., Hokanson, S. C., Maas, J. L., McCarthy, P. L., et al. (2004). Identification and detection of a virus associated with strawberry pallidosis disease. *Plant Disease*, *88*, 383–390.

Tzanetakis, I. E., Martin, R. R., & Wintermantel, W. M. (2013). Epidemiology of criniviruses: An emerging problem in world agriculture. *Frontiers in Microbiology*. *4*. http://dx.doi.org/10.3389/fmicb.2013.00119.

Tzanetakis, I. E., Wintermantel, W. M., & Martin, R. R. (2003). First report of *Beet pseudo yellows virus* in strawberry in the United States: A second crinivirus able to cause pallidosis disease. *Plant Disease*, *87*, 1398.

Wang, D., Coscoy, L., Zylberberg, M., Avila, P. C., Boushey, H. A., Ganem, D., et al. (2002). Microarray-based detection and genotyping of viral pathogens. *Proceeding of the National Academy of Sciences of the United States of America*, *99*, 15687–15692.
Wilson, S. C., Knight, V. H., & Barbara, D. J. (1983). Raspberry bushy dwarf virus and field infection of Malling Jewel. *Plant Pathology*, *32*, 357–359.
Wintermantel, W. M. (2004). Emergence of greenhouse whitefly *(Trialeurodes vaporariorum)* transmitted criniviruses as threats to vegetable and fruit production in North America. St. Paul, MN: The American Phytopathological Society. APSnet Feature, www.apsnet.org/online/feature/whitefly/.

INDEX

Note: Page numbers followed by "*f*" indicate figures and "*t*" indicate tables.

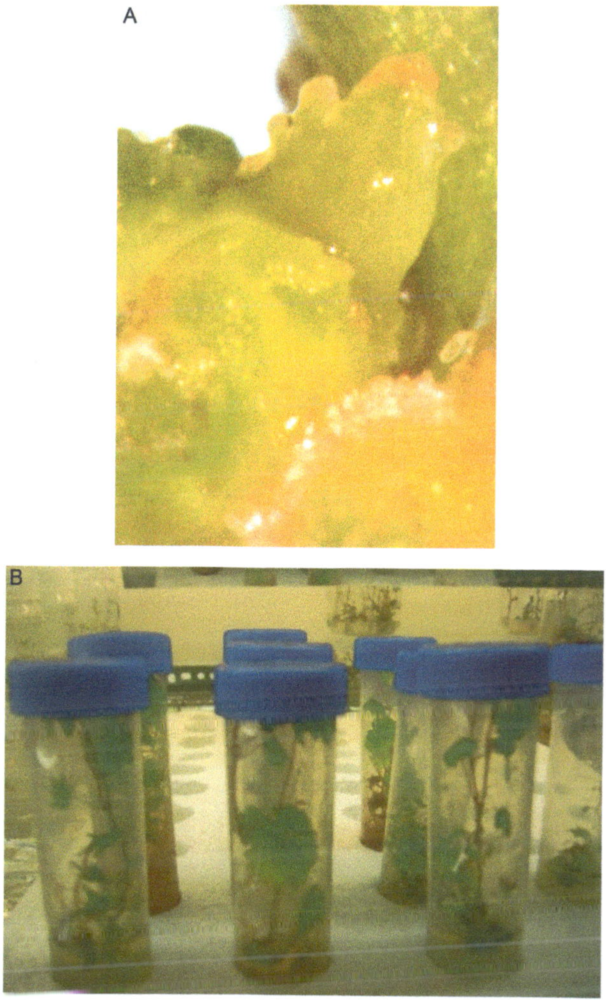

Christina Varveri *et al.*, Figure 1 Grapevine meristem culture *in vitro*. (A) Grapevine meristem and (B) rooted *in vitro* grapevine microplants in transparent polycarbonate tubes. *Minas, Papayiannis, and Kapari-Isaia (2007).*

Christina Varveri *et al.*, Figure 2 Potato meristem culture *in vitro*. (A) Aseptically prepared microcuttings for *in vitro* mass micropropagation and (B) culture with grown up and rooted microcuttings ready to be transferred for acclimatization *ex vitro*. *Minas, Gregoriou, Kapari-Isaia, and Papayiannis (2007).*

Figure 3 (See legend on next page.)

Christina Varveri *et al.*, Figure 4 Peach micrografting *in vitro*. (A) Peach meristems 0.2–0.5 mm long micrografted *in vitro*. (B) Micrografted peach plantlets in culture tubes ready to be transferred for acclimatization *ex-vitro*. *Kapari-Isaia, Kyriakou, Samouel, Papayiannis, and Gabriel (2009).*

Christina Varveri *et al.*, Figure 3 Citrus micrografing *in vitro*. (A and B) Meristems 0.2–0.5 mm long are aseptically isolated in a laminar flow cabinet using a binocular stereomicroscope to visualize the tissues. (C and D) Micrografted citrus plantlets in culture tubes. *Kapari-Isaia et al. (2002).*

Gad Loebenstein, Figure 1 *I. setosa* infected with *Sweet potato feathery mottle virus* (SPFMV), showing vein clearing.

Gad Loebenstein, Figure 2 Nursery of virus-tested stocks.

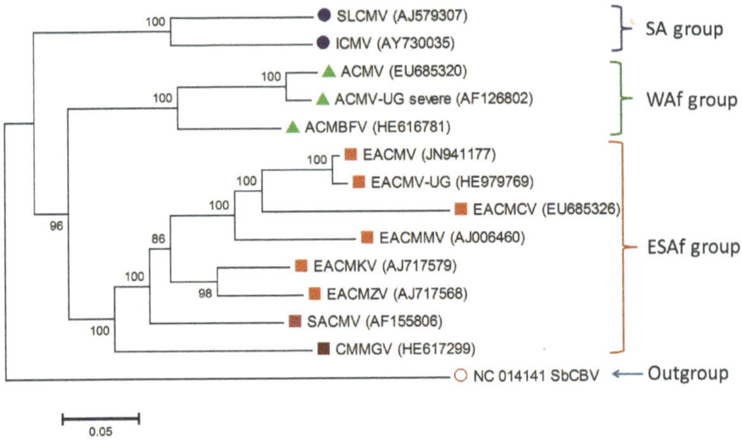

James P. Legg *et al.*, Figure 1 Phylogenetic relationships of the cassava mosaic geminiviruses. Phylogenetic relationship of 11 species of cassava mosaic geminiviruses based on the alignment of complete DNA-A using MEGA 5 software (Tamura et al., 2011). Sequences were aligned using the ClustalW algorithm, and the tree was constructed by the Neighbor-Joining method. The tree was rooted using *Soybean chlorotic blotch virus* (SbCBV) as an outgroup. The evolutionary distances were computed using the maximum composite likelihood method and are in the units of the number of base substitutions per site. The tree is drawn to scale, with branch lengths in the same units as those of the evolutionary distances used to infer the phylogenetic tree. The percentages of replicate trees in which the associated taxa clustered together in the bootstrap test (1000 replicates) are shown at the branch nodes. There were a total of 2627 positions in the final dataset. All positions containing gaps and missing data were eliminated from the dataset (complete deletion option). The NCBI GenBank accession numbers of 14 DNA-A sequences are indicated in parenthesis. ACMV: *African cassava mosaic virus*; ACMV-UG: ACMV-Uganda; ACMBFV: *African cassava mosaic Burkina Faso Virus*; CMMGV: *Cassava mosaic Madagascar virus*; EACMCV: *East African cassava mosaic Cameroon virus*; EACMKV: *East African cassava mosaic Kenya virus*; EACMMV: *East African cassava mosaic Malawi virus*; EACMZV: *East African cassava mosaic Zanzibar virus*; EACMV: *East African cassava mosaic virus*; EACMV-UG: EACMV-Uganda; ICMV: *Indian cassava mosaic virus*; SACMV: *South African cassava mosaic virus*; SLCMV: *Sri Lankan cassava mosaic virus*; SA: South Asian; WA: West African; ESA: Eastern and Southern African.

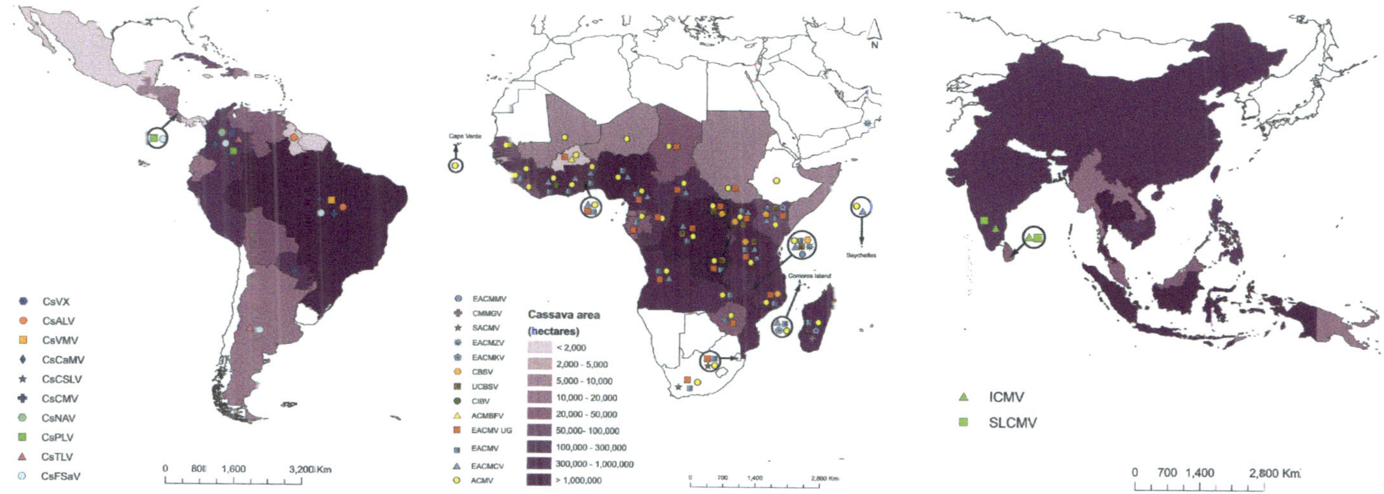

James P. Legg *et al.*, Figure 2 Global distribution of viruses affecting cassava.

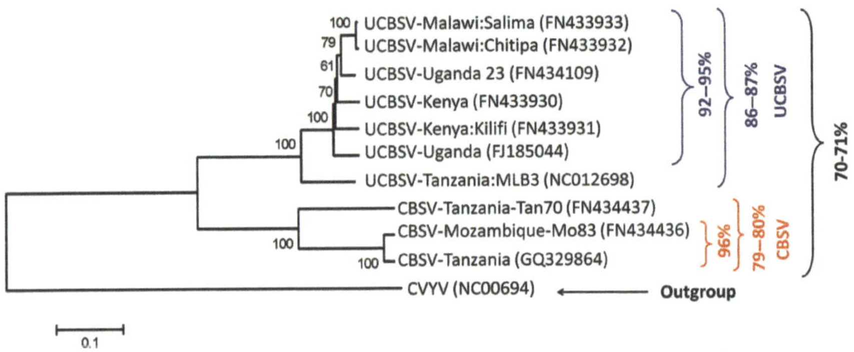

James P. Legg *et al.*, Figure 3 Phylogenetic relationships of the cassava brown streak viruses. Phylogenetic relationship of two cassava brown streak virus species based on the alignment of complete genome sequences using MEGA 5 software (Tamura et al., 2011). Sequences were aligned using the ClustalW algorithm, and the tree was constructed by the Neighbor-Joining method. The tree was rooted by using *Cucumber vein yellows virus* (CVYV) as an outgroup. The evolutionary distances were computed using the maximum composite likelihood method and are in the units of the number of base substitutions per site. The tree is drawn to scale, with branch lengths in the same units as those of the evolutionary distances used to infer the phylogenetic tree. The percentages of replicate trees in which the associated taxa clustered together in the bootstrap test (1000 replicates) are shown at the branch nodes. All positions containing gaps and missing data were eliminated from the dataset (complete deletion option). The NCBI GenBank accession numbers of 11 sequences are given in parentheses. Percentage sequence homologies within and between various clusters are indicated. CBSV: *Cassava brown streak virus*; UCBSV: *Ugandan cassava brown streak virus.*

James P. Legg *et al.*, Figure 4 Virus symptoms observed in cassava plants of the landrace Secundina (COL2063) infected by viruses detected in Latin America. Symptoms shown develop around 3-weeks after grafting on non-infected Secundina rootstocks: (A) uninfected control, (B) infected with CsTLV, (C) infected with CsCMV and (D) mixed infected with CsCMV, CsFSaV and CsTLV. Secundina plants single-infected with CsFSaV rarely show symptoms in leaves.

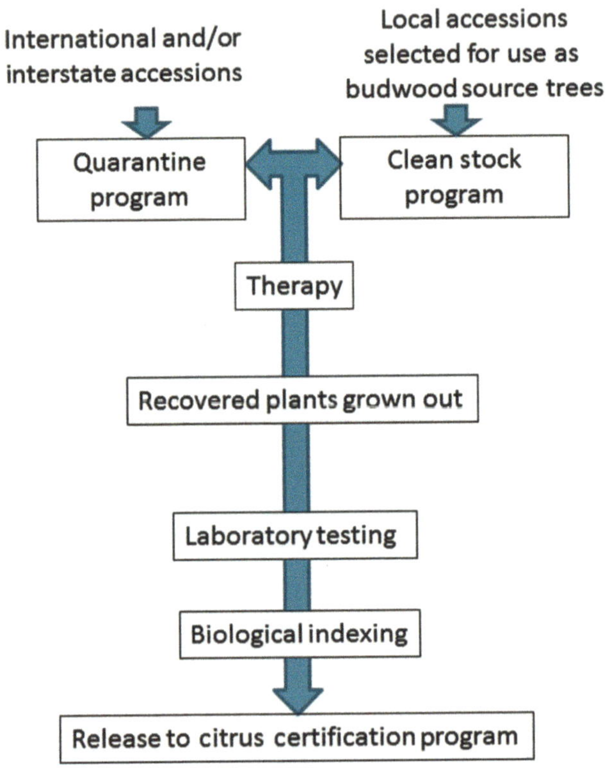

Richard F. Lee, Figure 1 Diagram showing the relationship between a quarantine program and a clean stock program and the general scheme of therapy and testing before incoming accessions may be released into the citrus certification program or made available to the citrus industry.

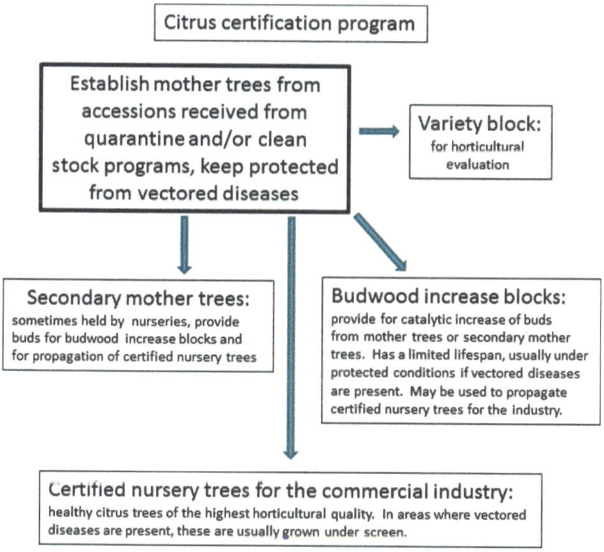

Richard F. Lee, Figure 2 Diagram of the general structure of a citrus certification program. The clean material received from the quarantine or clean stock program is used to establish the mother trees and maintained under conditions to protect them from exposure to vector-borne pathogens. Buds cut from the mother trees may be used to establish secondary mother trees, usually maintained under protected conditions in a commercial nursery, for budwood increase blocks to rapidly increase the number of buds available for propagation, or for direct propagation of certified nursery trees. A variety block is established from buds obtained from the mother trees to evaluate the horticultural trueness to type of the budline and to obtain production data.

Varvara I. Maliogka *et al.*, Figure 1 Leafroll symptoms on grapevine. (A) Mid vein reddening and curling of the leaves of an infected vine. (B) Close-up of a leafroll-infected leaf from a white-berried cultivar. (C) Close-up of a leafroll-infected leaf from a red-berried cultivar.

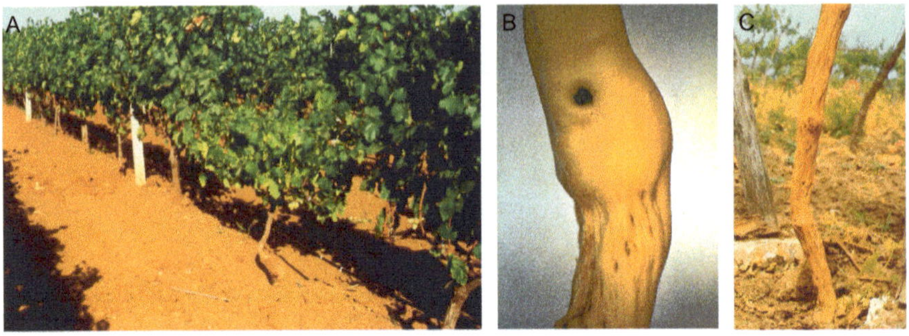

Varvara I. Maliogka *et al.*, Figure 2 Rugose wood symptoms. (A) Infected plant on the field. (B and C) Pitting and grooving on the stem.

Varvara I. Maliogka *et al.*, Figure 3 Symptoms of fanleaf disease. (A and B) Yellow mosaic on the leaves. (C) Canes from infected vines showing short internodes, double nodes, and zig-zag growth between nodes compared to cane from healthy plant (left). (D) Uneven development of the bunch on infected grapevine.

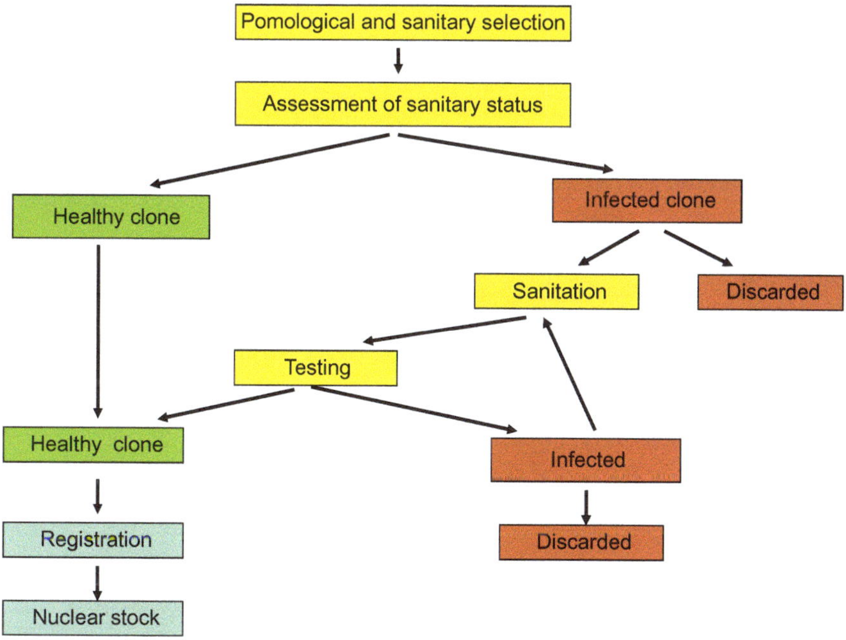

Varvara I. Maliogka *et al.*, Figure 4 Flowchart of the procedure implemented in Italy for the generation of certified nuclear stocks, in accordance with the certification program enforced in the European Union.

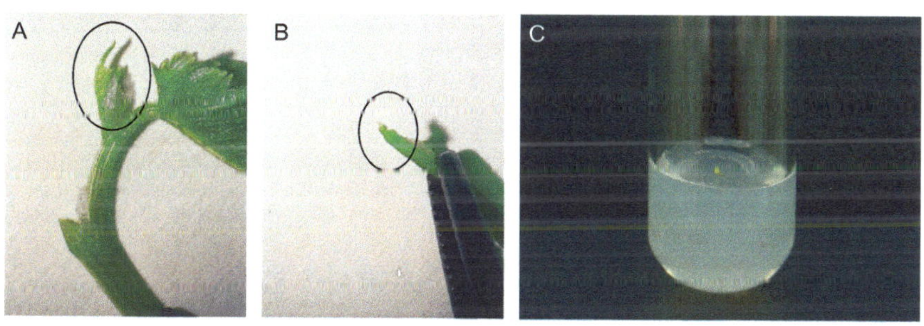

Varvara I. Maliogka *et al.*, Figure 5 Procedure of meristem tip isolation. (A) Apical bud before the meristem isolation. (B) Meristem tip after the removal of leaf primordia. (C) Apical meristem after a 4-week culture *in vitro*.

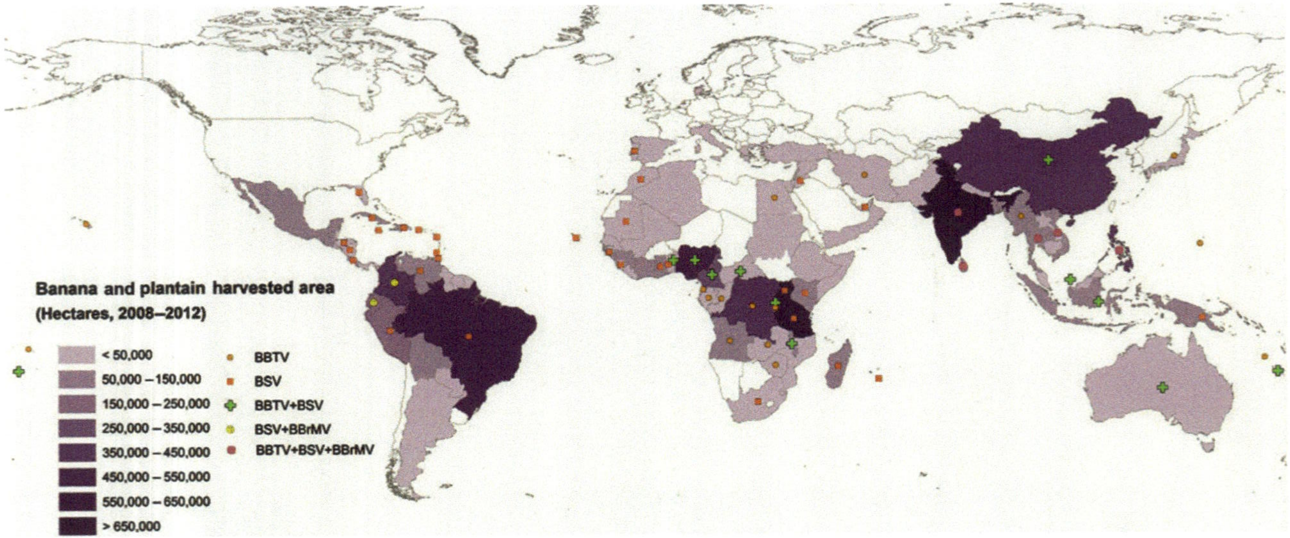

P. Lava Kumar *et al.*, Figure 1 Banana and plantain production in various countries and distribution of the three major banana viruses, BBTV, BSVs, and BBrMV. (Average production data for banana and plantain for the years 2006–08. http://faostat.fao.org/).

P. Lava Kumar *et al.*, Figure 2 Severe bunching symptoms (A) and green streaks (B) symptoms in BBTV infected plant; yellow streak symptoms on leaves of the BSV infected plant (B and C); and BBrMV symptoms on petioles (D) and bracts (E).

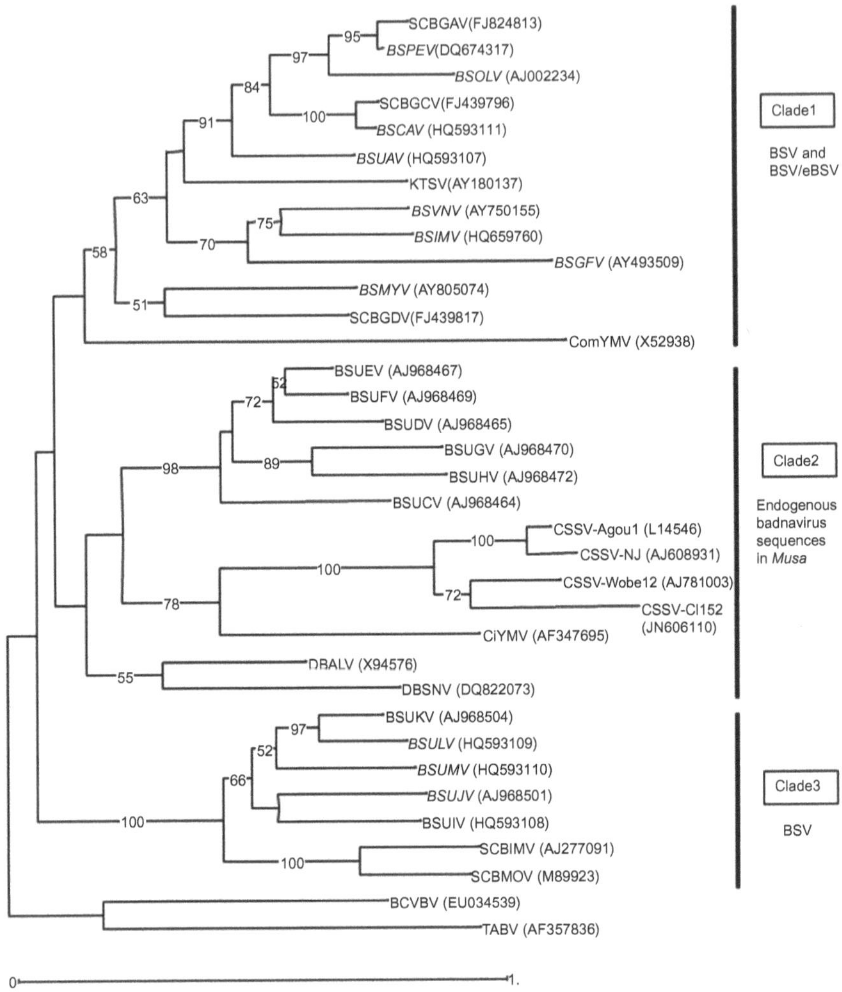

P. Lava Kumar *et al.*, Figure 4 Maximum likelihood phylogeny of badnavirus sequences based on alignment of a 540-bp fragment of the RT/RNase H viral region. Bootstrap values of 500 replicates are given when >50%. *Taro bacilliform virus* (TABV) and *Bougainvillea spectabilis chlorotic vein-banding virus* (BCVBV) are given as outgroups. Viral sequences isolated from *Musa* are in bold. BSV species where a full-length sequence is available are in italic. The scale bar shows the number of substitutions per base. The GenBank accession numbers of sequences are given in parenthesis. *Cacao swollen shoot virus*-CSSV-Agou1, *Commelina yellow mottle virus*-ComYMV, *Citrus yellow mosaic virus*-CiYMV, *Dioscorea bacilliform AL virus*-DBALV, *Dioscorea bacilliform SN virus*-DBSNV, *Kalanchoe top-spotting virus*-KTSV, *Sugarcane bacilliform MO virus*-SCBMOV, *Sugarcane bacilliform IM virus*-SCBIMV, *Sugarcane bacilliform Guadeloupe A virus*-SCBGAV, *Sugarcane bacilliform Guadeloupe C virus*-SCBGCV, and *Sugarcane bacilliform Guadeloupe D virus*-SCBGDV. *Adapted from Iskra-Caruana, Duroy, et al. (2014) and Iskra-Caruana, Chabannes, et al. (2014).*

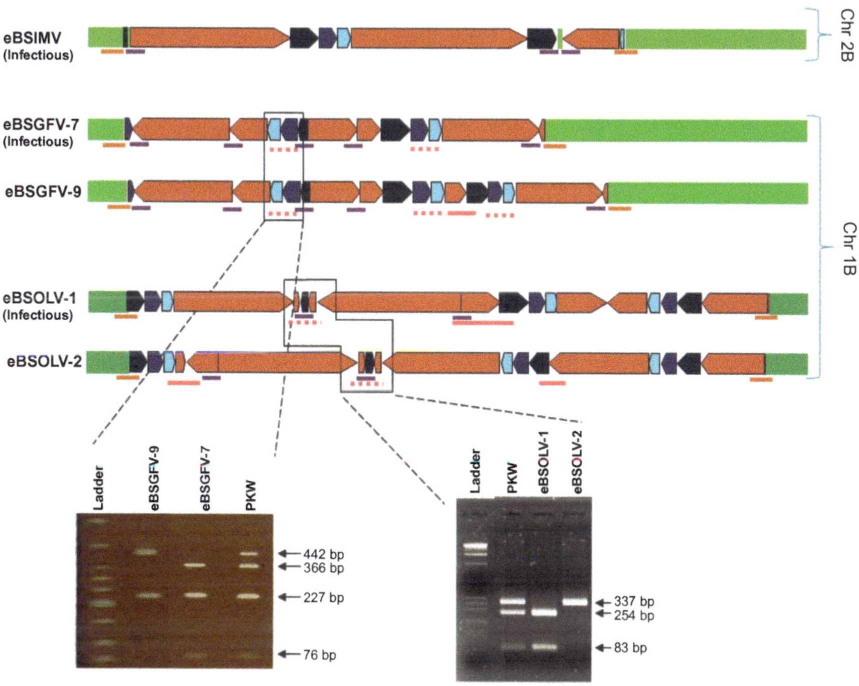

P. Lava Kumar *et al.*, Figure 5 eBSV genotyping markers for eBSIMV, eBSGFV, and eBSOLV in Pisang Klutuk Wulung (PKW). Banana genomic sequences of chromosome 1 and 2 are in green. Dark blue, light blue, red, and black arrowed boxes represent ORF1, ORF2, ORF3, and the intergenic region (IG) of the BSV genome. Arrows indicate the orientation of the viral fragment. PCR fragments are indicated by color lines. Orange, purple, and pink, lines correspond to *Musa*-eBSV junctions, internal eBSV rearrangements, and eBSV allele-specific markers. Two kinds of markers have been developed to specifically discriminate alleles of each eBSV species: pink -full lines represent classical PCR markers whereas pink -dotted lines represent dCAPS markers. Agarose gel images below correspond to eBSGFV (left) and eBSOLV (right) allelic patterns obtained with dCAPS markers. PKW exhibits both alleles for each eBSV species.

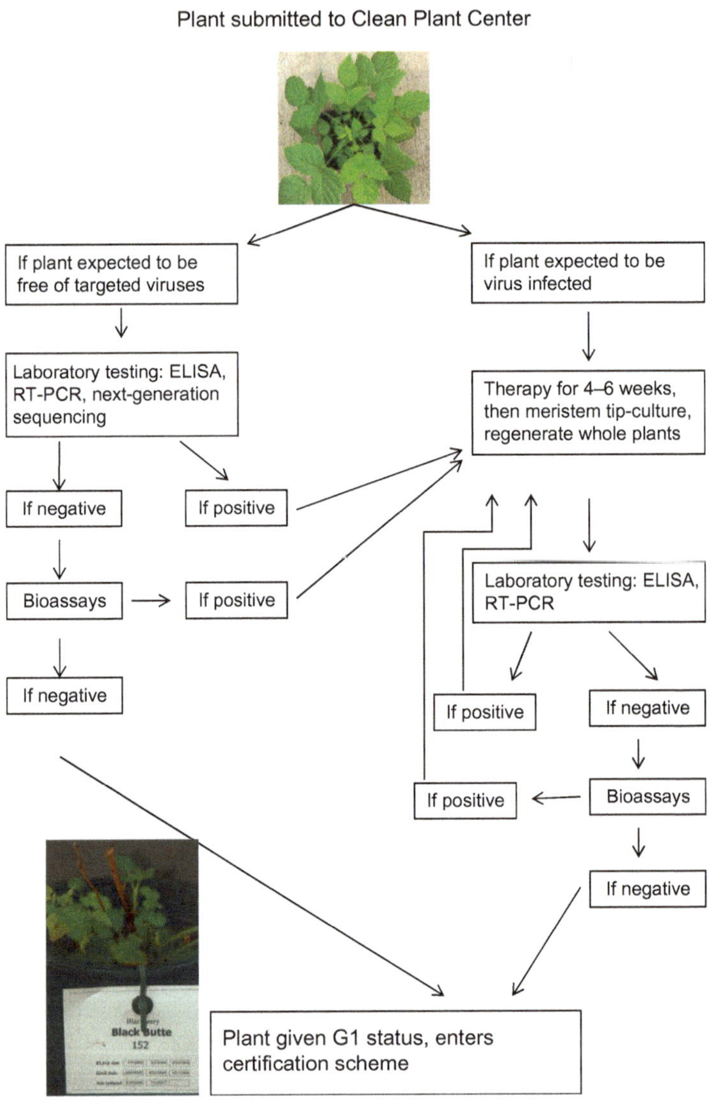

Robert R. Martin and Ioannis E. Tzanetakis, Figure 1 Path of a plant submitted to a Clean Plant Center from submission to obtaining G1 status.

Kolibri 360 GmbH
Ferdinand-Jühlke-Straße 7
99095 Erfurt, Deutschland
produktsicherheit@kolibri360.de